D0796372

# WRITINGS FROM THE WORKPLACE

## DOCUMENTS, MODELS, CASES

Carolyn R. Boiarsky

*Illinois Central College*

Margot K. Soven

*La Salle University*

**ALLYN AND BACON**

Boston London Toronto Sydney Tokyo Singapore

To our two Pauls, and to Ruthy and Michael for supporting yet one more writing project.

And to our teachers who encouraged us to be curious, supported our efforts to solve questions, and gave us the freedom to exercise our creative spirits.

*Editor in Chief, Humanities:* Joseph Opiela
*Editorial Assistant:* Brenda Conaway
*Marketing Manager:* Lisa Kimball
*Production Administrator:* Rowena Dores
*Editorial-Production Service:* Colophon
*Text Designer:* LeGwin Associates
*Cover Administrator:* Linda Knowles
*Composition Buyer:* Linda Cox
*Manufacturing Buyer:* Louise Richardson

 Copyright © 1995 by Allyn & Bacon
A Simon & Schuster Company
Needham Heights, Mass. 02194

All rights reserved. No part of the material protected by this copyright notice may be reproduced or utilized in any form or by any means, electronic or mechanical, including photocopying, recording, or by any information storage and retrieval system, without written permission from the copyright owner.

Boiarsky, Carolyn R.
   Writings from the workplace : documents, models, cases /
Carolyn R. Boiarsky, Margot Soven.
         p.     cm.
   ISBN 0-205-15012-8
   1. Readers—Technology. 2. Technology—Language—Problems.
exercises, etc. 3. English language—Technical English. 4. English
language—Rhetoric. 5. Technical writing. 6. College readers.
I. Soven, Margot. II. Title.
PE1127.S3B56 1995
428.6'0246—dc20
                                                      94-32252
                                                      CIP

 This book is printed on recycled, acid-free paper.

Printed in the United States of America

10  9  8  7  6  5  4  3  2        97

# Contents

# Preface

In the past few years, we have recognized the importance of providing our students with as thorough an understanding of the structure and style of technical documents as they have of sonnets, odes, and short stories. Results of research studies that indicate the significance of prior knowledge, including knowledge of story structures, have convinced us of the need for students to possess schema related to the story structures of technical documents. Furthermore, we believe it is as necessary for students to have a historical perspective from which to view technical documents as they have for studying fictional and poetic works. However, although some students may be familiar with such documents as business letters and memoranda, very few have read or written a project evaluation report or a proposal or a feasibility study.

*Writings from the Workplace* provides students with an opportunity to acquire schema for various technical documents by providing a variety of complete, actual documents similar to the kinds they may be asked to write once they enter the work force. These documents include recommendation letters and memos, sales proposals, environmental impact statements, project evaluation reports, and instructional manuals. In addition, *Writings from the Workpalce* provides an opportunity for students to learn how the context in which these documents are written affects the writing of the message as well as readers' interpretation of the message. They'll see how the executive summary in a report is prepared for the chief executive officer of a large company, and they'll discover how an engineer's failure to organize information appropriately may have cost him his job and the city of Chicago millions of dollars.

*Writings from the Workplace* provides exemplary models of technical genres and subgenres written in real-life situations so that students understand how the political, economic and social context of an organization affects the way a writer writes a document and a reader reads a document. By describing these real life situations, *Writings from the Workplace* provides an understanding of how writers base their rhetorical decisions on their audience, their communities, their purposes, and the situation in which they write. By providing not just an example of a letter of request, but the response to that letter, students discover how an audience responds to a text or fails to respond, as in the case of the Challenger accident. They'll see too why the audience fails to respond, as in the case of the Chicago flood. Thus, the text enhances students' awareness of the need to write reader-based prose and to follow appropriate conventions. In addition, by providing several examples of the same subgenre, the text helps students discover that even though documents in a particular genre follow a specific format and set of conventions, the writer must still invent the text for each document.

Furthermore, *Writings from the Workplace* acquaints students with technical "classics." They study such well-known documents as the report on Three Mile Island, the proposal to Congress to reduce lead poisoning, and the Chicago flood memo. They are also introduced to historical technical writers, such as Thucydides, Agricola, and Vitruvius. *Writings from the Workplace* allows students to see how the rhetorical aspects and conventions of these historical texts inform the technical documents of today. Students can follow trends in audience analysis and persuasion. Furthermore, because many of these documents concern the relationship of technology to the human condition, students discover the ethical implications underlying technical communications. Finally, students have an opportunity to study three specific cases: (1) the Challenger disaster, (2) the Three Mile Island nuclear accident, and (3) the Foley Square African Burial Ground.

## Objectives
By the conclusion of the course, students will have

- acquired schema for a variety of technical documents.
- acquired strategies for writing technical documents.
- learned the conventions of the major technical genres and subgenres.
- recognized that the audience and purpose for a document as well as the situation and community in which a document is written affect a writer's rhetorical decisions as well as a reader's interpretation of a document.
- perceived the ethical dilemmas that may be involved in writing technical documents.

## Organization
The book is divided into three sections. Part I: Writing and Reading Technical Documents discusses the various aspects of context that affect the writing and interpretation of a document. It provides strategies for writing within the various contexts.

Part II: The Documents provides a wide variety of examples of the major types of technical documents—correspondence, instructions, proposals, reports, and environmental impact statements—that students may write in the workplace. Each document is preceded by a brief description of the context in which it was written.

Part III: Case Studies allows students to follow the paper trail for three major events: the Three Mile Island nuclear accident, the Shuttle Challenger disaster, and the Foley Square African slave burial ground.

Each section contains a variety of examples, ranging in length, and including at least one historical example. Each chapter is introduced with a brief description of the genre. Suggestions for discussion and writing assignments are listed at the conclusion of each chapter.

## Special Features

*Integration of casebook, model, and process approach.* Allows students to acquire a schema for the format and conventions of various documents while simultaneously discovering how writers "invent" texts.

*Real world documents in actual situations.* Presents authentic documents and readers' responses to those documents in their real world setting.

- allows students to see how context affects the way a writer writes a document and a reader interprets a document
- reinforces the need for reader-based texts
- demonstrates invention and decision making in creating texts
- indicates the persuasive element underlying much of technical discourse
- points out the ethical implications of technical discourse

*Complete, full-length documents.* Contains a complete proposal, manual, and feasibility report, giving students an opportunity to see how the cover letter, executive summary, table of contents, appendix, figures, and tables reflect a document.

*Historical documents.* Provides a historical perspective of technical writing, allowing students to trace the origins of some of the conventions found in today's texts and thereby recognize how the documents of the past inform those of the present.

*Rhetorically based documents.* Places documents in the context in which they were written so that students can learn how audience, purpose, situation, and community affect the text.

*Ethical orientation.* Provides opportunities for discussing the ethical implications of a writer's decisions by placing each document in a political, economic, and social situation.

*Annotated documents.* Provides margin notes on the use of persuasion and other strategies to help students understand content, focus, and structure of a document.

*Pedagogical features.* Includes suggestions for discussion and writing assignments at the conclusion of each chapter.

### *The Beginning of an Anthology*

This book represents the authors' first attempt to create an anthology of technical discourse. We examined many works but settled on those included here as the most exemplary for our purposes. We welcome suggestions of other works that you would like us to include in later editions. We hope to expand on this selection in the future so that we can provide an exemplary book of technical discourse.

### *Acknowledgments*

This book could not have been written without the cooperation of the many people who spent time with us discussing the context surrounding the various documents and who made their own organizations' documents available for publication in our text. We wish to thank Stephen Hiles, American Electric Power Service Corp., Columbus, Ohio; Bailey Condrey, Council of Solid Waste Solutions, Washington D.C.; Albert Manville, II, Defenders of Wildlife, Washington, D.C.; Karen Florini, Environmental Defense Fund, Washington, D.C.; Howard Kreps, Environmental Defense Fund, Washington, D.C.; Richard Denison, Environmental Defense Fund, Washington, D.C.; Terri Capatosto, McDonald's Corporation, Chicago, Illinois; Marge Franklin, Franklin Associates, LTD., Prairie Village, Kansas; Marilyn Hailperin, West Jersey Health System Gibbsboro, New Jersey; David Jackson, Chicago Tribune Chicago, Illinois; Andrea Brands, City of Chicago, Chicago, Illinois; Diane Jacoby, New Mexico Environmental Law Center, Santa Fe, New Mexico; Naufumi Masuma, OMRON Electronics, Inc., Southfield, Michigan; Noboru Kato, Quality Spring/TOGO, Inc., Coldwater, Michigan; Alice Feldman, Caterpillar Inc., Peoria, Illinois; Dick Smalley, Balderson Inc., Wamego, Kansas; Michael Mudd, American Electric Power Service, Columbus, Ohio; Jay Feldman, National Coalition Against the Misuse of Pesticides, Washington, D.C.; George D. Lundberg, Journal of the American Medical Association, Chicago, Illinois; Arnold Relman, The New England Journal of Medicine, Waltham, Massachusetts; Peter Sneed, General Services Administration, New York, New York; Michael Duke, General Services Administration, New York, New York; John McCarthy, John Milner Associates, West Chester, Pennsylvaniz; Robert Neff, Vanderbilt University, Nashville, Tennessee; Bruce Kozarsky, Lin Pac Corp., Wilson, North Carolina; David Gibbons, Oil Spill Trustee Council, Juneau, Alaska; Mary Jo Zacchero, Department of Energy, Washington, D.C.; Eleanor Ledoux, Jenner and Block, Chicago, Illinois; Jody Howard, Advanced Technology Services, Peoria, Illinois; Merv Rennick, Caterpillar Inc., Peoria, Illinois; Harry Litchfield and Ed Wahlstrand, Deere & Company, Moline, Illinois; Linda Cooke, USDA Northern Regional Research Center, Peoria, Illinois; Robert J. Bruce, World

Distribution Services, Elk Grove Village, Illinois; Jay Miller, Illinois Department of Transportation, Springfield, Illinois.

We would also like to thank the librarians at both LaSalle University and Illinois State University, whose skills of ratiocination often resembled those of Sherlock Holmes as they helped us locate the various historical documents. We wish to express our appreciation especially to Margaret Ellen Wall, Head of the LaSalle Reference Department; W. Stephen Breedlove, Interlibrary Loan Coordinator at LaSalle; and Eithne Bearden, On-Line Coordinator at LaSalle.

Our gratitude must be extended to the many people who took time from their own work to read this manuscript and to offer their expertise to ensure that this text is both readable and teachable and that the discussions surrounding the documents reflect the most recent thinking in the discipline. To Paula Pomeranke, Illinois State University, Peoria, Illinois; Barbara Northrop, Johns Hopkins University Applied Physics Laboratory, Laurel, Maryland; Marianne Phillips, Air Products, Allentown, Pennsylvania; Olga Chesler Klein, Cherry Hill, New Jersey; Jane Allen, New Mexico State University; LynnDianne Beene, The University of New Mexico; Davida Charney, The Pennsylvania State University; Marc Glasser, Moorehead State University; Debra Journet, University of Louisville; Malcolm Richardson, Louisiana State University; Katherine Staples, Austin Community College; Erik A. Thelen, University of Wisconsin-Milwaukee; Thomas L. Warren, Oklahoma State University.

We would also like to recognize the support we received from our acquisitions editor, Joe Opiela, who believed this text could fill a void in the panoply of technical writing books currently on the market.

Finally, we wish to thank our students, whose requests to *see* complete documents spawned the idea for this text and whose discussions informed our writing.

# PART I

# WRITING AND READING TECHNICAL DOCUMENTS

- ■ Chapter 1: Technical Documents and their Contexts
- ■ Chapter 2: Discourse Strategies

Technical documents are written in an environment fraught with political, economic, and social consequences. The day-to-day business of an organization depends on the effective transmission of information within its walls and to its clients and customers outside it. While much of this information is related to routine business matters, some can have far-reaching effects. A company may decide to expand its environmental efforts because of a well-written report, and a state legislature may pass a bill to clean up local pollution because of a persuasive proposal. But million-dollar lawsuits for an injured employee or consumer may result from instructions that fail to carry adequate and appropriate safety warnings, and a company may fail to fix a product because of an improperly organized memorandum. Part I explains how the context surrounding a document affects the way it is written and the way readers interpret it.

In this section you will learn about the conventions and ethical responsibilities surrounding the writing of technical documents in the workplace. You will also study the various components of the context in which these documents are written and read—situation, purpose, and audience—and examine how these components affect the way in which a document is drafted and interpreted. Finally, you'll consider strategies writers use to meet their readers' needs while carrying out their own purposes.

# Technical Documents and Their Contexts

**H**undreds of thousands of documents are transmitted daily in U.S. industries. Regardless of their positions in the workplace, employees are continually involved in communicating technical information to their peers, subordinates, and supervisors. Recent surveys (Barclay et al. 1991; Kohl et al. 1993; Pinelli et al. 1993) indicate that those involved in the aerospace industry spend over 50 percent of their time reading and writing technical documents. These documents are closely related to the work the employees do and represent the social, political, and economic environment of their workplace.

In this chapter we will learn how the workplace affects technical documents. We will first study the conventions governing technical documents and the relationship between these conventions and the writer's task in creating texts that meet readers' needs. We will then consider writers' ethical responsibilities in writing a document. Finally, we will examine the context in which documents are written and read. We will learn how the context—the situation surrounding a document, the purpose for it, and the audience for whom it is written—affects writers' verbal and visual decisions, including decisions related to content, organizational pattern and sequence, point of view and focus, style, visual text, graphics, and layout.

## TECHNICAL DOCUMENTS

Technical documents comprise four major categories: correspondence, proposals, reports, and instructions. Each category has a variety of subcategories. Letters, memos, and electronic correspondence are among the subcategories of correspondence, and feasibility studies and environmental impact statements are among the subcategories of reports. These categories and subcategories are used for specific purposes and follow specific formats. Their purposes, formats, and overall organizational structures are governed by conventions that have developed over time and result from a document's use in the workplace.

## Conventions

Each of the various technical subcategories has its own conventions related to content, organization, style, and format. You will need to learn these conventions, just as you learned the conventions for thank-you letters and academic reports. For example, you know that social letters use salutations (e.g., *Dear Aunt Mary*) and that academic reports require in-text citations and lists of references.

Although each of the subcategories has a set of general conventions, there are many variations. For example, most proposals follow a problem/solution organizational pattern, but within that pattern is room for variation. A writer will not include a lot of information about a problem if readers are aware of it, but, if readers have limited knowledge of a problem, then a writer will provide more background data. A writer may begin with a problem and lead up to a solution if readers are expected to be hostile to a proposed project, but if readers are expected to be sympathetic to a problem, the writer will probably present the solution at the beginning and then explain how it solves the problem. Whether a writer begins a proposal with the problem or with the solution section depends on the readers' knowledge of the problem and attitudes toward it. Thus, different decisions create different texts and elicit different responses from readers.

### *Invention*

While remaining within the general conventions for a category, writers "invent" their text, basing their decisions on the context in which they write and in which their documents will be read. For example, whether Aunt Mary is a friendly, informal person or a prim matriarch will determine the content of your letter thanking her for the check she sent you for your birthday. Of course, your letter will include the information that you plan to use the check to buy a CD, but whether you tell her the kind of CD you plan to buy will depend on her knowledge of and attitude toward today's music. If Aunt Mary is familiar with the names of popular rock bands but thinks today's music is abominable, you probably won't tell her that you bought an album by the Grateful Dead, unless you want to take a chance that on your next birthday she'll send you a CD of Beethoven's Fifth Symphony instead of a check.

By studying readers' responses to the documents in this book, you will have an opportunity to discover how writers' misperceptions of a context, or their failure to consider aspects of the context in which they are writing or in which their documents are read, lead to inappropriate decisions that in turn lead to misunderstanding and miscommunication. You will also discover that when writers consider the context in which they write and in which their documents are read, readers can more easily understand and accept their messages.

## Legal and Ethical Responsibilities

Technical documents are usually considered impersonal, "scientific," and "objective." But most documents are related to human elements, regardless of the topic under discussion. A report on the health hazards of a potential leak in a landfill may be presented statistically in terms of the percentage of increased cases of cancer, but each one of those cases is related to a human being.

Writers need to keep the human element in mind as they write and to recognize that they have legal and ethical responsibilities to their readers. The legal responsibilities are fairly clear. They are designated in the Code of Federal Regulations, which must be followed by all industries contracted by the federal government. Court cases related to such aspects of industry as product liability and environmental cleanup establish precedents that must be followed.

Ethical responsibilities are less clear than legal requirements. As La Salle University Professor Marc Moreau explains,

> The ethical stance requires that we consider the welfare of others as well as our own. We have obligations to those who will be affected by what we do, including our peers, subordinates, employers, clients, and society at large. Therefore, before embarking on a course of action, its likely consequences for others should be anticipated. . . . When we act as members in an institution, we represent not only ourselves but our institution as well. Thus, a proposed action must be scrutinized for its consistency, not only with our commitments as individuals, but also with the established standards of the profession or industry we represent (Moreau 1993).

Thus, if writers disagree with the way in which management wants information to be presented in a document, they need to determine the ethical stance they should take. They need to weigh the consequences of the message as management would like it drafted with the consequences of the message if it were changed to reflect their beliefs.

For example, if a documentation writer for a software company believes that constant use of a computer may lead to carpal tunnel syndrome, then that writer should include a warning concerning this potential health hazard in the documentation. However, such a warning, according to present laws, could provide the basis of a lawsuit because it indicates that the manufacturer is aware that the product can be dangerous. There is less likelihood of a lawsuit if the writer omits the warning. The writer's decision is not easy. He must decide between responsibility to the consumer and responsibility to the software developer.

Moreau suggests that such a decision should not be the writer's alone but a joint decision between the writer and the management. "In an institutional setting . . . we must act as a collective body, not as independent agents" to reach consensus. "Unlike passive compliance (which can signal complicity in unethical practices) and unlike coercive or deceptive tactics (which usurp the autonomy of our fellow workers), energetic and thoughtful effort . . . can help us fulfill both our duty to oppose suspected wrongs and our duty to respect differences."

Today's technical writers could take a lesson in ethical responsibility from those of the past. For example, in *De Re Metallica* (1556), the standard text on the mining and working of metals (see Chapter 4), the author, Agricola, discusses procedures to ensure the safety of miners.

As you read the case studies in this textbook, you will have an opportunity to consider the ethical decisions made by various writers. Pay particular attention to the decisions made by the engineers involved with the Three Mile Island nuclear accident and the Challenger disaster, who repeatedly wrote memoranda warning of potential problems.

# CONTEXT

The context of a technical document involves three aspects: the situation, purpose, and audience.

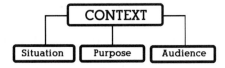

Let's examine how each of these aspects affects a text.

## Situation

The situation in which writers write documents and in which readers read documents can have a major effect on whether a message is understood easily and accepted by readers. Let's consider first how a writer's situation affects a document and then how a reader's situation affects its interpretation. Then we'll look at the economic, political, and social factors that further complicate communication between writers and readers.

### *The Writer's Situation*

In the workplace, documents are written by three types of writers: technical specialists, adjunct specialists, and technical writers. Writing is closely related to the specific tasks of technical and adjunct specialists. Technical specialists write about the specific content of their fields (e.g., chemistry, biology, electrical engineering, robotics, economics, archaeology, and psychology). They write to obtain information or materials for their projects, to respond to requests for information about their projects, and to communicate new information about their projects. Adjunct specialists, such as marketing analysts and business managers, write about the technical specialists' fields as they relate to their own areas of sales or budgets. They write about potential markets for the technical specialists' products and about financial matters related to the technical specialists' programs. The writing that technical writers do is closely related to work being performed by the technical and adjunct specialists. They write proposals to obtain support for the specialists' projects, and they write reports related to ongoing and completed projects.

Regardless of whether writers are technical specialists, adjunct specialists, or technical writers, the situation in which they write involves (1) helping readers achieve their purposes for reading a document, (2) providing for the needs of multiple or single readers, (3) collaborating with others to write a document, (4) adhering to the conventions of formal as well as informal organizational structures, and (5) engaging in day-to-day technical communication.

### Achieving Reader-Based Purposes

In the workplace, documents are written for specific readers and for specific purposes. These readers may be knowledgeable about the subject or they may know nothing about it. They may need to know the same information or they may each need different information. Regardless of who these readers are, they will all use the information to make a decision or to take action. As you read the case studies in Part III, you will have an opportunity to see how readers use and misuse the information they receive.

### Providing for Single or Multiple Readers' Needs

Documents may be read by a single reader or by multiple readers. When multiple readers are involved, their roles, positions, and fields may differ, causing their purposes and needs to differ. Creating a document that meets these various purposes and needs is one of the most difficult challenges in drafting a document.

The *Aqueducts of Rome*, one of the earliest examples of a report to be read by a diverse audience, was written for government officials who had little technical knowledge as well as for engineers (see Chapter 6). Frontinus provided readers with a series of cues to help them recognize which sections were appropriate to their particular needs. Thus, the politicians could avoid reading the technical sections that the engineers needed to study.

Readers in the workplace can be categorized as *primary, secondary, intermediary,* and *peripheral.* The primary reader is the person or persons for whom a document is being written. Secondary readers are people affected by a document, such as the marketing managers, comptrollers, and service representatives who need to design a marketing plan, develop a budget, or repair a piece of equipment. Intermediary readers are usually supervisors who are responsible for approving a document before it is sent to the primary reader. Peripheral readers, such as lawyers and news reporters, may read a document after the primary reader receives it.

An employee may write a memo to a subcontractor, the primary reader, requesting that a damaged piece of equipment be repaired. However, the employee's supervisor, an intermediary reader, may want to see the document before it is sent out to make certain the message is appropriate. If a document is important or a reader is difficult to work with, the supervisor's supervisor may also want to read the memo. The subcontractor who receives the memo may send it to a subordinate, who is a secondary reader, to carry out the request. If the subordinate gets hurt repairing the part, an attorney, who is a peripheral reader, may read the memo and use it as evidence in a lawsuit.

The number of readers and their range of roles, positions, and fields can affect the content, tone, and format of a document. For example, secondary readers who will carry out a project often need detailed information that primary readers don't require. To provide for the needs of both types of readers, writers may include the detailed information in an appendix so that the primary reader can skip it but the secondary reader can find it easily. As you study the documents in this book, you will find that some are read by a single reader but that most are eventually read by a multiple audience.

**Collaborating**   Often writers do not write a document alone; they collaborate with others. Sometimes the people involved write their own sections, or sometimes they get together to write a section. Often an editor or supervisor reviews a document and either makes changes or suggests revisions. These documents do not belong to a specific writer; they are owned by the group and by the organization. Often they are published without bylines (the names of the writers). Sometimes documents—especially letters and memos—are not even signed by the people who write them but by the writers' supervisors.

As you read the documents in this book, you will find some that are signed by the writer, some that have no byline, and some that list all of the people involved.

### Adhering to Formal or Informal Structures and Conventions

Organizations operate both formally and informally (Mathes and Stevenson 1991). Formal structures separate one company from another. They also divide a single company into departments, divisions, and sections, with each entity assigned a specific responsibility. Employees, too, are divided into separate categories and designated as hourly workers, supervisors, and managers. These structures can create communication barriers. Because of superstition ("Don't air your laundry in public"), the various entities often isolate themselves, tightly controlling information within their own boundaries. This was one of the problems in communication that you will read about in the case studies of the Three Mile Island (TMI) nuclear disaster and the Challenger accident. Personnel at Babcock and Wilcox, which manufactured the nuclear reactor at TMI, and at Morton Thiokol, which produced the solid rocket booster for the Challenger, didn't communicate their problems to the external organizations involved with the projects. When the Morton Thiokol engineers finally did communicate, they failed to provide sufficient background information for the readers to understand their message.

Informal relationships develop alongside the formal structures within an organization. Sometimes these create further barriers within divisions and sections; however, sometimes they cross barriers and create interdivisional entities, encouraging collaboration and allowing information to flow across formal structural lines. At Caterpillar Inc., for example, although the training division and the technical documents division are separate entities located in different geographical locations, their tasks are interrelated. The curriculum designers in the training division use the manuals developed by the technical documents division to develop the content for courses they present to the dealers. Until recently, this presented a problem when a new product was being introduced. Because there were times when the technical writers didn't complete the manuals until the product was ready to be shipped to the dealers, the training division was often unable to provide the dealers with training in the new equipment before they received it. To solve the problem, the technical writers agreed to provide the training division with early drafts of the manuals so that the trainers could develop their courses in time. In exchange, the training department, which used MacIntosh desktop publishing programs to create visuals, provided the technical writers with copies of their graphics for inclusion in the manuals. This interchange was based on an informal agreement between the members of the two divisions; no formal organizational chart indicated this relationship.

Just as formal and informal relationships can govern the communication of information within an organization, formal and informal procedures can affect the transmission of information. Memoranda may be transmitted by interoffice mail; by electronic mail (e-mail), which is sent through a computer network via a modem using telephone lines; by fax, which uses a facsimile machine to transmit messages over telephone lines; or by hand.

Conventions surround both formal and informal relationships and procedures. For example, e-mail is used only when a writer doesn't need an immediate response, but if a message is important, the writer needs to inform the reader via telephone that a letter is on the way.

As you read the description of the situation involving the Chicago flood, you will see how conventions involving the transmission of information can cause miscommunication. You will find that a similar problem occurred at Babcock and Wilcox when you read the case study about the Three mile Island nuclear accident in Part III.

**Engaging in Day-to-Day Communication**   Much of the writing done in industry involves short reports, memos, and letters, which are usually written as part of the daily routine that occurs in any organization. Most of the documents involved in the Three Mile Island nuclear accident and the Challenger disaster are brief and were originally written as routine correspondence. However, because of ensuing events, they eventually became key documents in major public investigations and court cases. In retrospect, they no longer appear routine but can be seen as documents containing urgent and important information. Some of them might have prevented the loss of life or millions of dollars in damages had they persuaded their readers to act.

Documents in the workplace are expected to be completed on schedule, and the deadlines are often tight, leaving writers little time to gather necessary information, let alone to plan and revise their drafts. Most of the documents in this text were written "under the gun," with tight deadlines.

Let's examine how a document was affected by the writer's situation.

### ◗◆ The Chicago Flood

On Monday, April 13, 1992, downtown Chicago came to a standstill. The Chicago River was flooding freight tunnels that had been dug underneath the city in the early part of the century. Water was seeping into the basements of office buildings in the city's financial district and into the retail area where department stores, such as Marshall Field's, were located. All buildings in the area were evacuated and traffic halted. It took the city weeks to pump all the water out. The cost in closed businesses and ruined inventory ran in the millions.

The cause of the flood was a leak in the wall of the tunnel abutting the river. Several weeks earlier Louis Koncza, the chief engineer for the Bureau of Bridges in the city's Department of Transportation (DOT), had sent a memo (Figure 1.1) to John LaPlante, DOT commissioner, notifying him of the leak and requesting permission to repair the walls. Why wasn't

Koncza
Kent
Naras
Chrzasc
Ociepka
DS/blm

RECEIVED
APR -3 1992
COMMISSIONER OF TRANSPORTATION
cc Johnson
Letter
Files

**INTER OFFICE CORRESPONDENCE**

CHICAGO DEPARTMENT OF TRANSPORTATION
BUREAU OF BRIDGES
CITY OF CHICAGO

Date 2 April 19 92

TO: John N. LaPlante ✓

FROM: Louis Koncza

SUBJECT: Freight Tunnel Repair

**Problem**

On March 13, 1992, City forces discovered a damaged section of concrete wall in the freight tunnel which passes under the North Branch of the Chicago River along Kinzie Street. The damaged wall area is approximately 20 feet long by 6 feet high. Some soil from beneath the river has flowed into the tunnel and this flow is slowly continuing.

**Background**

Investigation into the cause of the damage reveals that on September 20, 1991, new pile clusters were installed under a City contract to replace old deteriorated piles. These piles protect the Kinzie Street Bridge from river traffic. It appears that the added lateral soil pressure exerted by the new piles resulted in wall failure of the freight tunnel which is very close to the pile cluster.

**Recommendation**

This wall failure should be repaired immediately due to the potential danger of flooding out the entire freight tunnel system which is quite extensive. The City is currently receiving revenue by renting sections of the tunnel system to cable and fiber optic companies.

**Solution**

The most expedient and economic solution to this problem is to install 4 foot thick brick masonry bulkheads, keyed into the tunnel wall, on each side of the wall failure. Similar bulkheads have been installed many times in the past when CTA tunnels or building foundations were constructed through the freight tunnels.

The estimated cost of repair is approximately $10,000.00 and it will take City crews approximately two weeks to construct the bulkheads.

With your approval, work will begin as soon as possible.

Louis Koncza
Louis Koncza
Chief Engineer/Bridges

CHICAGO WORKS

**Figure 1.1 The Chicago flood memo.** Source: City of Chicago the wall repaired before the leak became a flood? Miscommunication between Koncza and LaPlante was one of the reasons.

Let's look specifically at the writer's situation. The Kinzie Street Bridge was only one of the many bridges for which the chief engineer was responsible. The problem had been brought to his attention by an employee of a local cable company that had planned to use the tunnels for stringing wire. The cable employee had gone into the tunnel in January to study the situation prior to sending people down to install the wire. When the employee found water and soil leaking into the tunnel, he notified his company, and the company notified the city. A city engineer investigated the tunnel on March 13 in response to the call from the cable company, which wanted to use the tunnel as soon as possible. After several meetings to determine what should be done and to estimate the cost for the repairs, Koncza wrote the memo as one of the many routine memos he had drafted over the years. His *purpose* was to obtain approval for repairing the wall. The memo was typed on a form for interoffice correspondence that included, in the upper right-hand corner, the names of six people other than Koncza's to whom documents in the department were routinely distributed. These people were often involved with some aspect of a project. Thus, Koncza had a *multiple readership.* In this case, one of his readers, Ociepka, had been the project manager for the installation of the pile clusters that had caused the wall damage. He would not be pleased to read that his project was responsible for the problem. Another reader, Chrzasc, worked under Koncza as his coordinating engineer and would be responsible for coordinating this project. Koncza needed to include enough details to provide Chrzasc with sufficient information to understand what he would need to do. Koncza penciled in at the upper left-hand corner the names of three other persons who needed to be informed of the repair work for the leak.

Because he was busy and because this was only one of several problems facing him, Koncza did not spend much time planning the memo. Nor did he take much time to revise it, other than checking that the facts were correct and then proofreading it for grammatical or spelling errors. Had he taken the time to plan or revise, he might have used a different organizational pattern, included additional information, or phrased the subject line differently to consider the reader's situation.

Koncza used the *formal office structure* to send the memo through interoffice mail to his supervisor. The situation surrounding a document involves not only the writing and reading of that document but also its transmission from writer to reader. The following newspaper article describes the conventions for transmitting interoffice correspondence in the city of Chicago's governmental offices and explains how Koncza's failure to adhere to these conventions caused his reader to fail to recognize the urgency of the message. ◆●

# In City Hall memos, everything is 'serious'

## By David Jackson

When he left work on Friday evenings, former city Department of Transportation Commissioner John LaPlante usually carried home with him a thick sheaf of memos warning that the city was about to fall apart.

Engineers wrote to him of leaky viaducts, inspectors sent ominous missives about sagging bridges, and urgent calls for spackle and bricks streamed in from across the city.

The memos rarely meant exactly what they said.

During the flood that paralyzed the Loop, Chicagoans were initiated into the arcane language of maritime engineering and learned to interpret the depth of piling clusters and the density of river silt.

And they were given an extraordinary glimpse into the subculture of the City Hall memo.

Now everyone knows that Chicago is governed not by people but by interoffice correspondence.

Like a second flood, paperwork courses through the corridors of City Hall, leaving a thick residue on every desk.

Following an elaborate protocol, memos are time-stamped, carbon copied, checked "for your action" and "please file," slipped into the manila envelopes that line the

> No one in City Hall starts a memo by saying, 'I've got a somewhat serious problem that you can get to on another day.'
>
> **Judson Miner**

city's metal filing cabinets, then left to disintegrate in darkness.

Interpreting these documents is a high art, like reading tea leaves.

"To determine the urgency of a memo, an individual has to read it very carefully," said former Mayor Eugene Sawyer

"Every letter starts, 'This is a serious problem,' " said former city chief attorney Judson Miner. "No one in City Hall starts a memo by saying, 'I've got a somewhat serious problem that you can get to on another day,' "

Take the memo written to LaPlante on April 2 by chief bridge engineer Louis Koncza. It has become famous, since it got LaPlante fired, and its words have been quoted often since the first soggy days of the flood.

Koncza wrote in his memo that a tunnel wall was rupturing, and "should be repaired immediately." If not, the city faced "the potential danger of flooding out the entire freight tunnel system."

There, spelled out in black and white, was a depiction of the catastrophe that lay ahead, a clarion call to act with urgency.

Sort of.

As a 30-year veteran of the City Hall bureaucracy, LaPlante knew that the memo carried another, subtler message—one that was only apparent to a seasoned insider.

On the upper left-hand corner of Koncza's memo was a time stamp showing it had been routed through the City Hall interoffice mail system, rather than delivered by hand, and so it did not reach LaPlante's desk until April 3, a day after Koncza sent it.

For the bureaucratic insider, the fact that Koncza sent his message through the sluggish office mail system was itself a signal that the tunnel repair job was not an emergency.

"It probably read like any other memo saying part of the city was about to collapse, which is something department heads are confronted with every day," said Michael Holewinski, an aide to former Mayor Harold Washington.

Only this time, part of the city did collapse.

Records released by the city since the flood show that Koncza's memo itself was no mere reminder, rattled off in a hasty moment.

(continues)

Before he typed it up for LaPlante, the chief bridge engineer or an aide first wrote out the text of the memo longhand, in a cramped, deliberate script.

Both the handwritten version and the final typed memo said the tunnel should be repaired immediately. But in the handwritten version, which LaPlante never saw, the word "immediately" was underlined.

And before even those simple words were committed to paper, there was a bureaucratic feat of Olympian proportions: the nearly three-month period in which word of the potential catastrophe was carried like a torch from engineer to engineer, making its way from the muddy banks of the river to the broad, busy desk of Commissioner LaPlante.

Only the torch was dropped time and again, flickered, and finally died.

One engineer wasn't reached for weeks because his telephone number had been changed in a City Hall reorganization designed to increase government efficiency. When the engineer finally descended into the tunnel, he couldn't find the breach in the wall.

So he waited three weeks, then

> **'You can't conceivably read all the stuff that's sent to you. You end up taking home a thick stack of these things on Friday and sitting by the garbage can, trying to catch up.'**
> **Michael Holewinski**

went back in, only to discover that the ruptured tunnel was so full of silt that it was impossible to walk through.

Then it took two more weeks and half a dozen meetings for as many city bureaucrats to decide that Koncza should draft his memo.

As Mayor Richard Daley said when he fired Koncza on Wednesday, if the chief engineer had really thought the tunnel damage was urgent, he should have acted, not written a memo.

"Bureaucratic paper shuffling in the face of such a monumental threat is unacceptable," Daley said.

Still, someone must give the bureaucrats their due.

For the workaday business of government, even the harshest critic of red tape concedes that the elaborate reporting procedures and formal layers of supervision have a purpose: They make government accountable.

City workers are required to send lots of memos so that lots of people know what they are doing, and how much money they spend doing it.

Of course, that sometimes means hiring more people just to read all the memos.

Holewinski said that when he was at City Hall, he had a staff of three aides as well as student interns who would "filter through every day's memos so that I could see the 5 or 10 percent that was most important, then pass on to the mayor the 1 percent he needed to see."

"The flow of paperwork is absolutely unbelievable," Holewinski said. "You can't conceivably read all the stuff that's sent to you. You end up taking home a thick stack of these things on Friday, and sitting by the garbage can, trying to catch up."

Or as former Public Works Commissioner David Williams put it: "The checks and balances of a bureaucracy can be its own worst enemy."

Source: Copyrighted April 26, 1992, Chicago Tribune Company. All rights reserved. Used with permission.

### The Reader's Situation

The situation in which a reader reads a document affects that reader's perceptions of the document. Readers are often busy and have little time to read. They read to obtain information to make a decision or take action or to learn what they need to do. They do not read to be entertained.

Readers often receive several pieces of correspondence at a time. They seldom have time to read letters and memos as soon as they are delivered. Usually they pile them in the "in" basket on their desks and look at them sporadically during the five minutes they are free between appointments, just before going home, or even the following morning when they arrive at work. Their reading may be interrupted by telephone calls or people stopping to talk. Longer documents, such as reports or proposals, usually do not get much more of their attention unless they are important. Readers may skim through these longer documents to determine whether they need to read them more thoroughly. If so, they may set aside time in their schedule or wind up taking the report home to read after dinner, before going to bed, or over the weekend. Most people don't want to be bothered by procedures; they scan them, if they read them at all. As for instructions, people read them as they work on a project, studying them just long enough to learn what to do.

Because readers read under these conditions, they are likely to miss information, misunderstand a message, or fail to comprehend a text unless the document is written clearly and succinctly, with markers for locating necessary and important information. As you read about the Chicago flood, the Three Mile Island nuclear accident, and the Challenger disaster, you will note that these conditions existed and were partially to blame for the ensuing problems. You will also note that when writers take care to overcome these conditions and to meet the readers' needs, as in the lead paint proposal and the Foley Square research design, problems are eliminated and writers achieve their readers' and their own purposes.

●◆    Let's look at the situation of the reader of the memo in Figure 1.1. The reader, John LaPlante, had assumed his position as temporary Commissioner of Transportation fairly recently. He continuously received memos describing construction problems and requesting approval to have them repaired. When he received this one on Friday afternoon, he gathered it up with the others and took it home with him. On Sunday afternoon he read through the stack of memos, including the one on the Kinzie Street Bridge. Having no secretary or typewriter available, he handwrote the response (Figure 1.2) on one of his memoranda forms, approving the repair and recommending that the repair job be put up for bid (a PSR [Project Specification Report] is a form for putting out a bid. For more information see p. 173.). He penciled in the names of several other readers who needed to be kept informed of the project. At some point, a copy of the memo was

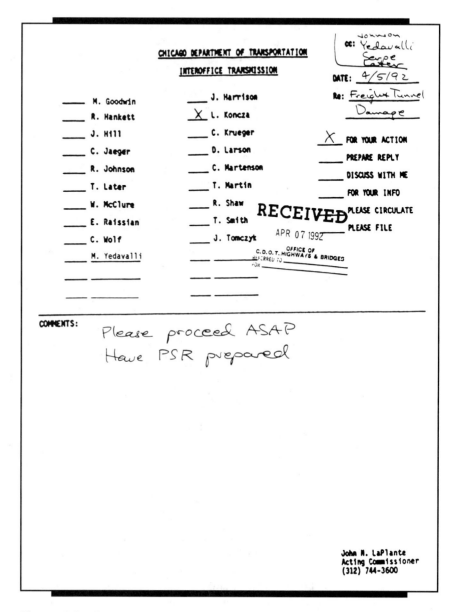

**Figure 1.2    Response memo.**    Source: City of Chicago.

sent to Sadowski, who was responsible for inspecting the newly installed pile clusters and who therefore had some responsibility for the damaged wall. Eventually both Sadowski and Ociepka were fired for failing to act quickly enough to repair the leak. Returning to work on Monday, LaPlante placed the memo in interoffice mail. It was delivered on Tuesday, as the stamp in the upper left-hand corner indicates. If Koncza's memo had not been one of many, if it had emphasized the urgency of the problem or detailed the possible consequences, or if LaPlante had taken the time to study and evaluate the message, he might have responded differently. But on a Sunday afternoon, he probably wanted to get through the paperwork so he could continue to enjoy his weekend.  ◆●

### Economic, Political, and Social Factors

The situation in the workplace involves three major factors: economic, political, and social.

These factors can affect the content, organization, point of view and focus, style, format, and visual elements of a document. For example, if employees are asked to write a description of their job in a department of a large corporation that has been losing money and laying off workers, they will include as many details about their responsibilities as possible to indicate to readers that their job is too important to be eliminated. However, if employees are simply changing jobs and a supervisor asks for a description to use in recruiting a replacement, the employees will probably provide only a *general* description of their major responsibilities.

These factors can also affect the way in which a reader interprets a document. For example, a consultant has used the following memo in a number of training sessions.

> In those establishments where suspension of labor is possible, direct those parties of management to the termination of the illumination.

The memo was actually written by a bureaucrat in the federal government during World War II and was revised by President Franklin Roosevelt to read, "When you're finished working for the day, turn out the lights." Participants at the training session who were employed by stable companies interpreted the memo fairly accurately. However, several participants from companies where layoffs were common or a strike was imminent invariably misinterpreted the message, thinking it related to being fired or laid off (*terminated*) or to going on strike (*suspension of labor*).

Technical writers of the past were as acutely aware of the context in which a message would be read as today's writers. Because in ancient times rationale for technology was based on the public good, writers had to be aware of the political, economic, and social situation if they were to persuade their readers to accept new technologies. As *Ten Books of Architecture* by Vitrivius in 27 B.C. (See Chapter 4) and *Aqueducts of Rome* (see Chapter 6) by Frontinus in A.D. 97 illustrate, recommendations for construction or repair were always accompanied by an effort to gain approval from those who kept the purse.

**Economic Factors**   The content of a document may be affected by economic factors. These factors may relate directly to an employee's personal life or to an organization's profitability. Because of the threat of a lawsuit and the potential of a monetary settlement, a company may decide to eliminate information or to include warnings about a process or piece of machinery. To persuade stockholders to continue to invest, a company may emphasize products that are successful in the marketplace and subordinate data related to those which are having problems.

•◆   Economic factors were very much present in the minds of both the memo writer and the reader in the situation involving the Chicago flood. The chief engineer included an estimated cost in his memo. He also alluded to the rental fees the city was receiving from cable and fiber-optic companies for their use of tunnel sections, implying that if the tunnel were not fixed, the city might lose its fees.

The reader's recommendation to put the job out for bid indicates his awareness of the need to limit spending.   ◆•

Economic awareness is not new. As the Secretary of Commerce under President Calvin Coolidge, Herbert Hoover demonstrated an awareness of economic factors even before the Great Depression. In his 1927 report "Relief Work on the Mississippi Flood and Principles of Flood Control" (See Chapter 6), he reassured President Coolidge that all programs could be completed within the budget and that some money would be left over to give to the Red Cross to continue operations.

**Political Factors**   The content and tone of a document may be affected by political factors. Competition and unwritten agreements between organizations, between divisions within an organization, and among personnel within a division often create hidden agendas—purposes that are not openly acknowledged. These purposes may cause writers to add information that readers may not need or want but that writers want readers to know. Or writers may exclude information that readers may want but that they don't want readers to know. Hidden agendas may also cause writers to create a certain tone that they want conveyed to their readers. If a supervisor believes that threatening an employee is the only way to get him to meet a deadline, then the supervisor may use the phrase *fails to meet deadlines* in a personnel evaluation. However, if the supervisor feels that the employee's improvement should be reinforced so that he will continue trying, then the supervisor may use the phrase *continues to have an opportunity for improvement in completing assignments in a timely manner* because it presents a less threatening tone.

•◆    Political factors were involved in the situation surrounding the Chicago flood memo. Mayor Daley, in the newspaper article on page 13, suggests that Koncza should have acted rather than written a memo. Office protocol, however, required that Koncza obtain his supervisor's permission before spending uncommitted departmental funds. If Koncza had acted without approval, his supervisor may have perceived it as an attempt to undermine his authority.

LaPlante was also aware of political factors when he read the memo. City politicians have been criticized in the past for giving jobs to their friends, for paying higher fees than necessary for a job, and for failing to provide jobs to minority- and women-owned businesses. The mayor, who was LaPlante's boss, was up for re-election at the time of this memo. By putting the job out for bid, LaPlante was making sure that the mayor's administration would not criticized.  ◆•

**Social Factors**    The content and tone of a document may also be affected by social factors. These factors help writers determine who their readers are and how their readers should be addressed.

Today's companies assume two major organizational patterns—hierarchical and collaborative. In a hierarchical pattern, managers and workers are at different levels and the roles they assume remain constant throughout a project. In a collaborative pattern, managers and employees work together in teams in which their roles may change as a project progresses.

The social relationships in a hierarchical organization, such as the city government of Chicago, are more formal than those in a collaborative one (Figure 1.3).

In a hierarchical organization, lines of command are maintained. Because supervisors are held responsible for their employees' actions, they usually check employees' written documents before sending them outside a division, and employees are expected to obtain approval before engaging in a project. Thus, Koncza needed LaPlante's approval before beginning work on the leak.

In a hierarchical organization, people often prefer that their subordinates in the hierarchy address them by their titles and surnames. The informal note in Figure 1.4 was sent to an engineer as a friendly warning, explaining that one of the conventions followed by employees of the city of

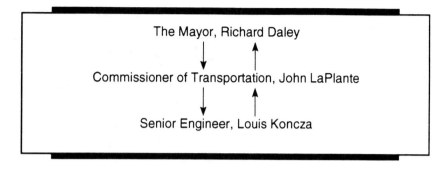

**Figure 1.3    The social structure of the city government.**

Chicago is the use of titles with surnames, especially when addressing someone at a higher level.

In a collaborative organization the emphasis is on equality in decision making. Thus, it is extremely important that all team members be kept informed of a project's progress. Members of a team often prefer to be addressed by their first names, without regard to their position in the hierarchy.

Regardless of the organizational pattern, personnel associated with a project want to be kept current on a project's status, just as all personnel within a division want to be kept informed of happenings within a division. If they are not, they feel personally excluded.

•◆    In the city of Chicago memo, the names of three people have been added to the distribution list to prepare them for the project and to make them a part of the project from the beginning.  ◆•

## Purpose

The writer has a purpose for writing a document; the reader has a purpose for reading a document. Both the writer's and the reader's purposes affect the document.

### *Writers' Purposes*

The writer's purpose may be to instruct, inform, or regulate the reader. In addition, the writer may want to persuade the reader to take action or to think in a certain way. For example, the vice president of an organization may have requested that the manager of the personnel department determine ways in which working conditions for physically disabled employees

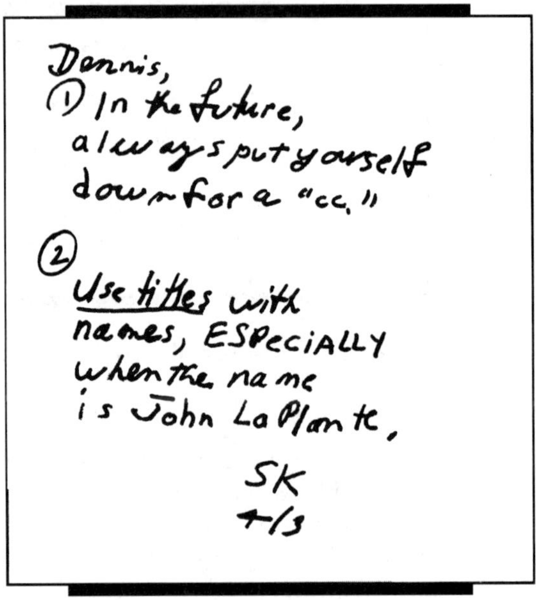
Dennis,
(1) In the future, always put yourself down for a "cc."
(2) Use titles with names, ESPECIALLY when the name is John LaPlante.
SK
4/3

**Figure 1.4 Note recommending conventions to be followed by a City of Chicago employee.** Source: City of Chicago.

can be improved. The manager may simply present the vice president with the suggestions listed at random. However, if the manager believes that some of the suggestions should be enacted immediately, she may list them in order of priority.

Sometimes writers' purposes are explicit; at other times they are implied. If the manager of the personnel department states that the first three recommendations should be enacted as soon as possible, the recommendation is explicit, and the vice president will be aware that the writer is trying to persuade her to respond in a specific way. However, the manager may believe that the VP is "going overboard," that such actions are too costly and, since they are not required by ADA (Americans with Disabilities Act), that they should not be enacted. Rather than state these beliefs explicitly, the manager may include cost estimates for each suggestion and introduce the costs with such comments as, "This will cost *at least* . . ." or "The cost of this is *over* . . . ," implying that the actions are too costly and therefore should not be carried out.

Koncza's purpose in writing the Chicago flood memo was to obtain the reader's prompt approval of his recommendation. Notice that he makes the request explicitly in the final paragraph. Because he receives La Plante's response relatively promptly, he can assume he has achieved his purpose. However, if he had indicated the urgent nature of the problem and the need for a prompt response at the beginning of the memo, he might have received a response sooner, and the response might have permitted him to go ahead without putting the project up for bid. By wording the subject line "Potential flooding of financial district and 'Million Dollar Mile' area of city," he would have indicated to the reader the seriousness of the problem. Furthermore, if Koncza had begun the introductory paragraph with a request for permission to repair the bridge, and if he had included not only the general consequences of the flood, but also the major details related to the area that might be flooded, LaPlante probably would have realized immediately that he had to make a decision, and he would have had enough details to determine whether the job should have been put up for bid. The introductory paragraph should have read as follows:

**Request**

**Effect of problem**

I would like permission to begin repairing a section of concrete wall in the freight tunnel that passes under the north branch of the Chicago River along Kinzie Street. If this wall failure is not repaired immediately, it may cause the flooding of the entire freight tunnel system, which is quite extensive and which runs under the financial district and the "Million Dollar Mile" section of downtown. Some soil has already flowed into the tunnel, and this flow is continuing. ◆●

### Readers' Purposes

The document must relate to readers' purposes if it is to be read and accepted by the reader.

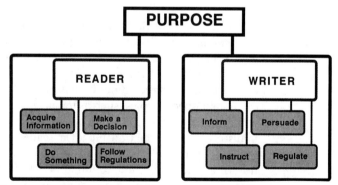

If users need to learn a software program to obtain architectural codes in different regions of the country, then the writer needs to write instructions so users can look up the codes; the reader doesn't need a description of how the program works. If a writer wants a reader to approve a recommendation, then the writer needs to request that the reader respond to the recommendation.

➡◇    LaPlante's purpose in reading the Chicago flood memo is twofold: (a) to obtain information about his area of responsibility and (b) to make a decision, if necessary, regarding that area. He learns about his area throughout the memo. The third paragraph implies that he needs to make a decision, which is specified in the final paragraph.

The purpose of his response in the form of the memo (Figure 1.2) is to inform Koncza that he has the approval to go ahead with the project and to persuade Koncza to put the job out for bids as quickly as possible. Three days later Koncza, responding to LaPlante's orders, sends a memo (Figure 1.5) requesting that the PSR (bidding form) to which LaPlante referred be processed.  ◇➡

## Audience

The most important factor affecting a text is the audience. We have already learned that readers' situations and purposes affect the way a message is interpreted. But there are other factors related to readers that also affect their understanding, acceptance, and response to a message:

- Readers' organizational communities
- Readers' professional communities
- Readers' reading processes
- Readers' thinking processes
- Readers' reading behaviors

### Readers' Organizational Communities

Readers' organizational communities are bounded by the organizations for which they work. Each organizational community has its own characteristics, standards, and conventions as well as its own vocabulary. In addition, each organizational community has an established structure in which readers assume specific roles, such as managers, assembly line workers, and data processors, and positions, such as hourly employees, professionals, and administrators. The relationship of a reader's organizational community to the writer's, the role a reader assumes in reading a document, and the position of a reader within a community affect the writer's choice of the content, focus, point of view, and style of a document.

**Readers' Relationships**    The contents, organizational pattern and sequence, point of view and focus, and style of a document may be affected by readers' relationships to their organizations. Readers may be *internal* or *external* to an organization. Internal readers are basically familiar with the product of an organization and the terminology related to that organiza-

cc: Yedavalli
Koncza
Kent
Naras
Malkos
Sadowski
Chrzasc
Bridge File
LPC:bim

**INTER OFFICE CORRESPONDENCE**

CHICAGO DEPARTMENT OF TRANSPORTATION
BUREAU OF BRIDGES
CITY OF CHICAGO

Date _____ 10 April _____ 19 __92__

TO:        John Hill, Assistant Commissioner

FROM:      Louis Koncza, Chief Engineer/Bridges

SUBJECT:   Emergency Repair of Freight Tunnel

Attached for processing are PSR-3 & 3A for the force account repair of
the freight tunnel which passes under the North Branch of the Chicago
River at Kinzie Street which was recently damaged.

We hereby request that a project number be assigned for this unfunded
emergency repair (Est. Cost $10,000.).

Louis Koncza
Chief Engineer/Bridges

Originated by:

Louis Chrzasc
Coordinating Engineer

**Figure 1.5a  A request memo with attachments.**  Source: City of Chicago.

tion. External readers, however, may not have this knowledge, and the
writer may need to include background and details an internal reader
would know. The writer may also need to avoid using certain terminology
or may need to define organization-specific terms.

CITY OF CHICAGO
**DEPARTMENT OF TRANSPORTATION**      PSR-3
PROJECT PROPOSAL AND APPROVAL

| | | |
|---|---|---|
| DATE: | _APRIL 6, 1992_ | DISTRIBUTION BY FINANCIAL ADMIN.: |
| TO: | _LOUIS KONCZA, CHIEF ENGINEER_ | |
| FROM: | _LOUIS CHRZASC COORDINATING ENG._ | ☐ HIGHWAYS & BRIDGES      ☐ PROJ. MC |
| PROJECT NO.: | _____ WORK ORDER NO. _____ | ☐ MASS TRANSIT      ☐ ACCTG. ( |
| PROJECT TITLE: | _FREIGHT TUNNEL REPAIR_ | |
| LINE REFERENCE 5-YEAR CAPITAL IMPR. No _____ | | ☐ NEIGH. IMPROV. & OPER.      ☐ _____ |

| PROJECT PROPOSAL | SCHEDULE | | | BUDGET | |
|---|---|---|---|---|---|
| | START | WEEKS | END | CDOT  COSTS | CONTRACT  COSTS |
| RIGHT OF WAY | _____ | ( ) | _____ | $ _____ | $ _____ |
| PRE-DESIGN | _____ | ( ) | _____ | $ _____ | $ _____ |
| DESIGN | _____ | ( ) | _____ | $ _____ | $ _____ |
| CONSTRUCTION | _____ | ( ) | _____ | $ _____ | $ _____ |
| FORCE ACCOUNT | _____ | ( ) | _____ | $ _____ | $ _____ |
| _____ | _4-1-92_ | ( ) | _7-15-92_ | $ _10,000.00_ | $ _____ |
| (OTHER) | | | SUB-TOTALS $ _10,000.00_ | | $ _____ |
| | | | TOTAL BUDGET $ _10,000.00_ | | |

SCOPE OF SERVICES:

_PERFORM EMERGENCY REPAIR TO THE FREIGHT TUNNEL_

_UNDER THE CHICAGO RIVER._

*Coordinating Engr.*

*Dennis Sadowski*

| CAPS      FUNDING | AMOUNT $ | PROJECT      MANAGER |
|---|---|---|
| | 10,000.00 | |
| | | |
| | | BUREAU/SECTION      HEAD |
| | | |
| | | OFFICE      HEAD |

PROPOSAL APPROVAL (Client Bureau Head)(By Commissioner, for unfunded projects)
The above project proposal, budget and schedule have been reviewed and are approved for the

DEPT. NO.: _____      BY: _____

DATE: _____      TITLE: _____

FORM CDOT 3/92

**Figure 1.5b**   *(continued)*

The terms *internal* and *external* are relative. Although a reader may work for the same company as a writer, the reader may be in a different division or in a different geographic location. Readers in one division or location may not know data or understand terms specific to another division or location, even though they work for the same company. When personnel

**DEPARTMENT OF TRANSPORTATION**
**WORK ORDER DESIGNATION**                    PSR – 3A

DATE _4 / 6 /92_

PROJECT NO. __-__-____          PROJECT TITLE: _FREIGHT TUNNEL REPAIR_

PROJECT MANAGER _D. SADOWSKI_ ___ EMP. NO. _713625_ __ R.U. __240__

| W.O. NUMBER | PROJ/RPTG/ OR APPR. | TITLE | CAPS FUNDING | BUDGET AMOUNT | START DATE | END DATE |
|---|---|---|---|---|---|---|
| 432 | | ENGINEERING SUPERVISION | | 1,000.00 | 4 1 92 | 7 15 92 |
| 433 | | TRADES | | 9,000.00 | 4 1 92 | 7 15 92 |

DEPT.# _8A_     ORGANIZATION #  _____          TOTAL     _10,000.00_

FOR USE BY ACCOUNTING

BUDGET REVISION AUTHORIZED _____        PROJECT MANAGERS SIGNATURE

FORM CDOT 3/92

**Figure 1.5c**   *(continued)*

with a nuclear utility were asked to define the term *dog house*, some said it related to an area under the reactor, and others indicated it was any small, enclosed area.

The primary reader in the Chicago flood memo in Figure 1.1 (LaPlante) is an internal reader; not only is he employed by the same organization as the writer (i.e., the city of Chicago), but he supervises the writer's division. Thus, Koncza doesn't provide a description or a map to specify where the Kinzie Street Bridge is or explain what sections of the city are located over the freight tunnels because LaPlante probably knows this information.

**Readers' Roles**   A document may also be affected by readers' roles. Readers assume a variety of roles—they may be decision makers or implementers, clients or subcontractors, users or producers. Readers in the role of decision makers need to know about various alternatives and the consequences of each alternative, but readers in the role of implementers don't need to know about the alternatives; they need only to know about the project on which they will work. Readers in the role of users need sets of instructions, not descriptive paragraphs, whereas readers in the role of learners may need a role play rather than a set of instructions.

➥    The primary reader of the Chicago flood memo in Figure 1.1 could have read the memo in the role of either a learner or a decision maker. Because the pattern Koncza used to organize the information in the memo does not include the request until the end of the message, LaPlante doesn't know he should be reading it in the role of a decision maker. Instead, he reads it in the role of a learner. Thus, rather than evaluating consequences and comparing alternatives as he reads, he simply studies the information to understand it. Once he discovers that he must make a decision, LaPlante must reconsider the information in terms of his new task. He may simply try to remember what he has read, or he may reread the message. The latter is preferable because he may not remember some of the data. However, it is also more time consuming and, because he wants to get through this task, he probably does the former.  ➥

**Readers' Positions.**    Documents are further affected by readers' positions. Readers' positions in an organizational hierarchy further affect a text. Readers may be above, below, or on the same level as a writer in the hierarchy. Traditionally, writers use a more formal, impersonal style in writing to people above them than they do in writing to their peers or their subordinates. However, this, too, is changing as an increasing number of companies adopt a "total quality management" approach and use teams of workers at various hierarchical levels to engage in decision making.

➥    The primary reader of the Chicago flood memo is above the writer in the organizational hierarchy. The people listed in both the upper right- and left-hand columns are on the same hierarchical level as the writer.  ➥

## Readers' Professional Communities

Readers are members of professional communities (e.g., medicine, oceanography, architecture, economics, civil engineering, engine manufacturing). A professional community may be as broad as engineering or as narrow as computer engineering for the manufacturing of engines. Although readers are experts in their own narrow areas, they are not necessarily experts in the writer's field.

Readers fall into three categories in relation to a topic: (1) experts, (2) generalists, (3) novices. Experts are those who have specialized in a field. Dr. C. Everett Koop, the former U.S. surgeon general, is an expert on public health issues. Generalists are those who have some knowledge in an area but have not specialized in it. For example, a marketing analyst in a pharmaceutical company is a generalist in pharmaceuticals. Novices are readers with little or no knowledge about a topic. An accountant who has been to a doctor to be tested for AIDS is a novice in terms of immunology and cytology.

Although readers are experts in their own fields, they may not be experts on a topic discussed in a document. A plastic surgeon who owns a twin-engine Cessna airplane may be an expert in plastic surgery, and she may be knowledgeable in operating a plane, but she is probably not an expert in avionics. Thus, letters to aircraft owners may be read by people who range from experts to novices.

Generalists and novices need background information that an expert would already know. They also need to have technical terminology defined or replaced by nontechnical words. For instance, when the members of the President's Commission on the Three Mile Island Nuclear Accident wrote their report for the President, Congress, and the general public, they knew they would be writing for readers who were novices in the nuclear field. They spent the first five pages providing readers with background information explaining how a nuclear reactor works so readers would understand what was happening when they described the problems that occurred. The commission members also defined such technical terms as *trip*—terms that are common knowledge to experts in the field but which mean nothing to the general public who are novices in the field of nuclear physics.

•◆  Although the commissioner who received the Chicago flood memo manages a department that is involved with engineering, the Commissioner was an administrator, not an engineer. He was a generalist rather than an expert in relation to the topic discussed in the memo. Michael Sneed, a columnist with the *Chicago Daily News,* suggests that LaPlante's lack of engineering knowledge may have been partially responsible for the extent of the damages. According to Sneed's source,

> The flood fiasco boils down to the work being placed in the hands of deputy transportation commissioner Ron Johnson and acting transportation commissioner John LaPlante, who are not engineers. When a leak occurred two weeks ago at the Kinzie site, they put the job up for bid—which takes six months. "And the only thing a small leak does is get bigger" (Sneed, May, 1992).

Perhaps Koncza should have recognized the commissioner's lack of knowledge and suggested that the city forego the bidding process and go ahead with the work. As we have already discussed, however, such a suggestion may not have been politically expedient. On the other hand, it was neither economically nor politically expedient to allow the leak to worsen.  ◆•

### Readers' Reading Processes

In addition to readers characteristics, readers' reading processes affect the writer's organization and sequence of a document. Readers engage in a three-step process—predicting, reading, and aligning—as they attempt to determine the meaning of a text.

Readers begin by *predicting* what they will read based on the situation in which they're reading, the cues they obtain from a document, their knowledge of documents, the topic under discussion, and so on. If they receive a letter, they will probably predict that they will read information from a client or customer. By glancing at the addressee, they may be able to predict the topic of the letter, or a subject line may provide them with a cue for predicting the topic. When they *read* the letter, if the first few sentences relate to the topic they predicted, their reading will be *aligned* with their predictions, and they will be able to read the text fluently, without stopping. However, if readers don't perceive that the first few sentences relate to the topic that they had predicted, then they will stop reading because the text is not aligned with their prediction. They may reread the sentences to try to find a relationship, or they may reconsider their prediction. Either way, they will not have fluency. Thus, a text needs to provide readers with accurate cues for predicting what they will read.

Tables of contents, chapter titles, and headings and subheadings provide readers with cues for predicting what they will read. In addition, a forecast at the beginning of a chapter, section, or paragraph helps readers predict what they will read. A forecast includes the organizing idea and a list of subtopics if several will be discussed. The writer must remember to discuss the subtopics in the text in the same sequence listed in the forecast so the reader doesn't become confused.

➽ Koncza's subject line in the Chicago flood memo provides the reader with a good cue for reading the memo. However, as we have already discussed, it does not provide the reader with the sense of urgency that succeeding events have indicated were needed. Furthermore, the first two paragraphs simply present information. LaPlante cannot predict the kind of a decision he will need to make until he reads the third paragraph. Once he reads the third paragraph, he may need to reread the details in the first two paragraphs to determine if Koncza's recommendations are valid. If the paragraph on page 21 had been the opening paragraph, then as LaPlante read the details, he would have been able to evaluate them in terms of the writer's recommendation. Knowing that he had to make a decision, he would have paid special attention to the information related to the consequences of his actions—that the tunnel could flood if the leak were not stopped and that if the tunnel did flood, it would affect two areas that have a major economic impact on the city. ◆●

### *Readers' Thinking Processes*

Reading is a thinking process. Readers *create* meaning from the information in a text. Comprehending a message is like putting together the pieces of a jigsaw puzzle to create a whole picture. Readers must put all of the pieces of information in a document together to create a picture of the writer's meaning. To understand a writer's message, readers relate the information in a text to their previous knowledge and experience, categorize related pieces of information into chunks, sequence information in logical order, and process information both verbally and visually.

In his memo in Figure 1.1, Koncza relies on LaPlante's previous knowledge and experience to help LaPlante understand and accept the recommendation for fixing the leak.

Related information is chunked. Koncza has kept all of the information related to the problem together and all of the information related to the solution together. Each paragraph contains a chunk of related information (e.g., the first paragraph is concerned with the damaged section; the second paragraph is concerned with the cause of the problem). The information is sequenced logically, following a problem/solution pattern. The problem section begins with the effects of the problem and is followed by the cause; the solution section begins with a general statement and is followed by the details.

### *Readers' Reading Behaviors*

The various styles and patterns readers use when they read need to be considered if writers are to be certain that readers locate and read the necessary information. Readers use a variety of reading styles, including skimming, scanning, searching, understanding, and evaluating (Huckin 1983). Readers usually simply scan a brief, routine memo or letter to pick out the specific information they need (e.g., the date and time of a meeting, the writer's specific request, the information they requested). They may skim through a brief report or scan a table of contents to learn the major areas covered in a document, or they may search through a report to locate information specifically related to their project or division. Readers will only spend the time reading to understand and evaluate a document if they need the information to work on a project of their own.

Readers also have different patterns for reading a document. They may skim or scan a document to determine if it is worth taking the time to read for understanding and evaluation. Some executives read only the executive summary before sending a report on to those under them who are responsible for the specific areas of the document. Sometimes executives read the summary along with chapters that interest them. Division managers may read only those sections specifically related to them, or they may also read the executive summary and even the introductory chapter, or they may read the entire report. If multiple readers receive a document, one may skim it and another may study it; one may read only the executive summary and another may read the appendices.

Because most readers are busy and are interested in reading only information that pertains specifically to their needs, they need to be able to locate information quickly and easily. Thus, the subject lines of letters and memoranda and the tables of contents, chapter titles, and headings and subheadings of reports, proposals, and instruction manuals help readers find the information they want. The information is even easier to find when these "jump out" of the text because they are in a larger type size, darker, in all capital letters, or in a different type style.

Readers can get brief overviews of a document without spending time on details simply by reading cover letters, executive summaries, or abstracts. Readers who want data that are not essential for an understanding of a document can find such material in an appendix. When readers assume a text is organized from most to least important, they can stop reading at any point and know they have read the most important information.

●◆ Let's consider LaPlante's reading behavior, based on the description of the situation in the newspaper article on pages 12–13. Because it is a Sunday afternoon and because he has a stack of memos to read, we can assume he will simply skim or scan a document to determine whether he needs to respond immediately or whether the matter can wait until he gets to his office. If he needs to respond immediately, he will probably scan it to determine what to do. Based on this scenario, it is a wonder LaPlante responded when he did to the flood memo, because neither the subject line nor the first paragraph give him any sense of urgency. Only because of his prior experience in reading many memos organized in this pattern does LaPlante read beyond the subject line and opening paragraph to find Koncza's warning of potential problems and his recommendation and request for preventive action. This was not the case in the Challenger disaster, in which the reader at NASA simply filed a letter that warned of potential problems with the shuttle because he didn't recognize the urgency of the problem (see the case study of the Challenger disaster). ◆●

## CHAPTER SUMMARY

There are four major categories of technical documents: correspondence, instructions, proposals, and reports. Each category comprises a variety of subcategories, and each subcategory adheres to certain conventions. Within the boundaries of these conventions, writers need to make a variety of decisions relating to content, organization and sequence of information, point of view and focus, style, graphics, visual text, and layout. Writers must also consider their legal and ethical responsibilities in determining the content of a document.

The context of a document affects the decisions writers make. The context involves the situation in which a document is written and read, the reader's and the writer's purposes, and the audience for whom a document is written. Economic, political, and social factors are aspects of a situation that affect a document.

A writer's situation involves helping readers achieve their purposes for reading a document. It may also involve multiple audiences; collaboration with writers, editors, and content specialists; adherence to the conventions of the organization; and routine communication.

Writers need to consider readers' organizational and professional communities as well as readers' reading and thinking processes and behaviors if they want readers to understand and accept their message.

## CHAPTER SUGGESTIONS FOR DISCUSSION AND WRITING

1. Someone once said, "Hindsight is wonderful." Based on all that we now know about the Chicago flood, if you had been Koncza, how would you have handled the situation once you knew about the tunnel problem?

2. Apply Moreau's explanation of ethics on pages 4 and 5 to Koncza's and LaPlante's situations. Before the disaster, what were their respective responsibilities to their supervisors and to the mayor? To the citizens? To the businesses that would be affected by a flood? Based on these responsibilities, discuss whether the project should have been put up for bid.

3. Discuss how Koncza might have revised his memo to ensure that the leak was fixed before it became a flood. Could the memo have been organized differently? If so, how? Should additional information have been included? If so, what information? Should the point of view and focus have been different? If so, what should they have been? Revise the memo to persuade the reader to fix the leak immediately.

4. Consider a problem you have had at some time with your car or with an appliance in your dormitory, apartment, home, or workplace or with a piece of equipment such as a CD player, a computer, or a video

recorder. If you had written a letter to someone to get it fixed, would you have written to the retailer? The distributor? The manufacturer? Describe the situation in which you think the recipient would have read your letter. How would your perception of the reader's situation affect the way in which you wrote your letter? How would it affect your decisions relating to the information you include, the organizational pattern you select, the sequence of the information you present, your focus, and your style? Keeping your decisions in mind, write the letter.

## Notes

Barclay, R. O., et al. 1991. Technical communications in the international workplace: Some implications for curriculum development. *Technical Communication* 38, no. 3 (August): 324–35.

Huckin, Thomas N. 1983. A cognitive approach to readability. In *New essays in technical and scientific communication: Research, theory, and practice,* edited by Paul V. Anderson, R. John Brockman, and Carolyn R. Miller. Farmingdale, N.Y.: Baywood.

Kohl, J. R., et al. 1993. The impact of language and culture on technical communication. *Technical Communication* 40, no. 1 (February): 66–79.

Mathes, J. C., and Dwight W. Stevenson. 1991. *Designing technical reports.* New York: Macmillan.

Moreau, Marc. 1993. What is ethics? In *Instructors's manual for technical writing: Contexts, audiences, and communities,* prepared by Carolyn Boiarsky. Rockleigh, N.J.: Allyn and Bacon.

Pinelli, T. E. 1993. The technical communication practices of Russian and U.S. aerospace engineers and scientists. *IEEE Transactions on Professional Communication* 36, no. 2 (June): 95–104.

Sneed, Michael. May, 1992. "Sneed." *Chicago Daily News.*

# DISCOURSE STRATEGIES

To meet their readers' needs and to achieve their readers' and their own purposes, writers use strategies that are based on the context in which they write and in which their documents are read. These strategies relate to persuasion, organization, visualization, readability, ancillary information, and navigation and location.

## PERSUASION

Almost all technical writing involves persuasion. The writer tries to persuade the reader to do something or to think in a certain way. Writers may try to persuade readers to follow instructions, to respond to a request, to approve a proposal, or to agree with the results of a research study. Sometimes writers' intentions are explicit; other times their purposes are hidden.

### Presenting Valid Claims

Regardless of whether the purpose is explicit or implicit, readers are more likely to be persuaded by arguments that reflect their previous knowledge, experience, and attitudes. For example, to persuade the Alaskan government to permit the aerial shooting of wolves, sportspersons claim that their actions are ethical because they help preserve the moose and caribou populations, a goal of the Alaskan government. If representatives of conservation organizations that want to halt the practice of aerial shooting, such as the Defenders of Wildlife, want to persuade the Alaskan government to pass measures to make aerial shooting illegal, they need to claim that research studies show the moose and caribou populations are declining because of factors other than wolves, thus appealing to officials' positive attitudes toward research studies and directly refuting their opponents' claim.

## Presenting Valid Evidence

Readers not only determine the validity of claims based on their past experience, knowledge, and attitudes, but they also determine the validity based on the evidence used to support the claims. If writers are to persuade them, then this evidence, like the claims, must be related to their prior knowledge and experience, values, and needs. The research cited as evidence to stop the aerial shooting of wolves in Alaska will need to have been conducted in Alaska if the readers are to accept it. If the research was conducted in other states, readers may counterargue that the conditions in Alaska differ and that the evidence therefore isn't valid.

Expert readers often judge the validity of a document by a writer's ability to refer to the appropriate authorities, follow the appropriate procedures, and use the technical terminology, conventional patterns, styles, and format appropriate to their professional communities. Each profession has a set of conventions that governs the way in which information is presented in writing if it is to be considered valid. These conventions are described in style guides developed and published by a community. The conventions cover everything from citing quoted information to writing an abstract to setting up a title page. The community of literary critics follows the conventions of the Modern Language Association (MLA), which requires that a writer cite an author's name and the page number of the material in referencing an excerpt or idea. The community of botanists follows the conventions of the American Psychological Association (APA) and requires writers to cite an author and the year an article or book was published rather than a page number in citing referenced material. In writing a research report, a writer needs to follow the style guide of the specific field on which the report is written if readers are to be persuaded to accept the results of the study. Many organizations print their own style guides, which can be often be obtained from their publication departments.

Before beginning to draft, writers try to determine readers' previous knowledge of their topic, experiences related to the topic, and attitude toward the topic. They then consider various arguments in an effort to determine which claims and supporting evidence would be most closely related to readers' concerns.

Let's study a document to discover the strategies a writer uses to persuade readers to think in a specific way. The text in Figure 2.1 is excerpted from a brochure designed by the Department of Energy to inform government officials and interested citizens about the Department's new program for "clean coal." The implied purposes of the brochure include persuading

Evidence →

Claim: A new coal age is upon us.

Graph has title.

Graph is labeled, numerical values are next to bars.

Claim (implied) coal is plentiful in U.S.

Evidence →

Claim: Coal provides more new energy to U.S.

Evidence →

Map and graph provide supporting evidence for the claim that U.S. coal is plentiful.

Map includes a legend.

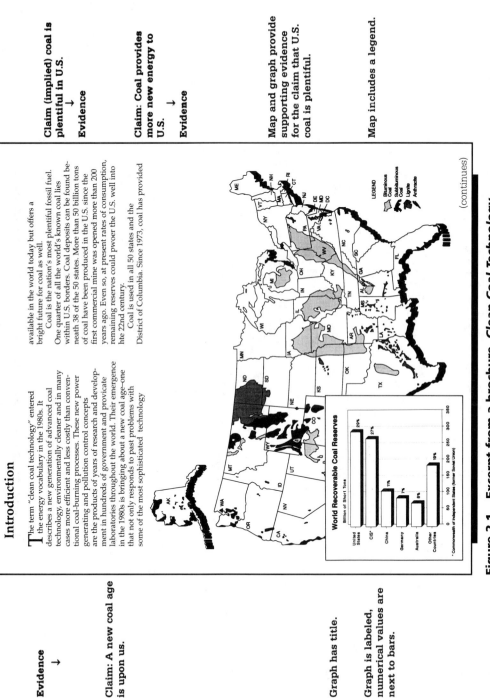

(continues)

**Figure 2.1  Excerpt from a brochure, *Clean Coal Technology* (March 1992).**

Source: U.S. Department of Energy.

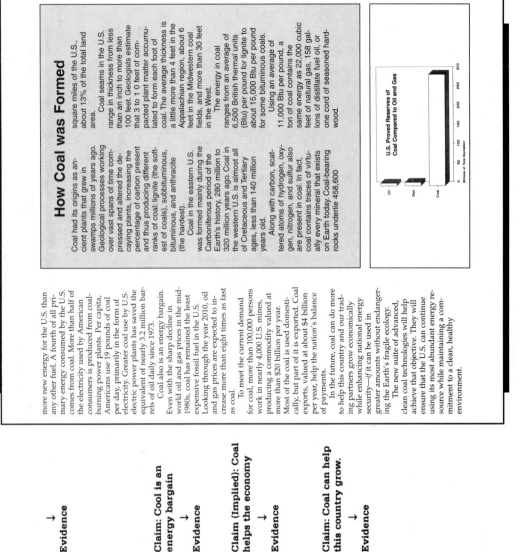

Evidence →

more new energy for the U.S. than any other fuel. A fourth of all primary energy consumed by the U.S. comes from coal. More than half of the electricity used by American consumers is produced from coal-burning power plants. Per capita, Americans use 19 pounds of coal per day, primarily in the form of electricity. Greater coal use by U.S. electric power plants has saved the equivalent of nearly 3.2 million barrels of oil daily since 1973.

**Claim: Cool is an energy bargain** →

Coal also is an energy bargain. Even with the sharp decline in world oil and gas prices in the mid-1980s, coal has remained the least expensive fossil fuel in the U.S. Looking through the year 2010, oil and gas prices are expected to increase more than eight times as fast as coal.

**Evidence** →

To meet the current demand for coal, more than 100,000 persons work in nearly 4,000 U.S. mines, producing a commodity valued at more than $20 billion per year. Most of the coal is used domestically, but part of it is exported. Coal exports, valued at about $4 billion per year, help the nation's balance of payments.

**Claim (Implied): Coal helps the economy** →

In the future, coal can do more to help this country and our trading partners grow economically while enhancing national energy security—*if* it can be used in greater amounts without endangering the Earth's fragile ecology.

**Evidence** →

The new suite of advanced, clean coal technologies will help achieve that objective. They will ensure that the U.S. can continue using its most abundant energy resource while maintaining a commitment to a clean, healthy environment.

**Claim: Coal can help this country grow.** →

**Evidence** →

## How Coal was Formed

Coal had its origins as ancient plants that grew in swamps millions of years ago. Geological processes working over vast spans of time compressed and altered the decaying plants, increasing the percentage of carbon present and thus producing different ranks of coal: lignite (the softest of coals), subbituminous, bituminous, and anthracite (the hardest).

Coal in the eastern U.S. was formed mainly during the Carboniferous period of the Earth's history, 280 million to 320 million years ago. Coal in the western U.S. is almost all of Cretaceous and Tertiary ages, less than 140 million years old.

Along with carbon, scattered atoms of hydrogen, oxygen, nitrogen, and sulfur also are present in coal. In fact, coal contains traces of virtually every mineral that exists on Earth today. Coal-bearing rocks underlie 458,600 square miles of the U.S., about 13% of the total land area.

Coal seams in the U.S. range in thickness from less than an inch to more than 100 feet. Geologists estimate that 3 to 1 0 feet of compacted plant matter accumulated to form each foot of coal. The average thickness is a little more than 4 feet in the Appalachian region, about 6 feet in the Midwestern coal fields, and more than 30 feet in the West.

The energy in coal ranges from an average of 6,500 British thermal units (Btu) per pound for lignite to about 15,000 Btu per pound for some bituminous coals.

Using an average of 11,000 Btu per pound, a ton of coal contains the same energy as 22,000 cubic feet of natural gas, 158 gallons of distillate fuel oil, or one cord of seasoned hardwood.

**U.S. Proved Reserves of Coal Compared to Oil and Gas**

Oil
Gas
Coal

Billions of Tons Equivalent
0    50    100    150    200    250    300

**Figure 2.1**   *(continued)*

readers to support the funding of the technology projects, to invest in companies using this technology, and to reinstate coal as a viable source of energy. Notice how the claim, that the "new generation of advanced coal technology is environmentally cleaner and in many ways more efficient and less costly than conventional coal-burning processes," is based on the readers' prior knowledge and attitude that coal is "dirty" and an environmental polluter. The claim is also based on readers' values for efficiency. In addition, the implicit claim, that using coal will help the American economy, is based on readers' values related to "buying American" and on their attitudes toward becoming more self-sufficient in terms of fuel supplies, toward providing jobs for more Americans, and toward reducing the deficit. To support these claims, the brochure includes statistics on the amount of coal used in the United States, the amount of oil saved by using coal, and the number of people involved in the coal industry. All of the evidence relates directly to readers' needs and values and is expressed in statistical terms that most readers consider valid. The graphics further emphasize the claims. The bar graph visually depicts the United States' potential for providing coal to the rest of the world. The map visually indicates that much of the United States can profit from the sale of coal because coal can be found in thirty-eight states.

# ORGANIZATION

The organizational pattern and sequence of information that a writer uses in communicating a message may be crucial in readers' attempts to read and understand a text. When important information is placed late in a text, a busy reader may never read it. When readers cannot follow a line of reasoning, or when their fluency is interrupted because they do not read the information they predicted, they may stop reading before they are persuaded by the writer's claims. When the organizational pattern does not reflect the purpose of a document, readers will not be persuaded to do or think as a writer wishes. If readers are to comprehend a writer's message, then they need to accurately predict what they will read. If they are to accept the message, then they must be able to follow the reader's line of reasoning. And if they are to be persuaded by the message, then they must focus on the writer's purpose and perceive the topic from the writer's point of view.

Writers use a variety of strategies to facilitate readers' fluency, comprehension, and memory and to persuade readers to respond appropriately. These strategies include ensuring that readers accurately predict what they will read, chunking related information, placing important information in a position of emphasis, and using an organizational pattern that reflects the writer's purposes.

## Ensuring Accurate Predictions

Organizing information from general concepts to specific details allows readers to predict what they will read and to understand the relationship between the details of a subtopic and the topic. A report, proposal, or instruction manual that uses this pattern begins with an introductory chapter that provides an overview of the following chapters. A chapter, letter, or memorandum begins with a paragraph that presents an overview of the following text, and a paragraph begins with a sentence that serves as an umbrella statement for all of the information in that paragraph. The overview, whether it is an entire chapter, a single paragraph, or even a sentence, indicates the main idea of the information that follows and presents the writer's point of view. When this overview introduces a chapter or section of a document, it is called a forecast. A forecast is composed of one or two sentences that summarize the topic and list the subtopics in the sequence in which they will be discussed. Thus, readers can use the forecast to predict what they will read. If the subtopics are discussed in the same sequence in which they are listed and if the details are related to the topic, then the information readers read will be aligned with their predictions.

Figure 2.2 is an excerpt from a letter written by the president of the Tennessee Sierra Club to persuade a member of the U.S. House of Representatives to answer questions concerning problems the club perceives with the proposed Superconducting Super Collider. The writer uses a modified analytical organizational pattern, dividing the letter into four categories, one for each of the four major issues. Each category follows a problem/solution pattern. The problem section is further subdivided into a cause/effect organizational pattern. The solution section is actually a set of questions, the answers to which provide the solutions to the problems.

Notice that the reader knows immediately why he is reading the letter because the purpose is stated at the very beginning. The brief paragraph on the background of the letter provides the reader with sufficient information to understand the remainder of the message. The forecast, which includes the sequence of topics to be discussed in the letter, helps the reader accurately predict the information that will follow.

The reader is also able to follow the writer's line of reasoning. A subhead provides the reader with an overview of paragraph four, the first paragraph to discuss the environmental issues. All of the following information relates to the topic specified in the subhead. In section two, *Irradiation of the Public and the Environment*, a forecast, which includes a sequence of the subtopics to be discussed, introduces the section just as the subsection *Ionizing Rays* is introduced by a paragraph that includes a forecast of the topic and the sequence of the two categories to be discussed.

SIERRA CLUB    Tennessee Chapter
10 May 1983

Honorable Bart Gordon
Representative Fifth District of Tennessee
1517 Longworth Bldg.
Independence and South Capitol St., S.E.
Washington, D.C. 20515


Dear Representative Gordon:

**States purpose at beginning so reader knows why he is reading the letter and what his purpose is in reading it (to respond with answers to questions).**

The following letter contains questions and comments about environmental problems that would result from the construction of a Superconducting Super Collider (SSC) in middle Tennessee. It is written on behalf of the Tennessee Chapter of the Sierra Club. The questions raised have yet to be addressed by local governments, the State of Tennessee, or by the U. S. Department of Energy (DOE).

**Provides background because of reader's situation. Reader may have forgotten meeting, since he has probably met with many people during the intervening time. Forecasts the main topics so the reader can accurately predict what he will read. Sequences categories from most to least important.**

To refresh your memory, I was in a group that met with you in your Murfreesboro office on 5 March 1993. At the end of the meeting, you volunteered to find answers to questions we had about the SSC. I understand other members in that group submitted questions to you some time ago.

The four environmental issues about which we are most concerned at present are the following: growth impacts on the area; irradiation of the public and the environment; disposition of the excavated limestone; and absence of a decommissioning plan. Each is dealt with below.

**Numerates items to signal reader that he is now reading about the first issue.**

**Introduces the first issue with a subhead to**

1. Growth impacts resulting from the SSC in Tennessee. The magnitude of the problem can be sensed by considering the influx of SSC work forces. The numbers were found in the

**Uses personal style (1st person). Addressed reader directly (2nd person).**

**Uses a modified analytical organizational pattern. The data is broken down into categories representing the four major problems, thus, reflecting the writer's purpose to persuade the reader to respond with the propoosed solution to these problems.**

PROBLEM

CLAIM
↓
EVIDENCE

"Not blind opposition to progress, but opposition to blind progress"

Recyclable

(continues)

**Figure 2.2    Excerpt of a letter following an analytical organizational pattern.**
Source: U.S. Department of Energy.

signal the reader he is now reading about the first issue. Underlines so reader easily sees the subhead. Repeats words from the forecast listing of the issue in the subhead so reader recognizes he is reading what he predicted.

State's brochure "The SSC for Tennessee." Initially a construction work force of 4,500, many with families, will invade the area. This will be followed in 6 years by a permanent work force of 3,000, most with families. This may involve a total of 10,000 new citizens in all. Many families of the new work force (1,000? 3,000?) will require new homes. (Also, more than 100 families who now reside in homes located over the SSC will lose them and must find new ones.)

We are told by local planners that many parts of the infrastructure, including waste disposal sites, sewerage systems, roads, schools, etc., are largely overburdened in may areas. The planners also complain of overloads. It is very expensive to upgrade and maintain the infrastructure. It is even more expensive to expand it in an environmentally sound way so as to avoid damage to local ecosystems, and to maintain open spaces, clean air and clean water. It is of interest that the State, in its brochure, has stated that "Open Spaces WILL NOT Be Destroyed." However, we have yet to find, in any of the documents, information detailing who will pay for the expanded infrastructure while preserving a clean open-spaced environment.

Each of the categories follows a cause/effect pattern. –{CAUSE}– Discusses problem in human terms. Uses emotionally-charged negative words which reader will probably perceive indicate writer is hostile to topic.

–{EFFECT}–

Speaks for Club (uses 1st person plural).

Relates argument to authorities accepted by reader's community as valid.

Chunks questions and comments in respective subcategories and provides subheadings to introduce them so reader knows what he is reading as he moves from subcategory to subcategory. Underlines subheads so reader can see them easily. Numerates questions so that reader will have no difficulty determining to what he should respond.

Questions on growth impacts of the SSC. 1. Who will plan and who will pay for expanding in an environmentally sound way, the new infrastructures necessitated by the constructions of the SSC? 2. Will the local communities be expected to realize enough funds in additional revenues to provide for the expansion? 3. Will the State be willing to underwrite an environmentally sound expansion of infrastructure? The State has already agreed to buy 16,000 acres of surface and subsurface rights and give them to DOE. 4. Or will DOE pay?

Comments. The Sierra Club believes that the growth impacts on local communities and the environment, due to SSC, is accepted by

SOLUTION (Expressed in the form of questions; i.e. the writer wants the reader to respond to him with the solutions).

**Figure 2.2**    (continued)

**Uses an outline format.**

**Helps reader accurately predict what he will read by providing a forecast of the main subcategories to be discussed and a forecast of the sub-subcategories to be discussed.**

**Provides a forecast.**

the State. With proper planning and funding the usual loss of open spaces and wildlife, characteristic of unplanned and underfunded development, can be mitigated if not avoided completely.

2. Irradiation of the public and the environment. Both DOE and the State of Tennessee have stated categorically that the SSC will be radiologically safe. As proof, both cite the exemplary radiological record of the Fermilab in Illinois. Fermilab is said to be much like what the SSC is to be in that both will have accelerators which accelerate protons and produce the same product after interacting with targets, beam abort dumps, or various ring components. The products are intense beams of subatomic particles, mainly neutrons and mesons as well as radioactive atoms, also called activation products or radionuclides. All are, or produce, ionizing radiations. In order to understand how Fermilab and SC could be as safe as touted the following publications were read: Fermi National Accelerator Laboratory Site Environmental Report for Calendar Year 1986, " Baker. Samuel I., May 1. 1927 (Fermilab 87/ 58, 1104.100, UC-41): "An Introduction to Radiation Protection for the superconducting Super Collider," Metropolis, Katherine (ED.), November 10, 1987, SSC-CR-1027....

It is clear from these two reports that there are three potential avenues by which the public and other living things may be irradiated both during the operation and following the final shutdown of the SSC. Irradiation may be by way of a. intense ionizing rays, b. airborne radionuclides, and, c. soluble or waterborne radionuclides. Background, comments, and questions about each follow.

a. Ionizing rays. There are two categories here. In the first, intense beams composed mainly of neutrons and muons are produced when the proton beams smash into the beam abort dumps or into solid targets, such as

(continues)

**EVIDENCE**

**PROBLEM**

**Cites authorities whom the reader's community considers valid.**

**CLAIM**

–{CAUSE}–

**Figure 2.2** *(continued)*

may be used in future experiments at the SSC. The second category is residual (or fixed source) radiation consisting mainly of gamma rays and given off by "activated accelerator components and shielding, mainly iron and concrete.

**Signals reader discussion will now concern first category. Allows reader to realize he will read what he predicted in the forecast. Achieves coherence by relating back to forecast.**

Consider the first category. There should be at least two beam abort dumps at the SSC, one for each beam of non-collided protons. Neutrons and muons emanating from these dumps would fan out under the I regions. Such neutrons and muons are very energetic and very penetrating. For example, at Fermilab the muons were detected at the site boundary which appears to be about three miles from the target source. The beams at SSC should be even more penetrating in that the protons will be accelerated to 20 TeV whereas those at Fermilab have a maximum energy of 1 TeV....

─{EFFECT}─

**Achieves interparagraph coherence and signals reader new discussion is starting which corresponds to the second category in the forecast.**

In the second category, residual ionizing radiation, resulting from activation products will be coming from SSC components such as beam pipes, magnets, detectors, cement rocks, cryostats, etc. The radiation of concern will be energetic and very penetrating gamma rays. The half lives of the activation products range from 54 days from beryllium to 5.3 years for cobalt-60....

**Achieves interparagraph coherence by making the section parallel to a section under the first category on growth impacts.**

_Questions on ionizing rays._ 1. What will be the individual doses of ionizing radiation to residents that live above or adjacent to the intense beams of neutrons and mesons originating from the beam abort dumps and/or targets (I and H areas)? 2. Will each individual resident in these areas be monitored continuously (such as by special film badges) for exposure to scattered neutrons, muons and their products? 3. Will above ground storage of discarded radioactive accelerator components occur at SSC?....

**SOLUTION**

**Follows outline format.**

b. _Airborne radionuclides._ Carbon 11 ($^{11}$C) and tritium ($^{3}$H) are reported to be the major airborne radionuclides at Fermilab. $^{11}$C is said to contribute the largest source of off-site ionizing radiation....

─{CAUSE}─

**Figure 2.2**  _(continued)_

Readers can also use a forecast as a frame in which to view the details of a text, just as those who put a jigsaw puzzle together use the straight-edged pieces to create a frame to help determine where to place the dark brown pieces for the barn, the gray pieces for the geese, and the light brown pieces for the horse. Readers, in trying to understand a document, must put all of the pieces of information together to create a picture. By viewing the pieces of information within the framework of the writer's point of view, readers are able to focus on the writer's purpose.

The forecast in Figure 2.2 establishes a frame for the reader. All of the information that follows relates to environmental problems. As early as the first paragraph, the reader can recognize that the writer views the aspects of the project negatively.

## Chunking Information

Because readers comprehend texts by relating various pieces of information to one another, related information should be chunked together. Thus, all background information should be in a single section or paragraph of a text. Notice in Figure 2.2 that the questions are chunked into one section and the comments into another. Notice also that all the information relating to the expected increase in population is chunked together in paragraph four and that all the information relating to open spaces is chunked together in paragraph five.

## Sequencing Information

Readers tend to remember information presented first or last in a manual, section, chapter, paragraph, or sentence. For this reason, important information should be placed at the beginning and end of these segments.

Important information within a paragraph should be placed in the independent clause of a sentence to emphasize it. Notice that in Figure 2.2, paragraph four *begins* with a claim and is followed by the supporting evidence. In paragraph five, the evidence is presented first and leads up to the claim at the *conclusion* of the paragraph.

## Reflecting Writers' Purposes

The organizational pattern should reflect the writer's purpose. If a writer's purpose is to *solve* a problem, then a problem/solution pattern may be appropriate, but if a writer's purpose is to recommend that action be taken to *prevent* a problem, then a cause/effect pattern may be more relevant. In the first situation, readers need to understand the problem in order to understand how the proposed project will solve it. In the second situation, readers need to recognize the effects the problem will cause if it is not prevented in order to understand the need for the proposed action.

In Figure 2.2 the writer's purpose is to prevent problems related to the Super Collider. Therefore, each of the four major sections uses a cause/effect organizational pattern. The writer discusses a potential problem and explains how the Super Collider may cause it. Many of the memoranda involved in the Three Mile Island accident and the Challenger disaster use inappropriate patterns, as you will discover when you read the case studies in Chapters 8 and 9, respectively.

# VISUALIZATION

Readers create meaning from visual as well as verbal information. They "look at" a document as well as read it. Because one-third of our memory system is devoted to visual images, visual representation of information can facilitate readers' thinking processes in comprehending written text. Charts and graphs, headings and subheadings, and the interplay of text and graphics on a page can help readers understand text, locate information, and read fluently.

Writers consider readers' previous knowledge, experience, and attitudes toward their topic to determine whether visual information may improve readers' comprehension of their message, enhance readers' memory of important points, and facilitate readers' search for information. Visual information assumes three forms: (1) graphics related to the verbal data discussed in the text, (2) visual markers integrated with verbal text, and (3) layout of the entire document and of the individual pages. Writers consider which form of graphics to use, which visual markers to use and where to place them, and a basic layout during planning. However, once the text is written, the decisions relating to the visual representations may need to be revised.

## Graphics

Graphics can assume either symbolic or realistic forms. Tables, graphs, charts, and diagrams provide symbolic representations of abstract or numerical concepts. Line drawings and photographs provide visual images of concrete items. Graphics facilitate readers' thinking processes and comprehension by providing additional information; indicating relationships between pieces of information; describing numerical data; clarifying, emphasizing, or summarizing numerical data; enhancing readability; improving aesthetic quality; and increasing reader interest.

### *Facilitating Readers' Thinking Processes*
Many readers tend to remember information better if they see it represented graphically. Furthermore, readers comprehend and remember concrete in-

formation better than abstract concepts. In addition, because people's perceptions are based on their prior knowledge and experience, no two people have identical perceptions. Graphics can be used to help readers perceive a mechanism, a process, or a relationship between entities in the same way as the writer does.

### Facilitating Readers' Comprehension

By replacing a reader's fuzzy image derived from a verbal description with a concrete visual representation, graphics can complement verbal text. Photographs and diagrams help readers actually *see* how phases, parts, or elements of an entity fit or work together. Charts and graphs help readers see relationships between numerical values.

The diagram on page 47 provides readers with a visual representation of a complicated process. The diagram indicates relationships between different aspects of a process and clarifies the description of the process described on page 46. If you turn back to Figure 2.1, you can see how a bar graph and a map describe and summarize numerical data. As you look at Figures 2.1 and 2.3, notice that the graphics include sufficient information for readers to understand them. All the graphics are labeled, the diagram in Figure 2.3 has a caption, and the bar graph in Figure 2.1 is titled. A legend is provided for the map in Figure 2.1, numerical values are stated on the bar graph in Figure 2.1, and the text of the report to Congress in Figure 2.3 specifically refers to the diagram. For both documents, the labels, captions, and legends are legible, the type size is sufficiently large and plain, and the words are surrounded by white space so they can be read easily. Finally, the pages are aesthetically pleasing to look at.

---

3.2.2  <u>Process Description</u>

The production of Self-Scrubbing Coal· involves the application of three different novel technologies that interact and make the process applicable to many medium- to high-sulfur eastern and midwestern bituminous coals. The integration of these technologies creates a synergistic effect, so that the efficiency and cost-effectiveness of the combination is superior to any technology applied alone. The three novel technologies are:

(continues)

---

**Figure 2.3    Excerpt from the *Comprehensive Report to Congress: Clean Coal Technology Program* (October 1992).** Source: U.S. Department of Energy.

- Advanced Coal Cleaning, which involves the application of a unique flowsheet design incorporating the use of modified, novel, fine coal classifiers and heavy-media cyclones to reduce the pyritic sulfur content of the coal by up to 90%.

- Magnetite Production, which involves the production of ultrafine crystalline magnetite for use in the advanced coal cleaning heavy-media processes by reducing hematite produced by spray roasting a solution of ferrous chloride in a restricted air environment.

- Sulfur-Capture Agents, which involves the addition of limestone-based sulfur-capture agents to the advanced clean coal product to provide increased $SO_2$ capture efficiency to as high as 70% when the coal is burned in existing pulverized coal boilers.

**Reference to diagram**

Figure 3 presents a block flow diagram of the CCCC process. The raw coal is first sized into an intermediate size fraction (1.5 in x 0.5 mm), a fine size fraction (0.5 mm x 0.105 mm) and an ultrafine size fraction (0.105 mm x 15 microns) with the fractions being processed in separate heavy-media cyclone coal cleaning circuits. The intermediate and fine-size coal cleaning circuits will be two-stage, with the capability of producing a low-gravity clean coal, a high-gravity refuse, and an intermediate-gravity middlings fraction, which contains coal particles with pyrite and other mineral matter locked in the coal matrix. The middlings fraction will be crushed or ground to a finer size to liberate the sulfur-bearing mineral matter from the coal matrix. The coal will then be processed in either the fine or ultrafine coal cleaning circuits to separate clean coal from refuse.

The effect of the cleaning process is to maximize clean coal recovery while simultaneously maximizing pyritic sulfur and ash rejection. If the composite clean coal can meet overall $SO_2$ compliance levels, then the product is ready for shipment as Carefree Coal·. If the sulfur content of the composite clean coal is too high (primarily due to the organic sulfur content), then before being blended with the other fractions, the ultrafine clean coal fraction is pelletized with enough sorbent to enable the clean coal to meet compliance levels. If this option is taken, then the coal product is called Self-Scrubbing Coal·. The reduced ash content of the clean coal allows the addition of relatively large amounts of sorbent without exceeding the ash specifications of the boiler or overloading the electrostatic precipitator (ESP).

**Figure 2.3**   *(continued)*

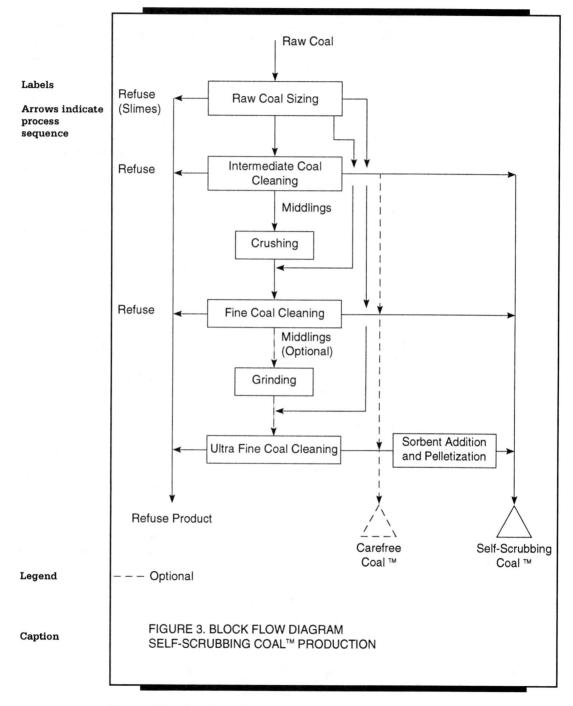

**Figure 2.3** *(continued)*

## Visual Markers

Visual markers are used to help readers comprehend information and read more fluently.

### *Facilitating Readers' Thinking Processes*

When people read a text, they often retain a visual image of the text; they can *see* the page in their mind. If important information is visually prominent, then readers will see it when they recall the visual image of the page. Important information can be made prominent by using spatial and typographical markers. Spatial markers include titles, headings and subheadings; margins and indentations; spacings; bullets and numeration; listings; and boxes. Typographical markers include capital letters and type styles, sizes, and fonts.

Spatial markers indicate related information. Titles, headings, and subheadings separate chunks of information as do margins, indentations, and spacings. Lists using numeration or bullets, and boxes enclose chunks of related data.

Typographical markers emphasize data. Verbal text that is larger or darker or that has a different type style or font stands out from the rest of the text. When spatial markers, such as titles, headings, and subheadings, are also darker or larger than the rest of the text, readers tend to remember the information they provide.

Consider how the writer uses visual text in Figure 2.4 to facilitate readers' thinking processes. Figure 2.4 contains a three-page excerpt from a report to industry, Congress, and interested citizens on the Department of Energy's Clean Coal Technology Demonstration Program. A reader should have no problem remembering the main topics, which are indicated by titles and headings, or the main subtopics, which are listed. The title and headings stand out. Their typeface is larger and darker than the text type. The two lists of information appear as separate chunks. The document designer has separated them from the main text by placing additional space above and below them and by indenting them further to the right. The bullets relate the items on the list to one another to indicate they are part of a single chunk. While the list on page two is comprised only of brief noun phrases that can be scanned quickly and are easily remembered, the items listed on page one contain a lot of information. The list is as easy to grasp as the one on the second page, because each item is introduced with a brief phrase in boldface type.

## Layout

The way in which a page of text is laid out affects readers' thinking processes and attitudes, and it facilitates their fluency.

Chapter heading. Large bold face type so reader can locate quickly.

Subheading (level 1). Bold face type. Begins with summary so readers can determine whether they want to read entire section for details.

3 columns, space between columns, ragged right margin facilitates fluency. Space between sections, visual marker so readers can *see* chunks.

Subheading serves as organizing idea so readers can accurately predict what they will read.

Lists. Items are bulleted so readers easily recognize each separate item. Items are introduced with boldfaced subheadings to help readers remember items.

Extra space around lists, hanging indentations help readers *see* lists as chunks.

Space between columns and sections, around lists, and following headings, subheadings, create airy appearance so readers are more motivated to read.

Page footer shows page number, chapter and date of publication for reader's quick location.

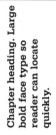

## 1. The Program

### Summary

The CCT Program has been identified in the National Energy Strategy (NES) as a major initiative whereby coal will be able to reach its full potential as a source of energy for the nation and the international marketplace. Achievement of this goal depends upon the development of highly efficient, environmentally sound, economically competitive coal utilization technologies responsive to diverse energy markets and varied consumer needs.

These technologies are being developed in the CCT Program and, when commercially available, will offer many options for addressing a wide range of energy issues that include air quality, acid rain, global climate change, power production, energy security, technology awareness, and international competitiveness.

The CCT Program, by minimizing or eliminating the characteristics of coal that have inhibited its use as a fuel, will make it one of the nation's most important strategic resources in building a more secure energy future.

### Role of the Program

The CCT Program is a technology development effort jointly funded by government and industry. In this program, the most promising of the advanced coal-based technologies are being moved into the marketplace through demonstration. The demonstrations are at a scale large enough to judge the commercial potential of the processes developed. The program seeks to make available to the energy marketplace a number of advanced, more efficient, and environmentally sound coal utilization technologies. The processes that will be commercialized are in recognition of the strategic importance of coal to the U.S. economy and the international marketplace. These are the ongoing technology development efforts that can resolve the conflict between increasing coal use and growing concern about the environmental impact of such use. The CCT Program is consistent with and directly supportive of overall NES goals:

• **Increase Efficiency of Energy Use.** Higher efficiency of energy use, where cost-effective, can help reduce energy costs to consumers, reduce energy demand, balance environmental concerns with economic development, and enhance energy security.

• **Secure Future Energy Supplies.** Even with advances in energy efficiency, the nation will likely need more energy in the future. All energy sectors will need to contribute to the achievement of our energy security, environmental, and economic goals.

• **Enhance Environmental Quality.** All energy options for the future must be sensitive to the environment—local and global. Energy and environmental goals can and must be achieved in mutually beneficial ways. Advanced technology and improved energy use practices can help to produce adequate supplies of affordable energy while maintaining the quality of the environment.

• **Fortify Foundations.** Expand the role that science and technology can play in achieving the objectives stand for energy security, economic growth, and enhanced environmental quality.

Specifically, the importance of the clean coal initiative and the CCT Program was further emphasized in the NES which noted the following:

If we as a nation are to benefit... from our enormous, low-cost coal reserves, a variety of efforts are necessary to (1) develop and demonstrate new "clean coal" technologies.

New clean coal technologies can substantially improve efficiency and reduce emissions from powerplants. Until they are proven at a commercial scale, however, their use entails more risk for utilities than conventional technologies. This additional risk could make it difficult for these new technologies to enter the marketplace quickly, especially given the tight deadlines of the Clean Air Act Amendments of 1990. The Clean Coal Technology Program, the single largest technology development program in the Department of Energy, is designed to help overcome this risk by offering the Federal Government as a financial partner in demonstrating worthy projects.

*Program Update 1991*  1-1

(continues)

**Figure 2.4  Excerpt from *Clean Coal Technology Demonstration Program: 1991 Program Update (February 1992)*.** Source: U.S. Department of Energy.

**Subheads can be found quickly by readers.**

By promoting the export of clean coal technologies the National Energy Strategy will also help other nations (especially in Eastern Europe and the developing world) to achieve common goals: a cleaner environment and less dependence on oil.

This clean coal initiative is to be accomplished on behalf of the United States, its people, and its private enterprise in the domestic and world energy markets by (1) sharing in the financial and technical risks associated with demonstrating innovative coal utilization technologies and (2) encouraging market acceptance by ensuring that key decision makers are informed as to the environmental, economic, and performance benefits of advanced coal technology options.

The innovative clean coal technologies being demonstrated offer tremendous potential as part of the solution to many complex problems that face the nation—and the world—in a rapidly changing arena dominated by energy, economic, and environmental issues. These issues include the following:

- Air quality
- Acid rain
- Global climate change
- Power production
- Energy security
- Technology awareness
- International competitiveness

*1-2    Program Update 1991*

## Air Quality

Passage of the Clean Air Act of 1970 and its subsequent amendments set the United States on a course that has committed billions of dollars to protecting the environment. The Environmental Protection Agency (EPA) has estimated that U.S. industry has expended well over $250 billion to control air emissions since enactment of air quality legislation in 1970.

Much of this has been spent by electric utilities, particularly those using coal to generate power. Since 1975, U.S. utilities have spent more than $100 billion for $SO_2$ capture alone. An average of $8 billion a year has been spent by coal-burning utilities to comply with air quality standards.

As a result of these air quality expenditures, $SO_2$ emissions have declined dramatically. Since passage of the Clean Air Act of 1970, nationwide $SO_2$ emissions from all sources have dropped more than 25 percent from their peak in 1973. Coal-fired power plants nationwide have reduced their $SO_2$ emissions by nearly 12 percent from their peak in 1977. While $SO_2$ emissions were falling, U.S. electric utilities were increasing their coal consumption by over 60 percent.

Unlike sulfur emissions, $NO_x$ emissions have risen only slightly over the last 15 years, having been generally constant at about 20 million tons per year. About half of all $NO_x$ pollution comes from automobiles and other vehicles, but coal-burning power plants also add to the problem. Power plants release about 25 percent of the total anthropogenic $NO_x$ emitted nationwide.

## Acid Rain

On November 15, 1990, Congress enacted the Clean Air Act Amendments of 1990 (CAAA of 1990). Under Title IV, emissions reduction targets were established for $SO_2$ and $NO_x$—pollutants identified long ago as acid rain precursors. These pollutants have been the focus of the CCT Program since the issuance of the *Joint Report of the Special Envoys on Acid Rain* in January 1986. Also introduced in the CAAA of 1990, are 189 toxic elements/compounds that must be addressed in developing compliance strategies for the post-1995 time frame.

The acid rain title of the CAAA of 1990 set emission reduction requirements on $SO_2$ to be met in two phases. Evidence to date suggests that the Phase I targets established for 1995 may be complied with by fuel switching; but the Phase II levels set for the year 2000 will require a technological solution. An equally important consideration is that the CAAA of 1990 set a permanent cap on $SO_2$ emissions beyond the year 2000.

To allow industry flexibility in meeting the $SO_2$ requirements, provision is made in the CAAA of 1990 to trade $SO_2$ allowances. $SO_2$ emission levels are established for each "affected source" (e.g., a power plant) by applying a formula that acts upon baseline levels existent in 1985 (or an equivalent). If an "affected source" exceeds allowed levels, measured in annual tons of $SO_2$, it must buy or have transferred within a utility the necessary allowances. Besides buying them, allowances can be generated by operating a plant below the allowed levels.

**List facilitates reader's memory.**

**List can be easily by readers.**

**Figure 2.4**   *(continued)*

**Figure 2.4** *(continued)*

### Facilitating Readers' Thinking Processes

When people read a text, they often create a visual image of the text in their minds. The page of text becomes a chunk of information that they "see" in their minds. By visually recreating a page in their "mind's eye," they can then remember information from that page. If a page is well-designed, i.e., it has an airy appearance, the various chunks of information are visually prominent, and headings and subheadings stand out as in Figure 2.4, then it is easier for readers to remember than a page that is dense with text and has no markers indicating various chunks of information.

### Affecting Readers' Attitudes

A well-designed page can affect readers' attitudes toward the content. Readers usually assume that dense print contains difficult concepts. When instructions appear dense, readers often assume the project will take a lot of time and be difficult to accomplish. Readers are likely to read a well-designed document from which they can quickly grasp the contents by scanning the headings and subheadings; they are more apt to set a dense document aside, thinking it will take too much time to read.

A balanced paged also looks easier to read than a cluttered one. When graphics are integrated with text, they should be balanced, as in Figure 2.5. Notice how the graph at the lower left-hand corner of the first page balances the box at the upper right-hand corner of the opposite page. There is also plenty of white space around both the graph and the box, so that the page does not appear cluttered.

### Facilitating Readers' Fluency

If a page is designed to correspond to readers' reading processes and behaviors, it can facilitate their fluency. When readers read, their eyes span a line of about sixty characters. By dividing a wide page into several short columns, document designers make it easier for readers to skim through text because they don't have to scan back and forth. Ragged right margins also make it easier for readers to scan text by helping them recognize which line they have completed reading. Plain type faces are easier to read, as are fonts with serifs (tails).

# READABILITY

A *readable* text is one that allows readers to read fluently through a text without stopping and which helps readers understand the writer's message easily.

Title

3 columns. Ragged right margins. Plain, serif type. Extra space balanced against space at bottom of opposite page (on page 54).

Caption

Graphic balanced against box in upper corner of opposite page.

Subhead

## Industrial Clean Coal Technologies

Coal consumption in nonutility markets has declined since 1950, even as the total use of coal expanded. Clean coal technologies could improve coal's environmental acceptability in the industrial and commercial sectors and perhaps increased its use.

Coal-burning electric power plants are no the only sources of pollutants from coal. Each year about 100 million tons of coal are used by factories and other industrial manufacturing facilities.

More than 9,000 industrial boilers today burn coal to produce steam for various manufacturing processes. Coal is also used to produce steel and cement, and it can be a valuable raw material for such products as perfume, dye, insecticides, and medicines.

Clean coal technologies are being developed for these applications. In some cases—industrial steam production, for example—scaled-down versions of utility clean coal systems, such as fluidized bed combustors, offer attractive options.

In fact, more than 100 process steam and small-power fluidized bed combustors are already operating in the U.S.; at least half of these units were added since 1983. Industrial-size fluidized bed combustors can be found today in paper mills, food processing plants, tire manufacturing factories, hospitals, and district heating systems.

Research is under way to further improve these systems, making them more practical and economic for smaller businesses and perhaps someday even for apartment buildings and homes.

But burning coal is not the only way to use this abundant resource.

Steelmaking and cement production are tow other applications that are being outfitted with special types of clean coal technologies.

### Steelmaking

An important use of coal in the industrial sector is to produce coke, which is used in smelting iron ore to make steel. Coke is made by a process called "carbonization" in which a blend of two or more bituminous coals is baked in the absence of air. The coke is then combined with iron ore and limestone in a blast furnace. The resulting carbon monoxide and heat reduce the ore to produce molten pig iron, essential to steel production.

The existing 30 coke oven plants in the U.S. emit about 300,000 tones of sulfur pollutants each year, along with airborne toxic chemicals such as benzene and other hydrocarbons. Many

(continues)

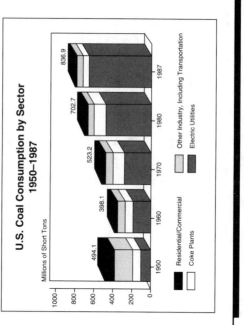

**U.S. Coal Consumption by Sector 1950–1987**

Millions of Short Tons

| | 1950 | 1960 | 1970 | 1980 | 1987 |
|---|---|---|---|---|---|
| | 494.1 | 398.1 | 523.2 | 702.7 | 836.9 |

Residential/Commercial
Coke Plants
Other Industry, Including Transportation
Electric Utilities

**Figure 2.5 Two pages from a brochure, *Clean Coal Technology* (March 1992).** Source: U.S. Department of Energy.

Space around title carried across this page.

Box.

Box title.

Extra space.

## Fill'er Up . . . with Coal?

America' transportation sector is the most vulnerable part of its energy economy. Nearly two-thirds of the oil consumed in the U.S. is burned in cars, trucks, trains and other vehicles. Alternatives such as compressed natural gas and electric vehicles are being tested in some urban areas, but a major shift from liquid transportation fuels will likely be slow.

Could coal be used instead of oil? Prospects for changing coal into a substitute fro petroleum have long intrigued coal chemists and engineers. The technology to accomplish this chemical transformation exists, but the drawback has been economic. Coal liquids historically have been too expensive to complete with natural crude oil.

Now that may be changing. During the 1980s, major advances were achieved in coal *liquefaction*. Scientists learned that by separating the coal-to-oil process into multiple

stages, they could squeeze 30% more liquids from the same amount of coal. They learned how to reduce construction and operation costs, how to operate at lower temperatures and pressures, and how to improve the performance of catalysts that accelerate the chemical reactions. They learned how to produce a higher quality liquid product that would be more valuable than a comparable quantity of raw crude oil.

The result? Today liquids can be produced from coal as low as $35 per barrel—almost half the cost of 15 years ago. In the future, new coal pretreatment steps, better solvents and improved process designs could lower costs to $25 per barrel. At these costs, the prospects for fueling tomorrow's vehicles with clean burning coal liquids could become much brighter, and America would have another option for reducing its need to imported oil.

coke ovens have no desulfurization equipment, while others use gas treatment processes that are more than 30 years old and rely on a cumbersome series of steps.

Modern-day clean coal technology can make coke plants both cleaner and simpler. In one clean coal technology project, ammonia will be captured from coke oven gas and used to scrub hydrogen sulfide from the gas. Then, using special catalysts, the ammonia is chemically changed into nitrogen and water vapor, and the hydrogen sulfide decomposed into elemental sulfur, a salable byproduct. More than 80% of the hydrogen sulfide and 98% of the ammonia can be removed, along with benzene and other pollutants.

### Cement Making

Cement is made by heating a mixture if limestone, clay, sand, and other minerals in a kiln until they fuse. More than 250 cement kilns have been built in the U.S. and along the St. Lawrence River in Canada. Because most of these kilns burn coal, they emit about 230,000 tons per year of $SO_2$.

One innovative clean coal technology uses the waste products from a cement kiln to reduce air pollutants. When the minerals in a cement kiln are heated, they release vapors containing sodium and potassium salts. These vapors later condense as a fine dust. Usually this dust had to be disposed of, but the clean coal technology can use it to absorb sulfur from the kiln's exhaust gases.

Sulfur-laden kiln gases are bubbled through a slurry made of the dust and water. Chemical reactions in the slurry remove at least 90% of the

sulfur pollutants, producing potassium sulfate which can be used as fertilizer. Additional process steps recover solid calcium products that can be reused in the kiln.

The result is a cement kiln that emits virtually no waste products other than distilled water.

**Figure 2.5**    *(continued)*

During the revision phase, writers review their texts to determine whether they have used strategies that create readable text. If they have not, they incorporate them. These strategies involve coherence (how the parts of a text hold together), sentence structure, and language (the vocabulary used).

## Coherence

When documents are coherent and cohesive, readers can follow the writers' line of reasoning and recognize the relationship between pieces of information. They can understand when information contrasts with other information, when information follows other information in a sequence, when information is the effect of previous information, and when information is expanding on previous information.

For a reader to perceive a document as a single, unified message, it needs to be coherent at three levels: global (between chapters and sections), interparagraph (between paragraphs), and intraparagraph (within paragraphs).

### *Global Level*

After they have completed a document, writers use a variety of techniques to hold the different parts together. These include the following:

1. Providing a table of contents that lists the chapters, headings, and first-level subheadings.

2. Providing section tables of contents with a more detailed list of chapters, headings, and subheadings if the document is long.

3. Beginning a chapter with chapter organizers that list or forecast the main topics of that chapter.

4. Paginating the text.

### *Interparagraph and Intraparagraph Level*

Strategies for creating interparagraph coherence are similar to those for creating intraparagraph coherence. To create a coherent document, writers must recognize how the various ideas they present are related to each other and to their main purpose. They need to begin to consider these relationships at the same time they begin to gather their data. If they are taking notes, they should indicate in a margin that a piece of information contrasts with another piece of information or that it serves as supporting evidence for it. Then, before they begin to draft, they need to sort their information, placing all related information together in categories and subcategories. For example, in writing a proposal, they should sort all of the information related to the problem that needs solving into a problem category, with the information related to each aspect of the problem serving as a separate subcategory.

But it is not enough for writers to simply draft these categories and subcategories. Writers need to signal the relationships among the various pieces of information and between the pieces and their purpose to their readers. They often use these signals subconsciously as they write. However, if they fail to do so, they need to insert them during revision. These signals include the following:

1. Repetition, restatement, or summarization of words, phrases, and concepts to indicate that the same topic is still being discussed.

2. Transitions and conjunctions to indicate that the relationship between the ideas is cause and effect, sequence, contrast, etc.

3. Omniscience to help readers understand the background or effect of an action.

4. Location of text in time or place so readers can recognize when changes occur.

Let's look at the excerpt from the *Comprehensive Report to Congress on the Clean Coal Technology Program* in Figure 2.6. Readers can put the pieces of information together easily since related information is placed in chunks. The three steps of the plan are chunked together as items in a list. All information concerning the bench tests is chunked together in the paragraph at the bottom of the first page just as all information relating to process optimization tests is chunked together in the first paragraph on the next page. Furthermore, all information relating to testing is kept together; the paragraphs about validation testing follow the paragraph about the optimization tests, which follows the paragraph about the bench tests.

Readers can quickly realize they are reading about the same topic as they move from paragraph to paragraph. In Figure 2.6 the names *Genesis* and *Duquesne* as well as the words *commercial* and *technology*, used in paragraph one, are repeated in paragraph two to provide interparagraph coherence. Furthermore, the concept of *cosponsor* and *partial ownership* discussed in paragraph one are restated as *joint venture* in paragraph two. Notice that paragraphs three, four, and five all repeat the word *test*.

**Paragraph #1**
**Background**

**List #1**
**Validation Steps**

**Paragraph #2**
**Forecast**

**Paragraph #3**
**Bench Text**

3.2  <u>Self-Scrubbing Coal· Technology</u>

3.2.1  <u>Overview of Technology Development</u>

In 1988, Genesis Research Corporation approached Duquesne Light to cosponsor development of an $SO_2$ emissions control technology, which Genesis had conceived. Duquesne Light contacted CQ Inc. (CQ), then the Electric Power Research Institute's (EPRI) Coal Quality Development Center, to perform an independent review of the proposed technology. A favorable review led to Duquesne Light's support of the effort in return for partial ownership of any commercial technologies that might result from the work. Duquesne Light and Genesis agreed on a three-step project for validation of the technology for Duquesne Light applications. The three-step plan included:

· Verification of Genesis's theories on fine-coal cycloning using ultrafine magnetite. This effort was conducted at CQ's demonstration-scale facility using a modified Krebs Heavy-Media Cyclone.

· Semi-continuous, commercial-scale testing of an integrated fine coal sizing, desliming, heavy-media cycloning, and media recovery unit, using specifically designed cycloning circuits installed in CQ's demonstration plant, including submitting samples of fine clean coal for pelletizing tests.

· Technology feasibility case studies.

Based on promising results from the experimental work and favorable economics evaluations, Duquesne Light and Genesis formed a joint venture (Custom Coals International), whose mission is to commercialize the technology. The following discussion summarizes the results from the three steps of process development and illustrates the readiness of CCI's technology for commercialization.

The initial bench-scale tests in 2-inch cyclones achieved greater than 90% rejection of the pyritic sulfur in Sewickley Seam coal and greater than 90%

(continues)

**Figure 2.6  Excerpt from *Comprehensive Report to Congress: Clean Coal Technology* (October 1992).** Source: U.S. Department of Energy.

Readers can also understand how each sentence relates to a topic within a paragraph. Sentence two in paragraph one repeats the words *Duquesne* and *technology* and restates the phrase *$SO_2$ emissions control technology* as *the proposed technology*. Sentences three and four also repeat

*Paragraph #4*
**Process Optimization**
**Tests**

retention of the coal's heating value.  Furthermore, the coal's $SO_2$ emissions potential was reduced from about 8 lb/million Btu to about 2 lb/million Btu.

In 1989, DOE conducted a series of process optimization tests on cyclone separations in ultrafine magnetite media.  The DOE results confirmed the Genesis/Duquesne Light results. The Genesis/Duquesne Light results also compared favorably with results obtained by Process Technology, Inc. using true heavy liquids (mixtures of methylene chloride and Freon).

*Paragraph #5*
**New Work**

Based on these promising bench-scale results, Genesis and Duquesne Light decided to move ahead with commercial-scale validation of the technology.  This work included testing larger diameter cyclones (6 and 10 inch), as well as integrating additional unit operations (i.e., coal sizing, desliming, magnetite production and recovery, coal dewatering, and coal pelletizing) into the testing scope.

*Paragraph #6*
**Test Validation**

The performance of key process steps of the CCCC process was validated at semi-commercial scale in late 1989 and mid-1990. The majority of the testing was again performed at CQ's facilities.  The key process steps which were validated include:

*List #2*
**Steps in Test Validation**

· Preparing ultrafine magnetite by more efficient methods than in earlier testing

· Desliming the less than 15-micron, high-ash material from the raw coal in a 10-inch diameter classifying cyclone

· Separately beneficiating two size fractions of the raw coal fines (600 x 100 microns and 100 x 15 microns) in a 10-inch diameter dense-media cyclone

· Separating and recovering the ultrafine magnetite from the various clean coal and refuse streams

· Dewatering the finest size fraction of clean coal (100 x 15 micron) in a high-G centrifuge

**Figure 2.6**   *(continued)*

the word *Duquesne* as well as the word *technology*. Sentence three also repeats the word *review* used in sentence two.

Furthermore, through transitions, readers can discover relationships between pieces of information. Sentence two in paragraph one uses the term *then* to indicate the next point in a sequence. In paragraph three, the term *furthermore* indicates additional information, as does the phrase *as well as* used in paragraph five.

Figure 2.7 shows a model of a coherent document in which all of the parts are related to create a single message.

## Sentence Structure

Sentence structure can affect readers' fluency and comprehension. To facilitate readers' reading processes, you should follow these guidelines.

- Include only a few concepts in a sentence.
- Follow the word order expected by readers, based on their previous experience with sentence structure.

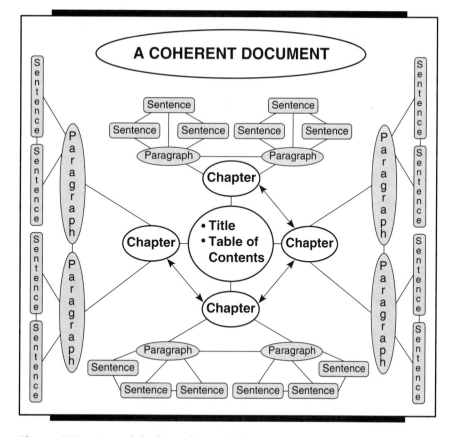

**Figure 2.7   A model of a coherent document.**

- Emphasize major concepts in independent clauses and at the beginning and end of a sentence.

When a sentence is dense, containing too many ideas, it becomes difficult for readers to understand the relationships between them. Read the following sentence.

In 1988, when Genesis Research Corporation approached Duquesne Light to cosponsor development of an $SO_2$ emission control technology which Genesis had conceived, Duquesne Light contacted CQ Inc. (CQ), then the Electric Power Research Institute's (EPRI) Coal Quality Development Center, to perform an independent review of the proposed technology.

This sentence contains at least eight propositions:

1. Genesis approached Duquesne.
2. The approach related to the cosponsoring of a project.
3. The project involved $SO_2$ emission control technology.
4. The company which conceived the technology was Genesis.
5. Duquesne contacted EPRI.
6. Duquesne requested an independent review.
7. The review concerned the $SO_2$ technology.
8. Duquesne requested the review when Genesis approached it.

By breaking the sentence into two separate sentences, the information can be made easier to read and comprehend. The preceding sentence is composed of two major concepts, each contained in a clause: (1) Genesis Research Corporation approached Duquesne Light Company (dependent clause) and (2) Duquesne contacted Electric Power Research Institute (independent clause). In the following example, the writer uses a separate sentence to discuss each concept. As you read them, determine whether the two are easier to understand than the preceding single sentence.

**Sentence #1**   In 1988, Genesis Research Corporation approached Duquesne Light to
prep.    phrase          subject                    predicate    object    infinitive
                              independent clause

cosponsor development of an $SO_2$ emission control technology, which
phrase        object of infinitive  prepositional phrase

**Sentence #2**   Genesis had conceived. Duquesne Light contacted CQ Inc. (CQ), then the
subject    predicate        subject        predicate    object    appositional phrase
dependent clause

Electric Power Research Institute's (EPRI) Coal Quality Development
Center, to perform an independent review of the proposed technology.
infinitive phrase   object of infinitive      prepositional phrase

These sentences, which appear in the first paragraph in Figure 2.6, also follow the word order readers expect (i.e., subject, verb, object). Notice how much more difficult it is to read these sentences when the words don't follow as expected.

**Expected order** Genesis Research Corporation which in 1988 had conceived the develop-
subject                      subject prep. phrase   predicate        object
independent clause                        dependent clause

ment of $SO_2$ emission control technology approached Duquesne Light.
prepositional phrase                       predicate     object
independent clause

**Words are out of** Duquesne Light contacted to perform an independent review of the
**expected order** subject          predicate      infinitive       object of infinitive  prepositional phrase
independent clause

proposed technology CQ Inc. (CQ), then the Electric Power Research
object            appositional phrase

Institute's (EPRI) Coal Quality Development Center.

If you read through the remainder of Figure 2.6, you will notice that sentences are not overly dense and that they follow the word order expected by readers. The average sentence is about twenty-two words, which is appropriate for a technical document.

During revision, writers need to review the density and structure of their sentences. If they find sentences that are too dense or which change the normal word order, they need to consider revising them. They can improve a dense sentence by creating several shorter sentences, each of which deals with only a few of the ideas in the denser one, and they can improve a sentence's complicated word order by rearranging it to follow the sequence expected by readers.

## Language

When readers are familiar with the language used in a document, their fluency and comprehension is facilitated. There is a saying, "One person's jargon is another person's technical terminology." A reader who is an expert expects to read the appropriate technical terms in a document. On the other hand, a novice, who is unfamiliar with the terms, won't understand the message unless the terms are defined. For example, nuclear engineers are familiar with the term *half scram* and expect writers to use the term in discussions of safety procedures. However, novices reading a brochure discussing procedures for maintaining plant safety will not understand a discussion that refers to a half scram. The writer will need to eliminate the phrase altogether and use a substitute, such as *a method for slowing down a nuclear reaction*, or the writer will need to explain the term in a parenthetical or appositional phrase. When unfamiliar words are used in a text, readers need them defined in terms of their past experience and knowledge. For example, the difference between a half scram and a full shutdown might be explained to a writing consultant in the following familiar terms:

> It's like the difference between a half-baked potato and half of a baked potato. A half scram is half of a baked potato. Some of the rods are inserted to lower the core's reactivity. In a full scram, all of the rods are lowered to stop the nuclear reaction.

Technical terminology is not the only terminology that causes readers to have difficulty reading and comprehending a text. Polysyllabic, Latinate terms that are not part of a reader's everyday vocabulary can cause readers to trip over words. Readers find familiar terms such as *help* and *deviation* much easier to read than synonyms such as *alleviation* and *anomaly*. In addition, because readers do not have the time or inclination to look up unfamiliar words, such as *anomaly,* they may not even understand or they may misunderstand a message. Language should not be used to impress readers but to express ideas. Readers are "impressed" when they can read a message quickly and easily because it is clear and simple.

Before drafting, writers determine who their readers will be. Then, considering readers' organizational and professional communities and social and personal characteristics, writers determine the most appropriate language to use.

The text in Figure 2.8 was included in a box in the brochure on clean coal technology that was developed for novices. Readers can quickly discover what a scrubber is; the information is appended at the end of the first sentence in the second paragraph. Furthermore, readers can easily picture the sludge resulting from a wet scrubber because it is compared to a commonplace object, toothpaste. They can also understand the consequences of sludge production because a description is included parenthetically at the end of the first paragraph in the wet scrubber section. The numerical data are also easy to comprehend. Although readers may not understand the significance of $300 per kilowatt of capacity, they can understand $150 million for a typical 500-megawatt plant, which is explained parenthetically. And although readers may not think a 1 percent to 2 percent reduction in electrical output seems like much, the parenthetical explanation following the numerical data sets them straight.

**The "Box on the Back End"**

Until clean coal technologies emerged, the flue gas scrubber, developed in the 1960s, was the only commercial technology capable of achieving the 70% to 90% $SO_2$ reduction required under the 1977 Clean Air Act amendments.

Scrubbers are actually complex chemical plants—separate gas processing facilities installed at the "back end" of a power plant leading to its smokestack.

As of 1988, 146 scrubbers have been installed at 82 of the 370 currently operable U.S. coal-fired power plants. Installation and operational costs for these scrubbers currently exceed more than $17 billion.

There are two categories of conventional scrubbers—wet and dry. Both remove only $SO_2$; neither reduces $NO_x$ emissions.

**Wet Scrubber**

Flue gases from the combustion of coal are sprayed with a slurry made up of water and an alkaline reagent, usually limestone. The $SO_2$ in the flue gas reacts chemically with the reagent to form calcium sulfite and calcium sulfate in the form of a wet sludge (having the consistency of toothpaste). Over its lifetime, a 500-megawatt coal-fired power plant will produce enough sludge to fill a 500-acre disposal pond to a depth of 40 feet (often creating a waste disposal problem).

Wet scrubbers are effective—removing 90% or more of the $SO_2$—but they are expensive to install, costing as much as $300 per kilowatt of capacity (or $150 million for a typical 500-megawatt plant). They consume 5% to 8% of a power plant's thermal energy to run pumps, fans, and a flue gas reheat system, thereby reducing electricity output by 1% to 2% (a significant reduction for a utility). They require large amounts of water, typically 500 to 2,500 gallons per minute for a unit of 500 megawatts.

**Dry Scrubber**

In a dry scrubber, the reagent slurry (usually lime) is injected in a finely atomized form, which is why these devices are also known as spray dryers. The droplets evaporate in the hot gas, leaving only dry particles for collection as waste. Although simpler in concept than the wet scrubber, the dry scrubber has not been as successful on high-sulfur coal due to the increased amounts of expensive reagents required to reduce $SO_2$ by 90%.

Extended definition of a scrubber, based on its function and location.

Explanation in terms of an object (a plant) with which readers are familiar. Explanation of significance of numbers.

Description uses an analogy to an object with which readers are familiar.

**Figure 2.8    Boxed excerpt from a brochure, *Clean Coal Technology* (March 1992).** Source: U.S. Department of Energy.

Language can affect readers' attitudes. People attach emotional responses to words. They react to terms such as *failure, lack,* and *problem* negatively, but they respond positively to words like *challenge* and *opportunity.* Technical documents should maintain a neutral tone.

When writers consider their readers' knowledge and experiences in relation to their topic, they also consider readers' attitudes toward their topic. If readers may respond unsympathetically, writers must be especially careful to avoid using terminology that may carry negative connotations. During revision, writers often look for emotionally charged terms or phrases, even if they expect readers to respond favorably to their topic. They usually change such terms or phrases to more neutral ones as a precaution against a potential negative response.

# STYLE

Present-day technical writing style is generally impersonal, neutral in relation to a topic, and succinct. Except for writing that is specifically aimed at the general public, it seldom includes narrative descriptions, philosophical digressions, or metaphorical language. Discussions that place the topic in its historical and social context, if included at all, are usually limited to introductory sections. The result of this type of writing is that readers acquire information about a topic without any humanistic base in which to place it. A memo issued during the Third Reich in Germany (see Chapter 3) demonstrates the extreme to which this style can be taken. Although such extremes may not occur today, environmental impact statements and other documents appear devoid of the human context in which they occur. The Morton Thiokol memo (see Part III) warning that a catastrophe could occur unless the problems with the O-rings are solved never explicitly links the catastrophe to the loss of human life.

There is some attempt today to provide a humanistic context so that readers can recognize the human elements involved in the various technical documents being written. However, at present, the trend is limited to a few writers.

The style for technical documents has evolved over the last 2,000 years. During the earliest years, when such writers as Vitruvius and Thucydides wrote, the social and political context of a document was interwoven with the scientific aspects. Writers included theoretical discussions to justify their ideas as well as to place their concepts in historical perspective. Thucydides' report on the plague in the fifth century B.C. is part of a historic treatise on the Peloponnesian war. He discusses the plague within the context of the war as a contributing factor to Greece's losses. And Vitruvius, in his directions on the building of Rome, not only provides instructions for constructing the walls of Rome, but also discusses the education of architects, which he claims should include instruction in history, philosophy, music, and medicine so they are wise and judicious.

During this early period, personal opinions were often interjected into technical discussions. In describing the course of the plague, Thucydides inserts his own thoughts concerning the ethical dilemma people faced in nursing the ill, knowing that to do so could cause them to catch the disease.

Because technical writing then incorporated writers' philosophical concepts along with their technical knowledge, the style was often personal: The writer would use first person and speak directly to the reader. Scientific discussions of mechanisms or research findings, even instructions, often assumed a narrative style as the author discussed the situation involved.

This style persisted through the nineteenth century. As late as 1859 William Rankine introduced his manual on the steam engine with an introductory chapter that discussed the nature of progress in the mechanical arts. In this century, however, technical writing tends to be impersonal.

Prior to drafting, writers determine the style for their document by considering their purposes, their readers' positions in the organizational hierarchy and cultural backgrounds, and the humanistic aspects of their topic. In most cases, writers select the conventional impersonal, neutral, and succinct style. However, as they consider their purposes, they may recognize that readers' responses are of utmost importance and that there are humanistic consequences. When these conditions exist, writers need to consider including less conventional style elements, such as narrative, descriptions using adjectives, metaphorical language, and philosophical digressions in an effort to persuade readers to respond appropriately to their message.

# ANCILLARY INFORMATION

Because multiple readers with different needs, purposes, and reading behaviors may read a single document, documents often include preview and supplemental sections. After writers have completed the main text of a document, they consider readers' reading behaviors to determine the types of ancillary documents readers need. Ancillary documents may provide readers with a preview of the information in a document so they can determine whether to read it, which parts they want to read, and whether they want others to read it. Ancillary documents may also provide supplemental information that readers do not need to know to understand the document or to be persuaded to respond appropriately, but that some readers may want to know.

## Preview Documents

Preview sections include cover correspondence or letters of transmittal, title pages, executive summaries or abstracts, and tables of contents and lists of figures and tables. These preview sections provide primary readers with a summary of the information to help them determine whether they need to read an entire document. Often a CEO reads only a cover letter and executive summary before sending a report to the vice president whose area is discussed in the report.

Cover correspondence provides readers with a brief explanation of the attached document to help them determine whether to peruse the document or send it to another reader. The letter includes the purpose for the document, a very brief summary of the topic to be discussed, and any personal comments concerning the document that might interest the reader or persuade the reader to accept a recommendation or support a proposal.

A title page helps readers orient themselves to the document. It provides them with the topic to be discussed, the date the document was written, the author and sponsoring agency submitting the document, and the name of the primary reader.

Executive summaries and abstracts provide readers with overviews of a document. A busy executive may read only an executive summary before sending the document to someone else to study it.

## Supplemental Documents

Supplemental sections include appendices, glossaries, and reference lists. These sections can contain information that primary readers may not need but that secondary readers do. They can also contain information that readers do not need to understand a message but that writers want their readers to know or that the readers, themselves, want to know. As you study the manuals, proposals, and reports in Part II, you will have an opportunity to see how writers include ancillary information with a document.

# NAVIGATION AND LOCATION

Because readers scan, skim, and search a text, they have to be able to navigate through a document and find information they need quickly and easily. They need to be able to locate the information at both the global and local levels. After writers have completed the main text and the ancillary documents, they need to provide readers with navigation and location aids.

## Global Level

At the global level, tables of contents, lists of figures and tables, and indices provide readers with the location of the information they want. Page numbers and headers that indicate the title of a chapter or section then help readers navigate through the pages of a report or manual to arrive at the information they want.

## Local Level

At the local level, visual markers provide location information. Readers can quickly spot chapter titles and headings and subheadings, especially those that stand out from the main text because they are larger, darker, or differ in style or font. Typographical markers, such as boldface type, and spatial markers, such as boxes, also help readers who are scanning a chapter or page. As you read the instruction manual in Chapter 4, the proposals in Chapter 5, and the reports in Chapter 6, you will have an opportunity to see how writers use these strategies.

## CHAPTER SUMMARY

### Persuasion

1. Appeal to readers' personal, social, and corporate needs, attitudes, and values to persuade them to do something or think in a specific way.

2. Base claims on readers' prior knowledge, experience, and attitudes.

3. Use evidence readers consider valid, based on their prior knowledge, experience, and attitudes.

4. Base evidence on attitudes and values considered valid by readers' communities.

### Organization

1. Organize and sequence information according to the readers' and your own purposes.

2. Adapt the point of view to that of the readers. Relate the focus of a document to the readers' purpose.

3. Help readers predict what they will read by using titles, tables of contents, headings and subheadings, and forecasts. Then align the information following these cues. Make certain information is sequenced in the same order it is presented in the forecast.

4. Chunk related information together.

5. Place main ideas in titles, headings, and subheadings, and place the organizing idea at the beginning of a document, section, chapter, or paragraph.

6. Place the main idea in the independent clause of a complex sentence.

7. Organize information efficiently for busy readers. Order text from most to least important by placing the most important information at the beginning of a document, section, or paragraph.

8. Begin with background, explanations, or descriptions to familiarize readers with a topic if they aren't knowledgeable in a subject.

9. Place the main organizing idea at the beginning or end of the introduction. The main organizing idea should reflect the purpose and should summarize all of the information that follows. There should be one and only one main organizing idea.

10. Include a forecast in the introduction. A forecast should include
    - the main organizing idea.
    - the sequence of categories if appropriate.

11. Sequence categories in a logical order, using one or a combination of the following organizational patterns: alphabetical, chronological, analytical, sequential, comparison/contrast, problem/solution, cause/effect, most/least effective, most/least important.

12. Break up long lists into manageable chunks.

### Visualization

1. Use graphics (i.e., tables, graphs, charts, and illustrations) to help readers visually as well as verbally comprehend your message.
   - Use tables to provide specific numeric data efficiently.
   - Use graphs to compare quantitative aspects of one or more topics.
   - Use charts to indicate hierarchical, directional, or time relationships.
   - Use illustrations to depict a mechanism or object or a process, procedure, or theory.

2. Make the text visually attractive.

3. Introduce text with titles, headings, and subheadings.
   - Include the organizing idea of a chunk as part of the title, heading, or subheading.

4. Indicate organization, hierarchical order, and sequence with headings and subheadings, typographical markers, numbering systems, indentations, and numeration.

5. Emphasize information using type styles, boxes, and lists.

6. Separate extremely important or extraneous information from the main text by placing the information in a box.

7. Use legible typography.
   - Use a type size that is easy to read. If you are using a typewriter or computer, use 10- or 12-point type.
   - Use a typeface with serifs (tails).
   - Do not use all capital letters for continuous text.
   - Do not use fancy typefaces. Roman type is very readable.
   - Use unjustified right margins for single-column pages; justify the right margin for multicolumn pages.
   - Keep line length within the readers' eye span.
     - Do not exceed forty characters and spaces of type.
     - Use columns if type is small or paper is wide. Two columns of approximately $3\frac{1}{2}$ inches on a paper $8\frac{1}{2}$ inches wide is a good width.

8. Create the impression of airiness by providing sufficient spacing in margins and between sections, paragraphs, and lines.

9. Keep paragraphs relatively short.

10. Help readers *see* chunks of information.
    - Use hanging indentation.
    - List items of information and steps in procedures and instructions.

## Readability

1. Use terminology, sentence structures, and document formats that are familiar to readers.

2. Use language that readers easily understand.
   - Use as few technical terms as possible, and define those you do use when writing for a novice.
   - Include details and define terms if readers may be unfamiliar with the information. Relate definitions to a document's purpose.
   - Avoid excessive or pompous language. Do not use polysyllabic words if a monosyllabic word is just as good.

3. Use sentences that readers can read fluently.
   - Avoid complicated sentence structure, but don't write simplistically.
   - Keep the text to an average of fifteen to twenty-five words per sentence.
   - Shorten sentences if you run out of breath reading them aloud.
   - Do not interrupt the normal sequence of sentence parts.

- Do not interrupt linguistic units.
- Vary the types of sentences (simple, complex, compound), sentence constructions (loose, periodic, cumulative), and placement of clauses and phrases.
- Do not overload sentences with prepositions.

4. Use neutral terms to avoid offending readers.

5. Keep documents short and to the point for busy readers.

6. Choose the language and tone carefully so you do not offend readers.

7. Indicate relationships between paragraphs and the main topic.
   - Indicate the relationship between each paragraph and the paragraph preceding it. Signal this relationship in the first sentence of each paragraph.
   - Indicate the relationship between the sentences in a paragraph and the main topic or organizing idea of that paragraph.
   - Indicate the relationship between each sentence and the one preceding it in each paragraph.
   - Use redundancy and restatement to indicate global, inter- and intraparagraph relationships.

8. Use redundancy and nouns rather than pronouns to provide unambiguous references to previous text.

9. Use transitions to indicate the relationship between ideas, especially if the relationship is sequential, contrasting, causal, or contingent.

10. Keep comparable items in a list and compound grammatical units (nouns, verbs, propositional phrases) parallel.

11. Keep elements that are being compared or contrasted parallel.

12. Emphasize aspects of your topic according to your purpose.

13. Use technical terminology for readers in your discourse community. Be careful that the language is not industry- or plant-specific and that all persons in the community define a term the same.

14. Interpret numerical data for readers.
    - Discuss data in terms of readers' needs and their numerical knowledge.

- Use percentages rather than whole numbers.
- Round off numbers when writing for generalists and novices.

### Ancillary Information

1. Provide cover letters or letters of transmittal, abstracts and/or executive summaries, and tables of contents with long reports so primary readers can get an overview of the main points without reading the entire document.

2. Include a glossary if a document includes many technical terms that some readers may not know.

3. Include a list of references if written sources are cited.

4. Provide an appendix to include information that is extraneous to the primary reader's needs, but which other readers may want or the primary reader may like to read.

### Navigation and Location

1. Create tables of contents, including headings and subheadings, and listings of tables and figures so readers can locate information that interests them. If a document is long, use section and chapter organizers, listing the contents that follow.

2. Number pages prior to main text with roman numerals. Use arabic numbers for the main text.

3. Use headers to indicate chapter titles on each page of a chapter so readers know where they are as they skim through a document.

---

## CHAPTER SUGGESTIONS FOR DISCUSSION AND WRITING

1. Read the text in Figure 2.3. Discuss its "readability." What terms don't you understand? What familiar terms could the writer have used instead? How could the writer have written the text so that you would understand those terms? Do the sentences flow smoothly? Which sentences do you think are too dense? In which sentences do you find that the word order is not as you expected? Rewrite the text so that readers like yourself can understand it, so that sentences are not too dense, and so that the word order follows an expected sequence.

2. Study Figure 2.4. Is it coherent? How does the writer help the reader make accurate predictions? Discuss how the various paragraphs in the first section, "Role of the Program," relate to the main idea. Discuss how each paragraph in this section relates to the previous paragraph.

3. What types of ancillary information does this textbook have? Why do you think the writers include the information? What strategies do the authors of this textbook use to help readers navigate through it to locate information? List the navigational aids, and explain how they help readers.

4. If you were a member of the Sierra Club and you received the report updating the clean coal technology program, could you locate information on the effect of the technology on acid rain by scanning Figure 2.4? How would you find the information? Write a summary of the section for inclusion in your local Sierra Club's newsletter.

# PART II
# THE DOCUMENTS

**P**art II contains examples of the four major categories of technical documents—correspondence, instructions, proposals, and reports. Environmental impact statements are a subcategory of reports, but they are discussed in a separate chapter because the conventions governing them differ from those governing other types of reports.

In this section you will learn the conventions governing these four categories and their subcategories. You will also have an opportunity to see how company-specific conventions and readers' needs may cause writers to deviate from the conventional organizational patterns and formats of specific types of document. In addition, you will have an opportunity to study how a writer's decisions concerning content, organizational pattern and sequence, point of view and focus, style, graphics, visual text, and layout are affected by the context in which various documents are written. You will then discover the effect of those decisions on readers' interpretations of a message because you will read the audience's response to many of the documents.

The historical contexts of these categories will also be examined so that you can learn how the rhetorical aspects and conventions of the past inform the documents of today. You will discover how considerations of audience have infused documents since Greco-Roman times and how writers in every age have had to balance the impersonal, stylized conventions of technical writing with a recognition that the content of their documents has a humanistic base.

# Correspondence

## INTRODUCTION

Letters and memoranda are a primary means of communication among people within an organization and between people in differing organizations. Industry relies heavily on correspondence to carry on its day-to-day business. When writers consider the audience's needs and situation, then readers can usually interpret a message as the writer intends and business goes on. But, as the memo relating to the Chicago flood demonstrates, when a document does not provide for readers' reading processes and behaviors, then readers often misinterpret or fail to understand a message, and business comes to a standstill. Eleanor Ledoux, one of the legal assistants with the law firm hired by the city of Chicago to handle the litigation resulting from the tunnel flooding, believes that most of the suits she has handled are the result of miscommunication. Such miscommunication is often caused by "third paragraph syndrome": placing the purpose and main organizing idea for a letter or memorandum at the end of the third paragraph instead of at the beginning of the correspondence. This malady can often be fatal or at least extremely costly, as you will see when you read the case studies of the Three Mile Island nuclear accident and the Challenger disaster. In this section you will have an opportunity to read examples of effective correspondence that have been written by writers who consider their readers' needs and successfully communicate their message to their readers.

Letters and memoranda are the most common documents written in the work force. Surveys have found that most college graduates write, on the average, at least one letter or memorandum a day that is related to work (Anderson 1985; Roth 1993). With the increased ease of sending messages via fax (facsimile machine using wireless transmission) and e-mail (electronic mail transmitted through computer networks), the average daily amount of correspondence per employee is increasing. Correspondence also increases in businesses that emphasize total quality management, which requires that divisions within a company maintain constant communication among themselves. Thus, employees need to acquire the skills necessary to write the various forms of correspondence so that they can communicate effectively in the workplace.

## Context

Letters are intended for readers who are *outside* a writer's organization; memoranda are for readers who are *within* a writer's organization. Because most correspondence involves only a limited amount of information and can therefore be written in a few short paragraphs, the writing of letters and memos may appear deceptively simple and writers may be tempted to draft their messages without doing much planning. The result is writer-based documents that fail to provide for the needs of the reader, causing miscommunication. Furthermore, without planning, writers often fall prey to third paragraph syndrome. Busy, impatient readers may fail to read as far as the third paragraph, or they may miss the statement of purpose as they quickly scan the text. No matter how routine or insignificant writers perceive their topic to be, they need to consider carefully what their readers know and what they need to know if their readers are to respond appropriately to a request or be persuaded to take action or think in a specific way.

### *Audience*

Correspondence is written to people in widely differing roles, positions, and communities. They may be scientists, engineers, fiscal officers, marketing salespersons, operators of technical machinery, design engineers, product developers, researchers, clients, customers, or subcontractors. They may assume the role of decision makers, learners, instructors, repairpersons, designers, or researchers. Some readers rank higher than a writer in an organizational hierarchy, some are at the same level, and others are subordinate. Some are experts in a writer's field, possessing the same background and familiar with the same vocabulary; others are novices, having little knowledge of the field or its terminology.

### *Purpose and Situation*

Correspondence is written for a wide range of purposes and in every kind of situation. An employee may need to draft a memo to remind other employees about an upcoming meeting, to request that a subordinate provide information about a project, or to persuade a supervisor to provide funding for attendance at a conference. A memo may be read solely by a supervisor, or it may be read by multiple readers, including the supervisor's supervisor, the employee's peers, or the supervisors' peers. Correspondence may be written when a great deal of competition exists among the employees or between the divisions in a company or when confidentiality must be maintained within a company.

## Style and Format

Correspondence assumes a variety of forms. Letters and memos may involve requests or responses, applications or complaints. They may be brief messages of one or two paragraphs, or they may be as long as ten or twelve pages and contain an entire report or proposal. Although much correspondence concerns details in the routine, day-to-day business of an organization, some correspondence contains requests and information of prime importance to employees, clients, customers, stockholders, administrators, and even the general public (Barabas 1990).

Letters and memoranda may include personal information and follow an informal style, or they may be impersonal and formal. The style depends on whether the writer is familiar with the reader; whether the reader is employed by the same organization as the writer and whether the reader is above, below, or parallel to the writer in the organizational hierarchy; whether the document may be used later in a legal case; and whether other readers may read the correspondence.

### E-Mail

With the past decade's introduction of the electronic transmission of mail, the style and format of correspondence is undergoing a transformation. Until a few years ago, letters were transmitted by U.S. Postal Service and memos by interoffice mail. Today, fax machines are used along with the U.S. Postal Service, and e-mail is quickly replacing the office "gofer" (mail carrier).

This change in the delivery medium is causing a change in the format and style of business correspondence. Letters and memos that are transmitted electronically appear to be more informal and personal, and contain characteristics of oral as well as written discourse. Studies show that these messages often have grammatical and spelling errors and fail to observe many of the rules of standard edited English for punctuation and capitalization (Sims 1991). This tendency sometimes leads to communication problems. Because writers may spend less time in considering their readers' needs and in making decisions related to those needs than they did when using pencil and paper or typewriters, the potential for miscommunication increases. This miscommunication may cause readers to miss meetings, get inappropriate information, and fail to act or think as writers intended. Such results cost an organization time and money.

In this chapter you will have an opportunity to study the types of routine correspondence that occur within and between organizations in the workplace. These memoranda and letters have been written for various readers and purposes in a variety of situations. In addition, the chapter includes a memorandum written in Germany during World War II. Like the others, it was written as a routine document. However, the subject of the memo is anything but routine, and the manner in which it is treated presents, for today's writers, a chilling warning to remember the human side of technical communication.

# MEMORANDA

The memorandum has become increasingly important as a means for people internal to an organization to communicate with one another. It has also become a highly political document, providing members of an organization with a means of documenting their ideas and concerns as well as for communicating them. The documents in this section represent the kinds of organizational situations in which memoranda are written.

## Context

Because readers of memoranda are internal to an organization, they probably know some of the technical terminology of the field in which the organization is involved as well as some of the background of a situation. However, because their positions in the organizational hierarchy, their professional communities, and even their geographical locations may differ from that of the writer, they probably do not know all of the background information, or all of the technical terminology, or all aspects of the content.

Keep in mind that readers of memos do not want to spend much time reading them. Most recipients look at the subject line of a memo to determine if the message is important. If they don't think so, they toss it in the waste basket. If it is of interest, they scan it; they seldom study a message.

## Conventional Format

The conventional format for a memo includes two sections, the heading and the body. The heading includes the date, the name of the recipient(s), the name of the sender, and the subject to be discussed. These elements are usually listed in the upper left-hand corner, where readers can find the information quickly by scanning down the page without having to read across the page.

The distribution list of other readers is included either under the subheading *TO:* along with the primary reader or under a separate subheading that may be to the right of the other subheads, under them, or at the end of the memo. The subheading should be designated as *cc, xc,* or *c.* The original designation *cc* was an abbreviation for *carbon copy,* but because almost no one makes an actual *carbon* copy anymore, the initials *cc* seem inappropriate. Some organizations are changing the designation to *xc* to indicate a Xerox copy, and others are simply using *c* to refer to copies. Readers quickly become accustomed to the abbreviation format as long as it is followed consistently.

Writers sign a memo by writing their initials next to their name in the heading.

| | |
|---|---|
| DATE: | xxxxxxxxxxxxx |
| TO: | xxxxxxxxxxxxx |
| FROM: | xxxxxxxxxxxxx |
| SUBJECT: | xxxxxxxxxxxxxxxxxxxxxxx |
| (Text) | xxxxxxxxxxxxxxxxxxxx |
| c: | xxxxxxxxxxxxxxx |

The body of a memorandum can assume a wide variety of forms, ranging from requests and responses to proposals and reports. The memo may use a block or semi-block format. In a block format, all lines are flush with the left margin; in a semi-block format the first line in a paragraph is indented. Formal memos usually line up the information next to the subheads in the heading, as shown in the preceding model, whereas informal memos may place the information directly after the colon in each subhead. Memos may or may not be written on letterhead. As you read the memos in this section, notice the different forms they assume.

## Examples

The memoranda in Figures 3.1 through 3.7 exemplify the kinds of routine internal communication that occur in the workplace and the variety of topics that are discussed. All of the memos are brief and do not exceed a single page so that readers can read them quickly. The memos in Figures 3.4, 3.5, and 3.6 are cover letters for longer documents. However, each one focuses on the attached information differently because their purposes and their readers' needs differ. The memo in Figure 3.7 is a brief report on a research project that is submitted monthly.

Figure 3.8 is the only memo that is not routine. This memo is written in response to a problem and could become evidence in a lawsuit. Figure 3.9 provides a different context for viewing memoranda. It is written as a routine document, but the content is anything but routine.

Look very closely at the documents in Figures 3.1 through 3.3 to see how the writers have followed or deviated from the conventions, how they present the information so the readers can comprehend their messages quickly and easily, and where and how they indicate to the readers their purposes and their main organizing idea. Then apply what you've learned to the documents in Figures 3.4 through 3.7. Because the memo in Figure 3.8 is not a routine memo, it is examined more closely, as is the memo in Figure 3.9.

### *Figures 3.1 through 3.3*

These documents are all written by Ron Rothstein, an engineer specializing in computers at Advanced Technology Systems (ATS), a small computer software company. Good morale exists among the employees, who have seen the company go from a small venture capital enterprise to a self-sufficient company. The company's administrators believe that the employees should be kept informed of the organization's progress. Their commitment to good communications has filtered down to the various divisions. Ron Rothstein tries to keep the various project managers in his division up-to-date on all projects being undertaken. Managers are responsible for keeping each other informed of developments in their respective areas. The memos in Figures 3.1 and 3.2 have the same readers. The purpose for the memo in Figure 3.3 is simply to provide personnel with information to assist them in their jobs.

The types of memoranda Rothstein writes range from a simple invitation to an announcement of a meeting change to a complicated discussion of requirements for a new software program. All of these memos will be read by multiple readers in the writer's own department and field.

**Conventions**   All the correspondence follows the general conventions for memos. Logistical information relating to the meeting follows an organizational convention. The information relating to the place, time, and date of the meeting in Figures 3.1 and 3.2 is formatted consistently.

**Content**   The two memos related to meetings provide little background information because readers already know it. The writer, however, takes pains to make certain that the memos are self-contained and that the readers do not need to refer to other documents for necessary information. The memo notifying employees of the meeting change provides the consultant's name and briefly summarizes the topic to be discussed to jog readers' memories. The memo also reminds readers of the previous meeting time and date. The memo issuing the invitation names the speaker and briefly summarizes the topic to be discussed but does not explain why the staff should be interested; they know why. The other two messages provide background information, with the response memo going so far as to provide definitions of terms. The definitions are necessary because the program is new, and the readers, even though they are in the same division, are not familiar with it. However, acronyms, brand names, and technical terms are used without any explanation through all of the memoranda because readers are members of the same general organizational and professional discourse communities.

**Organizational Conventions**   The memos follow the appropriate organizational and discourse conventions. The memo in Figure 3.3 is written in response to the reader's request for the writer to develop a piece of software to bridge two other software programs. Although the memo's primary purpose is to provide the requirements for the new program, the writer implicitly requests the reader's reactions to the program. The writer does not need to specify this request; it is part of the organization's conventions. The reader knows he is expected to respond. You can see the reader's response in the handwritten notes on the memo that he has returned to the writer.

**Style**   While the memos are relatively formal, the writer creates a personal and friendly tone by using first person plural to indicate that the reader is part of the writer's social group and second person to address the reader directly. Rothstein also includes an informal, brief note at the end of each memo.

**Visual Text**   As you look at the documents, notice the use of visual text. The memo relating to the meeting change uses asterisks to emphasize the topic and boldface type to emphasize the new time and date. Several memos use lists to facilitate readers' fluency and comprehension.

**Format**   The format adheres to the conventions for memoranda. It is in block style. Notice that the distribution list is designated by the traditional initials, *cc*. Uncharacteristically, the distribution list includes the telephone extensions of the readers in case they wish to discuss the document with one another. Because the list of readers is long, the heading appears excessively long. To shorten the length, the writer might abbreviate the word "extension" and placed the distribution list in a separate column to the right of the list of primary readers.

Standard block format.

Long list.

Includes extension numbers.

Distribution list.

Attempts to persuade readers to attend. Readers are familiar with products, terms. Speaks in persona of company.

Makes request. Uses language for formal invitation.

Lists information.

Speaks directly to readers. Closes informally.

---

## ADVANCED TECHNOLOGY SERVICES, INC.

**ATS**

*"Setting New Standards In Service Excellence"*

DATE:   December 1, 19XX

TO:     William Bakker    Extension:  15
        Jim Goble         Extension: 730
        Jody Howard       Extension:  43
        Bob Keime         Extension:  37
        James Larson      Extension: 787
        Avi Narula        Extension: 511
        Jack Rainey       Extension: 746

FROM:   Ron Rothstein    Extension: 33   *Ron*

SUBJ:   Primavera Project Management and Job Costing Software

cc:     Jim Brooks        Extension: 18
        Don Bowen         Extension: 31
        Bob Everts        Extension: 21

We are very interested in reviewing the Primavera project management software and its associated job costing capabilities. Primavera offers both DEC VAX and PC-based products. Primavera software is distributed in Illinois by Altus Management Services, Incorporated (FAX: 708-654-0210, TEL: 708-654-0080). Mr. Steve Secker, Account Executive at Altus will be coming to ATS to tell us about their project management software, job costing capabilities, and consulting services.

You are invited to attend the formal presentation which will be held as indicated below:

        Title:    Primavera
        Subject:  Project Management & Job Costing S/W
        Time:     1:30-3:30 p.m.
        Date:     Thursday, December 7, 1989
        Location: Lower Conference Room
                  ATS Main Building

See you there.

2000 E. WASHINGTON • EAST PEORIA, IL 61611 • (309) 698-5700 • FAX (309) 698-5714

---

**Figure 3.1   Invitation to a meeting.** Source: Advanced Technology Services. With permission.

### ADVANCED TECHNOLOGY SERVICES, INC.

**ATS**

*"Setting New Standards in Service Excellence"*

Standard format.

**DATE:** October 31, 19XX

List same as in
Figure 3.1.

**TO:**
| | | |
|---|---|---|
| William Bakker | Extension: | 15 |
| Jim Goble | Extension: | 730 |
| Jody Howard | Extension: | 43 |
| Bob Keime | Extension: | 37 |
| James Larson | Extension: | 787 |
| Avi Narula | Extension: | 511 |
| Jack Rainey | Extension: | 746 |

**FROM:** Ron Rothstein   Extension: 33

Emphasizes topic.

**SUBJ:** ***************MEETING CHANGE********************
********************************************************

Distribution list. Almost
same as in Figure 3.1.

**cc:**
| | | |
|---|---|---|
| Jim Brooks | Extension: | 18 |
| Bob Everts | Extension: | 21 |

Background. Speaks for
company by using 1st
person plural. Purpose.
Readers are familiar
with products so no
need to explain.

Our meeting on Thursday, November 2, with Tony Mayo, National Account Manager, Robbins–Gioia, Inc. has been changed to Tuesday, November 14.

Mayo will be coming to ATS to tell us about Robbins-Gioia's Job Costing software, Project Management software, and consulting services.

Makes request. Uses
formal language for
invitation. Lists
information in same
format as in Figure 3.1.

You are invited to attend the formal presentation which will now be held as indicated below:

| | |
|---|---|
| Title: | Robbins–Gioia |
| Subject: | Job Costing & Product Management S/W |
| Time: | 9:30–11:30 a.m. |
| Date: | **Tuesday, November 14, 1989** |
| Location: | Upper Employee Conference Room ATS Main Building |

Speaks directly to
readers.

In addition to the formal meeting, Mayo will be here in the afternoon on Monday, November 13, to meet with us and learn more about our requirements. I will contact you individually to schedule meetings on this Monday.

Same informal closing
as in Figure 3.1.

See you there.

2000 E. WASHINGTON • EAST PEORIA, IL 61611 • (309) 698-5700 • FAX (309) 698-5714

**Figure 3.2   Announcement of a meeting change.** Source: Advanced Technology Services. With permission.

**Readers' Response**   All memoranda were successful. Readers understood the messages easily. No one went to the canceled meeting, and everyone attended the rescheduled one. In addition, people were prepared for Ron's call to schedule individual meetings with the consultant. Readers were able to use the information in the memo in Figure 3.3 to avoid the pitfalls the writer warns about.

**Standard block format.**

**Provides background.**

**Describes information in attached document.**

**States purpose. Attempts to persuade readers to read and use the attached information.**

---

**ATS**                                    **ADVANCED TECHNOLOGY SERVICES, INC.**

*"Setting New Standards In Service Excellence"*

DATE:    July 17, 19XX

TO:      Jim Brook          Extension: 18

FROM:    Ron Rothstein      Extension: 33  *Ron*

SUBJ:    Microsoft Project

cc:      Bob Everts         Extension: 21
         Ron Isaia          Extension: 27
         Vince Neyens       Extension: 20
         Loren Sorensen     Extension: 15

When I began working on the response to the RFQ from General Motors, Central Foundry division, for the Integrated Casting Handling System upgrade, I decided to use the Microsoft Project software to organize the project plan, assign resources, and determine costs and schedules.

During this planning process, I encountered a variety of shortcomings with the Microsoft Project software. The attached pages summarize my experience with this software and provide a "checklist" that can be used to compare other project management software products to typical ATS requirements.

Many magazines have reported on the features of project management software but they typically do not list the product deficiencies and they certainly do not try to use the products in situations exactly like those encountered at ATS while responding to customer requirements.

My hope is that this information will help others at ATS become more productive by avoiding some of the pitfalls of software products. The attached comparison chart will help build an information base on project management software that ATS has used.

2000 E. WASHINGTON • EAST PEORIA, IL 61611 • (309) 698-5700 • FAX (309) 698-5714

---

**Figure 3.3   Evaluation of a software product for company use.**
Source: Advanced Technology Services. With permission.

### *Figures 3.4 through 3.7*

The memoranda in Figures 3.4 and 3.5 are related to the management of computer information. They are written by the assistant vice president in charge of the management information systems division of a national health service company.

The memo in Figure 3.4 was written after a meeting. The writer's boss realized that no one had taken notes during the meeting, and he needed a written report of the decisions. The content is culled from the writer's memory of the meeting and from the "sketchy notes" she took for her own purposes. The memo in Figure 3.5 addresses a large group of readers to provide them with information they will need to know at a future meeting.

The memos in Figures 3.4 and 3.5 follow a conventional format. Like Rothstein, Hailperin uses the initials *cc,* but unlike Rothstein, she places her distribution list at the end of the memo. She also emphasizes the subject of each document by using all capital letters.

Both of these memos, along with the memo in Figure 3.6, serve as cover correspondence for transmitting more detailed information. In each memo, the writer focuses on a different aspect of the attached document. The Figure 3.4 memo summarizes the attached information so that the primary reader does not need to read the enclosed details if he doesn't want to. In the Figure 3.5 memo, rather than summarize the attached document, Hailperin explains how the information will be used so that readers will study and evaluate the document in terms of its projected use. The memo in Figure 3.6 neither summarizes the attached information nor explains its use but simply describes the information included.

All these memos indicate that documents are enclosed with the designation *attachment* at the end of the memo so readers know that additional information is included. The writer at the electric company, J. J. Chambers, places the designation before the distribution list, whereas Hailperin places it after the distribution list.

The memo in Figure 3.7 transmits a monthly research report. Notice that the headings in the memos in Figures 3.6 and 3.7 deviate from the conventional format. In Figure 3.6 the subheads *FROM* and *TO* form a second column. In Figure 3.7 the subhead *VIA* is added to designate the writer's supervisor, an intermediary reader. Furthermore, the date is placed in the upper right-hand corner in Figure 3.7. With subheads in the right half of the page, both memos force readers to scan across a page rather than simply move straight down it.

The memoranda in Figures 3.6 and 3.7 use letterhead; the others do not.

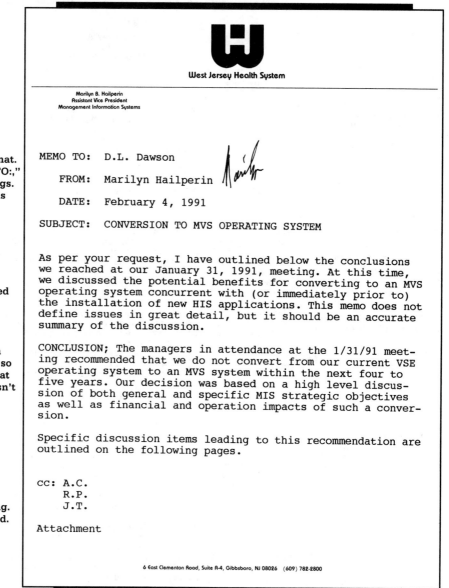

**West Jersey Health System**

Marilyn B. Hailperin
Assistant Vice President
Management Information Systems

Standard block format. Places date after "TO:," "FROM" subheadings. Aligns colons. Spells out word "Subject." States purpose.

MEMO TO:   D.L. Dawson

  FROM:    Marilyn Hailperin

  DATE:    February 4, 1991

SUBJECT:   CONVERSION TO MVS OPERATING SYSTEM

Summarizes attached information.

As per your request, I have outlined below the conclusions we reached at our January 31, 1991, meeting. At this time, we discussed the potential benefits for converting to an MVS operating system concurrent with (or immediately prior to) the installation of new HIS applications. This memo does not define issues in great detail, but it should be an accurate summary of the discussion.

Provides conclusion reached at meeting so manager knows what results are and doesn't have to read details inside if he doesn't want to.

CONCLUSION; The managers in attendance at the 1/31/91 meeting recommended that we do not convert from our current VSE operating system to an MVS system within the next four to five years. Our decision was based on a high level discussion of both general and specific MIS strategic objectives as well as financial and operation impacts of such a conversion.

Specific discussion items leading to this recommendation are outlined on the following pages.

Copies go to other managers at meeting. Information attached.

cc: A.C.
    R.P.
    J.T.

Attachment

6 East Clementon Road, Suite A-4, Gibbsboro, NJ 08026   (609) 782-2800

**Figure 3.4   Cover memo summarizing attached documents.** Source: West Jersey Health Systems. With permission.

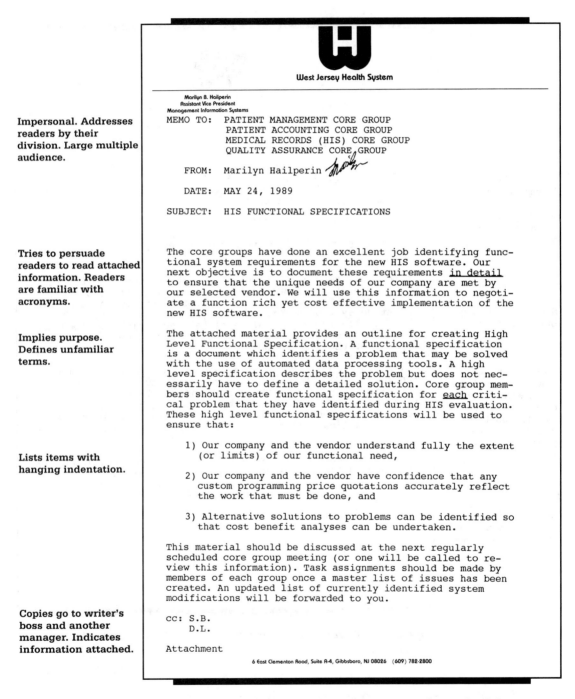

**West Jersey Health System**

Marilyn B. Hailperin
Assistant Vice President
Management Information Systems

MEMO TO:   PATIENT MANAGEMENT CORE GROUP
           PATIENT ACCOUNTING CORE GROUP
           MEDICAL RECORDS (HIS) CORE GROUP
           QUALITY ASSURANCE CORE GROUP

FROM:   Marilyn Hailperin

DATE:   MAY 24, 1989

SUBJECT:   HIS FUNCTIONAL SPECIFICATIONS

The core groups have done an excellent job identifying functional system requirements for the new HIS software. Our
next objective is to document these requirements in detail
to ensure that the unique needs of our company are met by
our selected vendor. We will use this information to negoti-
ate a function rich yet cost effective implementation of the
new HIS software.

The attached material provides an outline for creating High
Level Functional Specification. A functional specification
is a document which identifies a problem that may be solved
with the use of automated data processing tools. A high
level specification describes the problem but does not necessarily have to define a detailed solution. Core group members should create functional specification for each critical problem that they have identified during HIS evaluation.
These high level functional specifications will be used to
ensure that:

   1) Our company and the vendor understand fully the extent
      (or limits) of our functional need,

   2) Our company and the vendor have confidence that any
      custom programming price quotations accurately reflect
      the work that must be done, and

   3) Alternative solutions to problems can be identified so
      that cost benefit analyses can be undertaken.

This material should be discussed at the next regularly
scheduled core group meeting (or one will be called to review this information). Task assignments should be made by
members of each group once a master list of issues has been
created. An updated list of currently identified system
modifications will be forwarded to you.

cc: S.B.
    D.L.

Attachment

6 East Clementon Road, Suite A-4, Gibbsboro, NJ 08026   (609) 782-2800

**Marginal annotations:**

Impersonal. Addresses readers by their division. Large multiple audience.

Tries to persuade readers to read attached information. Readers are familiar with acronyms.

Implies purpose. Defines unfamiliar terms.

Lists items with hanging indentation.

Copies go to writer's boss and another manager. Indicates information attached.

**Figure 3.5   Cover memo explaining purpose of attached document.**

Source: West Jersey Health Systems. With permission.

**Uses letterhead.**

**Indiana & Michigan Electric Company**

# Innovations

DATE:       January 11, 1988              **FROM:** J.J. Chambers

**Deviates from
conventional heading.**

SUBJECT:  Proposed Operating Idea-Innovation          **TO:**      T.R. Winenger
          G. O. T& D 289 "Cost Cutting Procedure
          for Tower Light Projects" by N. E. Weaver

**States purpose.**

The attached proposed Operating Idea was submitted as an advance copy in
September, 1986. We indicated that, based on a one year field test, we
would formally submit the idea at a later date. The field test has been
completed and the results are quite favorable. We would like to formally
submit this article as an Operating Idea.

As we have previously discussed, I have included reviewers' comments with
the prepared article. I have also provided the one year inspection results
and some cost estimates prepared by the author. Additionally, I have included some questions with answers provided by the author. You should
already have photographs. If not, please let me know.

**Secretary's initials.**

JJC/dd

**Indicates enclosures.**

Attachment

**Places distribution list
at bottom.**

c: P. F. Carl
   W. L. Cross
   W. G. Dillehay
   R. E. Gifford
   A. H. Potter
   T. L. Cory
   T. K. Nussbaum

**Figure 3.6    Cover memo describing attached documents.** Source:
American Electric Power Service Corporation. With permission.

## *Figure 3.8*

The memo in Figure 3.8 (page 90) is written by an engineer with a private
company that was subcontracted by the City of Chicago's Department of
Transportation. The memo is written by the engineer to his supervisor to
provide information concerning a broken cable. The writer has a hidden
agenda in writing this memo; he does not want to be blamed for causing the
cable to break. In addition, he is aware of the economic and political consequences involved in the situation. If his workers are found at fault, his

Logo.

Includes persons'
positions. Includes
unconventional
category (via) to
indicate writer's
supervisor who reads
memo before it goes
out.

Brings reader up to
date on proposal.

(1) Changes to proposal.

(2) Present status.

a. Testing to data.

(1) Results of testing
    so far.

---

UNITED STATES GOVERNMENT

## *Memorandum*

LIBRARY OF CONGRESS

TO    : Frazer G. Poole       **DATE:** March 20, 1972
        Assistant Director for Preservation
VIA  : John C. Williams *JW*
        Research Officer
FROM : George B. Kelly, Jr.
        Chemist

SUBJECT: One Month Review of Research Proposal No. 2: Vapor Phase
           Methods for Deacidification and Buffering Paper

At present there is no correction to the accepted research
proposal which was approved February 4, 1972.

The proposal has been modified by the addition of ethylene or
propylene oxides as possible deacidifying agents as outlined in a
memo from G. B. Kelly to J.C. Williams, "Suggested Additions to
Research Proposal No. 2, dated February 9, 1972."

The present status of the Research Proposal No. 2 is in the
initial phase. Literature on epoxides has been acquired and
samples of volatile epoxides and diepoxides have been requested,
but not yet received. Most of the vacuum pump, etc., has been
received, so that the actual impregnation and testing can begin
as soon as the chemical samples are received.

A preliminary test of the effectiveness of an epoxide-amine
system on paper permanence was carried out by the liquid-phase
impregnation of paper with the water soluble epoxy-amine polymer
prepared by John Williams as reported in his memo to Frazer G.
Poole, Subject: Rebinding Brittle Books-memo of January 15 continued, dated February 15, 1971.

Accelerated aging tests indicated that although the treatment raised the pH of the paper and did provide some protection
against acid build-up on aging, the paper was severely yellowed
by the aging. Some protection against degradation of the paper
was obtained, but the protection was minimal, as the folding
endurance of the treated paper was only a little better than that
of the untreated paper after aging the equivalent of 70 years.
These results are somewhat encouraging in that some of the desired results were obtained, but point out some of the side effects and difficulties that will have to be overcome in the development of a vapor phase deacidification treatment.

BUY U.S. SAVINGS BONDS THROUGH THE PAYROLL PLAN

---

**Figure 3.7 Memo including monthly report.** Source: Library of Congress.

**b. Present development efforts.**

In the development of the vapor phase treatment, the initial efforts will be directed towards evaluation of pro-pylene oxide as a deacidifying agent and toward finding a non-yellowing, volatile, amine-epoxy system for impregnation.

*[signature]*

**Figure 3.7** *(continued)*

company will be responsible for paying for the repairs. The company may also be blocked from receiving another contract from the city. This is known as a CYT (cover your tail) memo. The reader will use the information in the memo as the basis for a letter to the city, defending the company's position that it is not at fault (see Figure 3.21).

As you read the memo, determine the claims the writer makes to defend his argument, and then consider whether the evidence he uses to support his claims will appear valid to the city.

## *Figure 3.9*

This memo, on page 91 and known as the *Geheime Reichssache* [secret Reich business] memorandum, is a historical document written during World War II in Germany.

One of the aspects identifying technical writing and separating it from other types of written discourse is its "objective," scientific, and impersonal style. Recently, technical documents have been criticized for their lack of humanism. Critics argue that technical writers should be responsible for pointing out unethical or dangerous aspects associated with the topics on which they write.

Perhaps nowhere are the extremes to which this style may be carried more apparent than in the communications that circulated in the Third Reich. The deleterious effects resulting from such dehumanism are only too well known from the pictures at Dachau and Auschwitz.

Plans for the extermination of the Jewish population were worked out at the Wannssee Conference on January 20, 1942. Six months later the memo in Figure 3.9 was sent from "Just" to Walter Rauff, a leader of one of the "Special Action Groups" organized by Himmler to carry out the execution of Jews during the early stages of the "Final Solution."

Although firing squads were initially used to carry out the executions, in 1942 Himmler ordered that gassing vans be used to execute women and children, because, he claimed, these techniques were more "efficient,

Heading in all capital letters. Deviates from conventional heading by using a second column and including an additional item, "Contract No." Right column makes it difficult to scan heading. Nonconventional subheading (Copies To) for distribution list. Semi-block style.

Provides background, using a chronological organizational pattern. Lack of extra space between paragraphs makes it difficult to read. Bases evidence on authority (Lendohl) readers will accept.

Provides detailed data.

Emphasizes information.

States claim #1. Evidence.

Claim #2. Evidence.

Presents argument (organizing idea) that crew not responsible for damaged cable. Reader would have wanted to know this information earlier. Nonconventional placement of signature (initials usually placed next to name in heading).

---

## INTEROFFICE CORRESPONDENCE

TO:    WAYNE VALLEY         DATE: SEPTEMBER 27, 1991

FROM:    PETE McCLAIN         CONTRACT NO. : 32514

SUBJECT:  DAMAGED SUBMARINE CABLE         COPIES TO:
          CHICAGO AVENUE BRIDGE


ON WEDNESDAY, AUGUST 21, I WAS NOTIFIED BY THE CITY ENGINEER THAT ONE OF THE SUBMARINE CABLES MUST BE DAMAGED. THE BRIDGE WAS FULLY OPERABLE AT 10:OO A.M., MONDAY, AUGUST 19, WHEN WE MOVED OUR EQUIPMENT THROUGH TO START WORK ON THE SOUTHWEST CLUSTER.

WE REMOVED THE EXISTING SOUTHWEST CLUSTER, STARTING MONDAY AT 12:40 P.M. AND FINISHED ON TUESDAY AT 9:40 A.M. WE STARTED DRIVING THE NEW CLUSTER AT 10:45 A.M. TUESDAY AND FINISHED DRIVING AT 12:45 P.M. ON WEDNESDAY.

ON THURSDAY, AUGUST 22, THE CABLE WAS DE-ENERGIZED BY A CITY ELECTRICIAN. A DIVING SURVEY WAS MADE BY LENDOHL MARINE CONTRACTORS, INC. THE CABLE WAS FOUND TO BE DAMAGED BELOW THE MUD LEVEL (SEE ATTACHED "DIVER INSPECTION 8/22/91"). THE CABLE WAS PULLED FROM THE MUD AND INSPECTED BY CITY PERSONNEL. THE ARMOR WAS UNWOUND, FRAYED AND RUSTED AWAY IN THE VICINITY OF THE DAMAGED END. THE END WAS CUT BACK A FEW INCHES TO CLEARLY IDENTIFY THE CONDUCTORS. THE CABLE CONTAINED THREE (3) #2 CONDUCTORS. THE CABLE WAS LEAD SHEATHED WITH SINGLE ARMOR WRAPPING.

THE DIVER WAS ABLE TO RETRIEVE OTHER <u>PREVIOUSLY DAMAGED</u> NEOPRENE WIRES FROM THE SAME LOCATION. IT SHOULD BE NOTED THAT ALL SORTS OF SCRAP LITTERED THE BOTTOM AROUND THESE CLUSTERS; STEEL LADDERS, CAST IRON RADIATORS, BROKEN PILES, ETC.

I DON'T BELIEVE THE DAMAGE WAS CAUSED BY DRIVING A PILE ON THE CABLE BECAUSE TOO MUCH SLACK REMAINED FROM THE ABUTMENT TO THE MUDLINE. DRIVING A PILE ON THE CABLE WOULD PUSH THE CABLE INTO THE BOTTOM UNTIL NO SLACK WAS AVAILABLE, THEN THE CABLE WOULD BREAK.

IT IS MY OPINION THAT THIS CABLE HAD SOME DEGREE OF PREVIOUS DAMAGE OR DETERIORATION AS THE ARMOR WAS RUSTED AWAY AND FRAYED, DISTURBING THE BOTTOM (AND ALL THE TRASH IN THE VICINITY). EITHER BY REMOVING OR DRIVING PILES COULD HAVE CAUSED THE FINAL BREAKAGE OF THE UNPROTECTED PORTION OF THE CABLE.

THE NEW CLUSTER WAS DRIVEN 3.5' TO THE SOUTH TO PROVIDE A SAFETY MARGIN AWAY FROM THE CABLES. THE CABLES ARE ON A SLOPE THAT SLOPES DOWNWARD AWAY FROM THE ABUTMENT TOWARDS THE CLUSTER. REMOVING THE CLUSTER COULD HAVE ALLOWED THE CABLE TO MOVE FURTHER DOWN THE SLOPE.

I FEEL WE TOOK PRECAUTIONS BY RE-LOCATING THE NEW CLUSTER. THE DAMAGE WASN'T CAUSED BY IMPROPER POSITIONING OR NEGLIGENCE.

PETE MCLAIN, SUPERINTENDENT

---

**Figure 3.8   CYT memo.** Source: City of Chicago.

humane." Just writes to his superior to request permission to make "technical changes" to these gas vans.

Just is well aware of his audience's attitude toward the situation. As we know from accounts of the Holocaust, many officers carried out their grisly tasks only by denying the humanity of the victims. This memo is an extreme case of using language in a way that deliberately masks the ethical issues. Notice how the writer uses technical, object-oriented terms to make the reader forget that human beings are the subject of the memo. The writer never uses terms such as *people, persons, men, women, children*. He talks, instead, of *ninety-seven thousand* being processed, van *loads*, the number of *pieces*, and the *merchandise*. The focus is on mechanical and procedural aspects, such as vehicle stability, lighting, and cleaning.

As you read, consider the ethical responsibility of the writer.

---

Geheime Reichssache (Secret Reich Business)
Berlin, June 5, 1942
Changes for special vehicles now in service at Kulmhof (Chelmno) and for those now being built

Since December 1941, ninety-seven thousand have been processed (*verarbeitet* in German) by the three vehicles in service, with no major incidents. In the light of observations made so far, however, the following technical changes are needed:

The vans' normal load is usually nine per square yard. In Saurer vehicles, which are very spacious, maximum use of space is impossible, not because of any possible overload, but because loading to full capacity would affect the vehicle's stability. So reduction of the load space seems necessary. It must absolutely be reduced by a yard, instead of trying to solve the problem, as hitherto, by reducing the number of pieces loaded. Besides, this extends the operating time, as the empty void must also be filled with carbon monoxide. On the other hand, if the load space is reduced, and the vehicle

(continues)

**Figure 3.9    Geheime Reichssache memorandum.** Source: Lanzmann, Claude. *Shoah: An Oral History of the Holocaust.* Copyright © 1985 by Éditions Fayard and Claude Lanzmann. Reprinted by permission of Georges Borchardt, Inc., for the author.

is packed solid, the operating time can be considerably shortened. The manufacturers told us during a discussion that reducing the size of the van's rear would throw it badly off balance. The front axle, they claim, would be overloaded. In fact, the balance is automatically restored, because the merchandise aboard displays during the operation a natural tendency to rush to the rear doors, and is mainly found lying there at the end of the operation. So the front axle is not overloaded.

2. The lighting must be better protected than now. The lamps must be enclosed in a steel grid to prevent their being damaged. Lights could be eliminated, since they apparently are never used. However, it has been observed that when the doors are shut, the load always presses hard against them [against the doors] as soon as darkness sets in. This is because the load naturally rushes toward the light when darkness sets in, which makes closing the doors difficult. Also, because of the alarming nature of darkness, screaming always occurs when the doors are closed. It would therefore be useful to light the lamp before and during the first moments of the operation.

3. For easy cleaning of the vehicle, there must be a sealed drain in the middle of the floor. The drainage hole's cover, eight to twelve inches in diameter, would be equipped with a slanting trap, so that fluid liquids can drain off during the operation. During cleaning, the drain can be used to evacuate large pieces of dirt.

The aforementioned technical changes are to be made to vehicles in service only when they come in for repairs. As for the ten vehicles ordered from Saurer, they must be equipped with all innovations and changes shown by use and experience to be necessary.

Submitted for decision to Gruppenleiter II D, SS-Obersturmbannführer Walter Rauff.

Signed: Just

**Figure 3.9**   *(continued)*

**Suggestions for Discussion and Writing**

1. Because the distribution list containing the names of the readers to receive copies of the messages are placed in a list along with the other information at the top of the memo, the headings in the ATS documents in Figures 3.1 and 3.2 appear to take up a great deal of space (one-half the page). Try changing the layout of one of these documents so that the heading doesn't look so long and so that it is easier for readers to locate the subject line as well as the main body of the memo. First try to improve the layout by borrowing from the letter format and placing the distribution list at the end of the memo. Next try to improve the layout by placing the list in a separate column about halfway across the page from the *TO* heading. After you have tried each of these revisions, determine which layout is best for transmitting the memo to all the readers. Why do you think it is easier?

2. Unless readers are expected to be hostile to a topic or a request or proposal, the purpose for a message should be placed at the beginning of a document so readers know why they have received it. Review the eight memos. Have the writers indicated the purpose at the beginning? Revise those memos that do not. Do you think your revisions make it easier for readers to read and understand the message?

3. Identify the conventional features of the Geheime Reichssache memo. How is this memo similar in format and style to the contemporary memos in this section?

4. What do we learn about the pitfalls of technical writing from the Geheime Reichssache memo? How does the impersonal tone dehumanize the people affected by this memo? How does it manipulate the reader's response? Rewrite the memo, replacing the dehumanizing terminology with human terms. How does the revision affect the tone of the memo? How might it affect the reader's response?

5. Linguists believe that the language we use affects the way in which we think about things. How might Just's use of language in Figure 3.9 affect the way in which the reader thinks about the people in the vans?

   Recently there has been a great deal of discussion of gender-specific language. By using the terms *mailman, Congressman, chairman,* and *mankind,* we cause people to perceive that these positions are for men, not women. For this reason, our language is changing to accommodate a perception that both genders are capable of assuming these positions. The terms *mail carrier, Congressional representative, chairperson* or *chair,* and *humankind* have replaced the gender-specific ones. Can you think of other terms and phrases related to women, minorities, and people with physical or mental disabilities that dehumanize or degrade them or otherwise provide unacceptable perceptions of them? What are these terms and phrases, and how might we change them?

6. The following comments were made by a survivor of Chelmno, a German concentration camp during World War II.

> When we built the ovens, I wondered what they were for. An SS man told me, "To make charcoal. For laundry irons." That's what he told me. I didn't know. When the ovens were completed, the logs put in, and the gasoline poured on and lighted, and when the first gas van arrived, then we knew why the ovens were built. . . . When they were thrown into the ovens, they were all conscious. Alive. They could feel the fire burn them.

> Source: Lanzmann, Claude. *Shoah: An oral history of the holocaust.* Copyright © 1985 by Éditions Fayard and Claude Lanzmann. Reprinted by permission of Georges Borchardt, Inc., for the author.

Contrast the tone of this survivor's account with Just's discussion of the vans. What do we learn from this account that makes Just's memo seem all the more ethically irresponsible?

# ELECTRONIC MAIL

The two memos in Figures 3.10 and 3.11 are routine memoranda, but they are transmitted by e-mail rather than interoffice mail. One was sent to a reader in the same building, while the other was sent across the ocean to a reader in another country.

The memo in Figure 3.10 was sent from an employee in one division of John Deere to an employee in another division. Both the writer and the reader are located in the same building. Notice that the document follows the conventions for memos and resembles the ATS memoranda in Figures 3.1 through 3.3. However, this memo is written in all capital letters, as that is how the e-mail program transmits its messages, and the memo does not contain a distribution list.

The memo actually has two purposes. It provides a written record of a telephone discussion so that the reader has something to which he can refer if he wants to check on the topics involved. It also provides the writer with a record of the discussion. If the reader fails to follow his suggestions, he has a written record to prove that it is not his fault; he gave the reader the suggestions. And if the results of the suggestions receive special commendation, he has a record to indicate that he, too, should be commended, perhaps even more so than the reader since it is his idea. The reader appreciated receiving the memo and used it in planning a meeting.

The memo in Figure 3.11 involves intercultural correspondence between employees of the same company but in two different countries and of two different cultures. The writer adheres to the conventions of a memo. Like the message in Figure 3.5, this memo is a follow-up, in this case of a previous e-mail message. In the previous message, the writer thought he had made it clear that the parts for the reader's division were to be ordered from the United States. However, he had recently received a message indicating that the reader planned to order parts from Japan. Therefore, to ensure that this time there is no misunderstanding about the reader's plans, the writer inserts questions in this message that he wants the reader to answer.

The purpose of this memo is to persuade the reader to order parts from the U.S. firm rather than from a Japanese vendor, although the reader would prefer the Japanese firm. The reader easily understood the memo, but additional correspondence occurred as the writer and reader negotiated the purchase of the parts.

**Follows conventions of regular memoranda:**

- **Heading**

- **Style**

- **Format**

**Starts out by picking up previous conversation. Impersonal.**

```
JOHN DEERE DAVENPORT WORKS
TRAINING AND DEVELOPMENT GROUP

TO: M. E. ADDINGTON

FROM: R. E. WAHLSTRAND

DATE: 11 MAY 1990

SUBJECT: ORGANIZATION ISSUES

PER OUR CONVERSATION, HERE ARE SOME ISSUES WHICH YOU MAY WANT TO RAISE
WITH YOUR PEERS AND AT THE SAME TIME BE PROACTIVE.
THESE ARE ISSUES WHICH I BELIEVE ARE OPEN (AS A RESULT OF MY ONE-ON
ONE CONVERSATIONS WITH MIKE'S STAFF AND GENERAL SUPERVISORS) AND HAVE
POTENTIAL FOR SIGNIFICANT IMPACT ON OUR BUSINESS AND LONG-RANGE PLANS.
DEALING WITH THESE COULD ALSO PRESENT AN OPPORTUNITY FOR THOSE WHO HAVE
BEEN TRAINED IN IT TO USE THEIR "DECISION FOCUS" SKILLS:

1. WHAT PRODUCTS WILL DAVENPORT WORKS MANUFACTURE INTO THE 21ST CENTURY?
   (THIS NEEDS THOROUGH DEFINITION.)

2. SHOULD VEHICLE PAINTING BE PART OF OUR BUSINESS?

3. WHAT WILL THE FIRST LEVEL OF MANAGEMENT BE SUPERVISING WAGE EMPLOYEES?
   IF THE ANSWER IS SUPERVISORS, WHAT WILL THEIR ROLES BE? (LIKEWISE, THIS
   NEEDS THOROUGH DEFINITION.)
```

**Figure 3.10 E-mail follow-up memo summarizing information discussed during a telephone call.** Source: John Deere Company. With permission.

<table>
<tr><td>

Left side of heading is
conventional; right side
is related to E-mail
addresses.

</td><td>

```
Date:    06/16/88
From:    SUTHERLAND WAYNE      SUTHEMB   -ADCCHOST
  To:    M. OTABE              PETERWA   -NPEO
         M. SASAKI             66ENG     -NPEO
         TWEED JOSEPH          66PURCH6  -ADCCHOST
Subject: SOUND TEST EQUIPMENT
```

</td></tr>
</table>

**Picks up conversation.**
**Truncated sentence.**
**States purpose of**
**memo.**

    As I mentioned in my last E-mail message, I have
sent complete set of prints on system hardware. You
probably won't receive until next week. I believe
these prints will provide the detailed information you
require. I would also like to clarify what I feel are
various options relative to the procurement of this
equipment.

**States position right**
**from beginning.**

    I believe you know that I feel the "least risk"
path is to allow ATS to assemble, de-bug, and ship
complete system to Sagami. At the time of installation, either Tweed or myself would be on site to help
with system installation and training. With that said,
let's take a look at the alternatives.

Please Note - Item numbers listed related to prints
sent by plant mail.

**Inserts truncated**
**explanatory note. Lists**
**items in order of**
**importance. Uses item**
**number from previous**
**discussion.**

BOOM (Item 4) -
    Initially you requested that we consider using the
boom that was used for the seminar. I believe you have
accepted our recommendation that this boom is unacceptable and that ATS will provide you boom. Although
we can provide all documentation required for you to
have the boom made locally, I believe we are in agreement that the procurement time would be too lengthy.

**Provides support for his**
**position. Uses 2nd**
**person and 1st person**
**plural to include reader**
**in his decision.**

We do not have a verbal order with the boom supplier,
but we can rescind the order for Sagami. I should mention that the Boom print you will receive is for East
Peoria. We will merely need to make minor modifications to adapt for Sagami 5.5M radius.

    From whom will you obtain the Boom?

**Requests specific**
**response to ensure**
**reader will adhere to**
**his decision.**

FOOT AS. (Item 2)
    I believe we are in agreement that the Foot AS. will
be provided by us. This has been the longest lead time
item and I don't believe it could be procured in Japan
in an acceptable time frame. It is for this reason
that we have given our supplier a "firm" order on this
piece of hardware. I do not see an acceptable alternative.

**Goes on to next item.**

    From whom will you obtain the Foot As?...

**Figure 3.11 E-mail follow-up memo providing additional information**
**related to topic discussed in previous memo.** Source: Caterpillar, Inc. With
permission.

The e-mail software program on which this memo was written allowed the message to be transmitted in both capital and lowercase letters, making the message easier to read than the Deere memo. The heading contains a nonconventional item, the readers' codes, in the upper right-hand corner. Notice that in this message the writer's sentence structure is clipped (articles and pronouns are omitted in paragraph one). It also includes informal speech patterns (e.g., "With that said").

As you read, compare the conventions in these e-mail memos with the previous memos.

---

**Suggestions for Discussion and Writing**

1. The memo in Figure 3.12 provides a set of instructions.

   Even though Howard's readers are experts in this field, she might have used additional visual cues to facilitate her readers' use of the information. Try to revise the memo by chunking related information and each step separately. List the steps in sequential order, and number each step sequentially. Single-space the lines within each step, and double-space the lines between the steps. Use hanging indentation so that the number of each step juts out into the left margin and is easy to see (See the list in Figure 3.5).

   A tool to help readers understand their alternatives is a decision chart. You may want to use such a chart for step one. In the following sample decision chart, readers can look down the left column to find the choice that matches their space unit. They can then enter the appropriate command by following the direction in the accompanying right column. You may want to set up a chart to help readers determine what to do according to the drive they are using and the subdirectory to which they are installing the program.

| IF SPACE UNIT IS TO BE: | THEN ENTER: |
|---|---|
| Tracks | TRKS |
| Cylinders | CYL |
| Blocks | BLKS |

SOURCE: Caterpillar Inc. Developed by Bud Wilson.

```
                        M E M O R A N D U M


DATE:       March 1, 1990

TO:         Systems Engineering Employees

FROM:       Jody Howard  J H

SUBJECT:    Misc. Time Tracking Information


On your time log diskette this week you will find a new tt.exe
program.  A few small bugs have been fixed.  These have only
aesthetic impact and do not affect the function of entering time
data.  New job number and RFQ number help screens are available in
the files job.lst and rfq.lst respectively.  The last new item is
a utility written by Paul to allow you to view a breakdown of your
time data by job number, activity, and category for a chosen time
period.  The program is called ttmate.exe.  To install the new
programs on your hard disk insert the time log diskette into drive
a of your computer and type the following command:

            C> a:install a: c:\setimer      <enter>

If you are installing the programs from a drive other than drive
a: then replace "a" in the command above with the drive letter you
are installing from.  If you are installing the programs to a
different subdirectory other than c:\setimer, replace "c:\setimer"
in the command above with the path of the subdirectory to install
to.

The TTMATE.EXE utility can be used to help you determine how many
hours per specified time period you have spent working on a job.
The program is able to use the old job numbers (January 1990 –
February 14, 1990) or the new job numbers (After February 14, 1990)
as long as you specify old or new when you run the program as
discussed below.  To avoid having to remember both old and new
numbers, I have a program that will recreate an individual's
HISTORY.TM file using new job numbers starting January 1, 1990.
See me if you want one.

To run the program, just type TTMATE and press ←⏎.  By default,
the time check program will process the TIME.LOG file in the
\SETIMER directory using new job numbers.

To override the default parameters, the following is the command
line syntax:

            C> ttmate filename type

where filename is the path and name of the file to analyze (usually
TIME.LOG or HISTORY.TM) and type is the job number type (either OLD
or NEW)
```

**Figure 3.12 Memo containing instructions.** Source: Advanced Technology
Services. With permission.

2. Study the memos in Figures 3.5 and 3.11. Does the tone of each seem
   formal or informal, personal or impersonal, friendly or businesslike, authoritative, gracious, hostile, pleading, scientific, persuasive, subjective, or objective? What strategies do the writers use to create the tone

of their memos? Do they use first person plural to include the reader in their social group? Do they use second person to address the reader directly? Do they make any personal references related to the reader or themselves? Do they use informal phrases or slang? Do they refer to people by their first or last names?

---

# LETTERS

Letters, unlike memoranda, are written to readers outside an organizational community. Letters have been a means of communicating information both socially and in business since Greco-Roman times. In fact, during this early period, the distinction between public and private discourse was blurred as the great value placed on friendship between two people permeated every aspect of public and private life. Letters written for political and commercial purposes were written in a style that simulated a relationship akin to friendship, even when the correspondents were not close friends.

The following letter was written by Apollonius in the second century A.D. to Dioscurides. Apollonius was probably a tax collector, and Dioscurides and Polydas were his field men. The letter, written to persuade Dioscurides to collect the accounts owed to him, adheres to the conventions of friendship although the letter is being written by a superior to his subordinates and although Apollonius is irritated with his reader, who has apparently failed to collect these accounts.

Apollonius to his very dear friend Dioscurides greeting.

As I asked you when I saw you to proceed vigorously with the collection of the accounts on your list, so I now ask you still more to apply yourself to the collection with urgency of the substantial accounts, since we have been compelled to pay everything into the bank. I have found that Polydas has collected nothing and I have remained here of necessity.

Farewell.

To Dioscurides from Apollonius his friend.

Source: Reprinted by permission of Westminster Press.

The remnants of these flowery greetings remain today in the correspondence of certain European and South American cultures. A typical letter of response from France begins with the phrase "I have the honor of receiving your letter" and concludes with the closing, "Please accept my most distinguished sentiments." A letter from Argentina begins with the conventional opening, "I have the pleasure of addressing you" and concludes "I send you my most distinguished sentiments."

In the fifteenth century, the middle classes began communicating their commercial transactions through written documents. Their letters became short and curt, with the English reducing the first two sections to a brief formula and virtually eliminating the gradations between social positions. As individuals' public and private lives moved apart, the business letter became increasingly impersonal and assumed the form in which it is presently written.

## Context

Because readers of letters are external to a writer's organization, they may not be as familiar with a topic as someone within an organization. They may need background information. Unless they are experts in a writer's field, they probably won't be familiar with the technical terminology or the technical aspects of a project. Most readers are busy. They look at the subject line and the signature to determine if a message needs to be read immediately. If they think the subject is of interest or the writer is important, they will read the first paragraph. They seldom read beyond that unless they need additional information or are motivated to do so.

## Conventional Format

Letters are written either in semi-block style (with indented paragraphs) or block style (no indentation). Block style is considered more formal than semi-block.

The conventional format (opposite) includes the heading, inside address, subject line, salutation, message, closing, and signature. It may also include the typist's initials or the word-processing number, a distribution list, and a designation of enclosures at the conclusion of the letter. The subject line, borrowed from the memorandum, is a new addition to a letter. It provides readers with a quick overview of the subject to be discussed so readers can determine whether they want to read it, pass it to another reader, put it aside to be read later, or throw it away.

Like the memo, a letter may assume a wide variety of forms, including requests, responses to requests, to proposals, and reports.

## Examples

The letters in Figures 3.13 through 3.19 exemplify the wide range of purposes for which letters are written. All of the letters involve some aspect of persuasion. The writers are attempting to persuade readers to think in a certain way or to take action. The letters in Figures 3.13, 3.14, and 3.19 request readers to take action. The letters in Figures 3.17 and 3.18 are responses to requests by the reader.

The letters in Figures 3.13 and 3.14 are impersonal. They are sent to multiple readers, and the writers address their readers as a group rather than as individuals. They also omit a conventional personal closing, although the letter in Figure 3.13 ends by thanking the readers. Furthermore,

| | |
|---|---|
| **Heading** | High Technology Assoc.<br>6677 Mayfair Street<br>Mystic, CT<br>July 2, 1994 |
| **Inside address** | Mr. Tom Morton<br>Software Products, Inc.<br>450 Brookfield Park<br>Lombard, IL 60677 |
| **Subject** | Evaluation of Software Package #244 |
| **Salutation** | Dear Mr. Morton: |
| **Message** | XXXXXXXXXXXXXXXXXXXXXXXXXXXXXXXXXXX<br>XXXXXXXXXXXXXXXXXXXXXXXXXXXXXXX<br><br>XXXXXXXXXXXXXXXXXXXXXXXXXXXXX.<br>XXXXXXXXXXXXXXXXXXXXXXXXXXXXXXXXXXX<br>XXXXXXXXX.<br><br>XXXXXXXXXXXXXXXXXXXXXXXXXXXXXXXXXXX<br>XXXXXXX.XXXXXXXXXXXXXXXXXXXXXXXXXX<br>XXXXXXXXX<br><br>XXXXXXXXXXXXXXXXXXXXX. |
| **Closing** | Yours truly, |
| **Signature** | John Mahoney |
| **Typist's<br>initials/<br>processing no.** | I.A. |
| **Distribution<br>list** | c: J. Edwards<br>    P. Smith |
| **Enclosures** | enc. List of applicants |

Letterhead.

# Germantown Hills Fire Department

R. R. #2, Metamora, Il 61548
Chief Wayne Cox

August 26, 1991

Inside address.

Emergency 116 Ambulance Service
Metamora, IL 61548

Salutation. Impersonal style. Addresses readers as a group by their role. Provides background. Establishes purpose.

Dear EMTs:
I want to address a serious safety situation that is occurring in the auto accidents we answer together. People from your department as well as mine are getting into wrecked automobiles without any protective gear on. To ensure the safety of our people from this date on, I am going to reinstate the basic safety rules we, in the fire and rescue service, have had set down for us for many years. These include the following:

Lists procedures, using hanging indentation.

1. Everyone who approaches or enters an automobile that has been in an accident must wear protective clothing.

2. Upon the arrival of my department anyone who is in or near the automobile without protective clothing will be asked to step back or leave the car.

3. After the victim or victims have been extricated, they will be moved to the ambulance gurney and turned over to the other EMTs for further evaluation and transport.

4. At all times we will provide protection with a charged fire line and any stabilization necessary to protect all rescuers.

5. We will try to set up a safety zone around the car and maintain it to keep spectators out of the work area.

Relates to readers' values.

I am not initiating these changes for any reason other than the personal safety of all the responding emergency personel. We all work together for the good of our community and I feel a friendship and mutual respect for all our EMTs and firemen, and I will not allow any of them to be injured or killed engaging in an unsafe act while I am chief of the fire department.

Courteous closing.

If you have any questions on this subject feel free to call me and I will attend your meeting to address these concerns. As I stated earlier my people will be ordered to back away from the scene also if they do not have the proper protective gear on. Thank you.

Signature and position.

Wayne Cox, Chief

**Figure 3.13    A letter of request.** Source: Germantown Hills Fire Department, Peoria, Illinois. With permission.

the letter in Figure 3.14 omits the personal signature. Instead, it gives the name of the division in the company.

The remaining letters in Figures 3.15 though 3.21 are sent to a single reader. These letters are more personal than the previous ones. The writers of these memos address their readers by name. However, except for the letters in Figures 3.18 and 3.20, in which readers are addressed by their first name, the writers are not very familiar with their readers and address their readers by their title and surname. They also include conventional personal closings, with the letters in Figures 3.15 through 3.18 using "Sincerely," and the letters in Figures 3.19 and 3.20 using "Sincerely yours" and "Very truly yours," respectively.

Except for one letter, all of the letters are written on letterhead. All use block style. Although the letters in Figures 3.15, 3.16, and 3.17 are several pages long, the others are brief and do not exceed a single page.

We'll look closely at several of these letters in terms of the conventions they use. Then you can examine the others in light of what you learn.

### Figure 3.13

The chief of a fire department in a small town wrote this letter to emergency medical personnel in an adjoining town to persuade them to follow appropriate emergency procedures. The letter includes a list of the procedures to be followed. Although the writer and readers are in separate professional communities (the writer is a member of the community of fire fighters and the readers are members of the community of medical practitioners), their communities overlap as both are members of emergency teams involved with car accidents. Thus, the writer does not need to provide a great deal of detail to describe appropriate procedures; the readers know the details. The single salutation addressing the readers by their role rather than by their individual names is appropriate because the entire group is being addressed as a unit.

Although the letter is impersonal, the message is friendly. The writer addresses the readers directly, using second person and, by using first person plural in the second to final paragraphs, he includes himself as part of the readers' social group. His argument is also persuasive because his claims relate to readers' attitudes as well as prior experience. He stresses that he is enforcing the rules for the readers' own welfare and that these rules have been in effect for a long time.

### Figure 3.14

The manager of customer services at a large manufacturing plant wrote this letter to owners of the plant's aircraft to persuade them to have their planes inspected. Although the owners may have some knowledge of their planes,

June xxx, 19xx

**71**
**Cessna**

Dear Cessna Owner:

Several improvements have been made to the exhaust system
of Pressurized Centurions which are designed to extend the
service life of the exhaust system on your airplane.

It is suggested that you discuss these improvements which
have been covered in our Service Information Letter SE79-
52, Revision #1, with your Dealer at your first opportunity. Your dealer can inspect your airplane and determine
which of the improvements would be beneficial to you.

Your affected aircraft serial and registration number is
shown on the mailing label.

Cessna Single Engine
Customer Services

**Figure 3.14   A letter of request.** Source: Cessna Aircraft Company. With
permission.

they may not be familiar with the mechanical aspects. Doctors and lawyers
may own planes just as they own cars. They may know how to operate
them but not necessarily how to repair or modify them. Thus, the letter
does not include mechanical details of the improvements.

This letter is an impersonal form letter; the same message is sent to all
readers, who are addressed in the salutation by their role, not their individual names. Unlike the situation in Figure 3.13, in which a single letter is sent
to an entire group, in this situation a letter is being sent to each individual
owner. Because readers paid a high price for the writer's product, many
probably feel they should receive personal attention. Thus, the impersonal
form letter is inappropriate. Whereas, previously, it would have been too
time consuming to personalize each letter, many of today's word-processing
software programs usually include a mail merge program that makes it possible to individualize a large number of letters in a relatively short amount
of time. The writer could have personalized each letter by addressing each
owner by name. Furthermore, the writer could have facilitated the reader's
task by using the software program to include the owner's individual air-

craft serial and registration number in the letter rather than making the owner obtain the information from the label on the envelope. Placing necessary information on the envelope not only requires readers to turn to another document, but may require them to retrieve the envelope out of the waste basket where they pitched it after opening it. In an even worse scenario, they may not be able to retrieve it if the trash has been collected since they threw it away.

### Figures 3.15 through 3.17

The following three letters exemplify the kind of correspondence that may occur between organizations involved with the same project. The written discussion between the organizations suggests a conversation, with one party asking questions or making suggestions and the other party responding. This conversational format is typical of a type of correspondence that often occurs between organizations and the agencies regulating them. Because of the recent emphasis on environmental, health, and safety regulation, this type of correspondence is becoming increasingly common.

The communication between these two organizations occurs in writing rather than over the telephone because the topic, remediation of a toxic spill, involves legal procedures. The letters constitute a form of contract between the two parties. Furthermore, by putting in writing the appropriate procedures to be used, both parties avoid the misunderstandings that can occur from a telephone conversation, which leaves no written record if a question concerning the information comes up at a future time.

Environmental Science & Engineering, Inc. (ESE) had been asked by a firm located in Springfield, Illinois, to clean up a toxic spill. ESE wrote a corrective action plan describing how it would clean up the area. The Illinois Environmental Protection Agency (IEPA) had to ascertain that these plans were in line with the state's environmental guidelines before it could give ESE the go-ahead to implement them.

The first letter (Figure 3.15) is IEPA's response to ESE's corrective action plan. Because the plan did not include all of the information readers needed to determine whether to approve the plan, the agency is requesting additional information. The second letter (Figure 3.16) is ESE's response to the IEPA letter. In it ESE provides the information IEPA requested and requests approval to implement the plan. The final document (Figure 3.17) is the IEPA response to ESE's request to begin operations. Although the agency accepts the information, it places limitations and qualifications on ESE's implementation process.

Each letter begins with a statement of purpose and the main organizing idea so that the reader knows immediately why he has received the letter. All three letters list items, using numeration and hanging indentation, so readers can scan the information quickly and easily. The second letter

(Figure 3.16) uses a question and answer format to provide information the reader needs. This strategy can be used in response letters when readers have requested information related to several issues. By repeating the items from the IEPA request in the ESE response letter, the writer facilitates the reader's fluency; the reader does not have to refer back to the original letter to learn what the questions were.

None of these letters uses a subject line, but all use a reference line. This line lists documents that have been transmitted between the two organizations in relation to the topic being discussed. In this case, they each list the corrective action plan that ESE filed with IEPA.

The letters are formal and impersonal. Because the readers for the documents are both experts, the language is technical. All three letters conclude with a courteous offer to provide responses to questions.

Illinois Environmental Protection Agency P.O. Box 19276, Springfield, IL 62794-9276

Mary A. Gade, Director 2200 Churchill Road, Springfield, IL 62794-9276

217/782-5504

November 23, 1992

Mr. Michael J. Hoffman, P.E.
Environmental Science & Engineering, Inc.
8901 North Industrial Road
Peoria, IL 61615

**Cites previous document to which this letter refers.**

Refer to: 1671205168 -- Sangamon County
Springfield
Superfund/Technical Reports

**Formal salutation.**

Dear Mr. Hoffman:

**Refers to previous communication. States purpose of this letter.**

The Agency has reviewed the report submitted 13 March 1992 titled "Corrective Action Plan" for the above referenced facility. Based on the Agency's review, please provide responses to the following comments:

**Lists items. Uses hanging indention.**

1. Provide data on the hydraulic conductivity of the Vandalia Till and Silty Loess present on-site. Does the Vandalia Till cover the entire site?

2. Provide information on the off-site extent of the groundwater plume and soil contamination to the east of the site.

3. Discuss how the vertical extent of contamination in and below the Vandalia Till will be investigated.

4. The Agency will require any new monitoring wells to be constructed of stainless steel, type 304 or 316, in the saturated zone.

5. All decontamination rinsates, drill cuttings, and development waters are to be contained in drums, on-site, until contents are proven below cleanup objectives.

6. Placing contaminated soils on poly sheeting would create a hazardous waste pile. Soils may be placed in drums or in covered roll-off boxes until such time as disposal can be accomplished.

7. Discuss, in further detail, the proposed bio-remediation of the site.

8. Permission from the local Publicly Owned Treatment Works (POTW) will be required prior to any treated discharge from the site. A sewer connection permit may be required from the Bureau of Water.

9. Figure 5.2 (trench cross section) is missing in the Corrective Action Plan.

If you have any questions or comments, please feel free to call or write me at the above number and/or address.

Sincerely,

Stanley F. Komperda
Environmental Protection Specialist

**Figure 3.15 Letter of request in a two-way communication via letters.**

**Cites original document to which this letter refers.**

**Formal salutation.**

**Purpose.**

**Repeats item #1 from November 23 letter.**

**Provides the information requested.**

**Repeats item #2 from November 23 letter.**

**Provides the information requested.**

---

**ESE** Environmental
Science &
Engineering, Inc.

A COLCORP Company

8901 N. Industrial Road     (309) 692-4422
Peoria, Illinois 61615-1589   Fax (309) 692-9364

January 20, 1993

Mr. Stanley F. Komperda
Remedial Project Management Section
Division of Remediation Management
Illinois Environmental Protection Agency
2200 Churchill Road
Springfield, IL 62794-9276

In Reference to:     1671205168 -- Sangamon County
                     Springfield
                     Superfund / Technical Reports

Dear Mr. Komperda:

This letter is in response to a request by the IEPA dated November 23, 1992 for additional information on the voluntary cleanup efforts underway at the          facility in Springfield, Illinois. The following paragraphs address the items which were specifically requested by IEPA:

1)   Provide data on the hydraulic conductivity of the Vandalia Till and Silty Loess present on-site.

     Hydraulic conductivity tests will be performed at each of the four existing monitoring wells (MW-1, MW-2, MW-3, and MW-4). Recovery rates from baildown tests will be collected using a Hermit data logger and analyzed using the method described by Bouwer and Rice (1976). Existing well and boring logs presented in the previously submitted Corrective Action Report show the presence of a till surface at less than a 20 feet. Additional information concerning the location and extent of the underlying till will be provided by additional soil borings discussed under item 3 and from published investigations of the Illinois State Geological Survey.

2)   Provide information on the off-site extent of the groundwater plume and soil contamination to the east of the site.

     Provided authorization can be obtained from the City of Springfield, two borings will be placed on the east side of          Street in the City's right-of-way. Continuous samples will be collected from each borings which will be screened and analyzed to evaluate the extent of off-site migration. If contamination is suspected based on PID or visual evidence, borings will used for installation of monitoring wells.

---

**Figure 3.16   Letter of response to a request in a two-way communication via letters.** Source: Environmental Science and Engineering, Inc. With permission.

2

3)    Discuss how the vertical extent of contamination in and below the Vandalia Till will be investigated.

Existing information on distribution of petroleum compounds within the soil column will be supplemented by four additional borings to be advanced to the top of the underlying till surface or to a maximum depth of 30 feet. Completed borings not used for monitoring well installation will be grouted closed. Borings will be sampled continuously using a CME continuous sampler. Soil borings will be screened at two foot intervals using a Photoionization Detector (PID). Sample intervals showing the highest PID reading for each boring will be submitted for laboratory analysis of BTEX and PNA values.

4)    The Agency will require any new monitoring wells to constructed of stainless steel, type 304 or 316, in the saturated zone.

As required by IEPA, any new monitoring wells will be installed using 304 or 316 stainless steel materials in the saturated zone in accordance with IEPA monitoring well requirements.

5)    All decontamination rinsates, drill cuttings, and development waters are to be contained in drums, on-site, until contents are proven below cleanup objectives.

All drill cuttings from borings performed for this investigation will be containerized and disposed along with soil which is to be excavated. Based on a telephone conversation between Mr. Komperda and myself on December 14, 1992 this requirement will be unnecessary provided TCLP benzene analysis shows that soil benzene does not exceed hazardous waste limits. Prior to excavation, screened intervals from soil borings showing high PID readings will be analyzed for TCLP benzene to assure that excavated soil does not exceed the TCLP benzene hazardous waste criteria of 500 µg/L. A maximum of two soil samples collected from soil borings will be submitted for TCLP benzene analysis.

6)    Placing contaminated soils on poly sheeting would create a hazardous waste pile. Soils may be placed in drums or in covered roll-off boxes until such time as disposal can be accomplished.

Analysis for TCLP benzene will be used to determine the need for temporary storage in covered containers. Soils having TCLP analysis exceeding hazardous waste criteria will be handled in accordance with hazardous waste requirements.

7)    Discuss, in further detail, the proposed bio-remediation of the site.

Proposed site remediation includes the installation of a down-gradient trench to collect shallow groundwater and prevent off-site migration. Petroleum compounds will be removed from the site as groundwater containing these compounds is collected by the trench. Diesel fuel contaminants are slow to

(continues)

**Figure 3.16**    *(continued)*

3

migrate with groundwater. Remediation of the site by using groundwater collection and treatment alone will involve collection of groundwater for an extended period. Site cleanup time can be reduced by increasing the groundwater recharge at the site and as a result the rate at which groundwater flows to the trench. In addition to simply increasing the flux rate of groundwater, infiltrating water used as recharge could also be used to accelerate naturally occurring degradation of petroleum compounds within the soil by acting as a mechanism to supply nutrients and oxygen to soil bacteria.
          will evaluate the cost-effectiveness of applying bio-enhancement or bio-remediation technology to supplement their groundwater collection and treatment activities.

8)    Permission from the local Publicly Owned Treatment Works (POTW) will be required prior to any treated discharge from the site. A sewer connection permit may be required from the Bureau of Water.

      Acquisition of required permits will be undertaken prior to system operation. Contacts made with the Springfield Sanitary District indicate that discharge of groundwater through their sanitary system is feasible.

9)    Figure 5.2 (trench cross section) is missing in the Corrective Action Plan.

      The missing Figure 5.2 is enclosed.

**Requests approval of plan.**

IEPA approval of the previously submitted Corrective Action Plan (CAP) and this work plan will allow                    to proceed with the investigation outlined here and the appropriate permit applications to begin implementation of the CAP.

**Gracious closing.**

Should there be any questions or comments on the information contained in this letter or the remedial action plan, please contact myself or Chris Everts here at ESE, Inc.

Sincerely,

ENVIRONMENTAL SCIENCE & ENGINEERING, Inc.

Michael J. Hoffman, P.E.
Senior Environmental Engineer

enclosure

**Figure 3.16**    *(continued)*

**Refers to previous letter.**

**Approves plan if following items are incorporated.**

**Lists items. Uses hanging indentation.**

**Makes request to visit site.**

**Provides guidelines for future work.**

Illinois Environmental Protection Agency     P.O. Box 19276, Springfield, IL 62794-9276

February 9, 1993

Mr. Michael J. Hoffman, P.E.
Environmental Science & Engineering, Inc.
8901 North Industrial Road
Peoria, IL 61615

Refer to:  1671205168 -- Sangamon County
                 Springfield/
                 Superfund/Technical Reports

Dear Mr. Hoffman:

The Agency has reviewed Environmental Science & Engineering's written response letter dated January 20, 1993 regarding the Agency's November 23, 1992 comments.  The Agency finds the work plan proposal to be acceptable pending incorporation of the following comments and understandings:

1.  No work may occur until a site specific health and safety plan (SSP) has been developed and submitted for Agency review.  The SSP should be developed to protect the health and safety of personnel involved in the project. The plan should address not only on-site workers, but also off-site public concerns.  The SSP shall be consistent with all applicable State and Federal laws and regulations, including but not limited to "Department of Labor, Occupational Safety and Health Administration-Hazardous Waste Operations and Emergency Response" as outlined in 29 CFR Part 1910, vol. 51, No. 224, December 1986, p.  45654.

2.  Soil samples from the off-site borings will be submitted for laboratory analysis if PID readings (above background) or visual evidence are present.  The boring(s) will be developed into a monitoring well(s) if this evidence is found to be present.

3.  The Agency recommends delaying the installation of the proposed groundwater trench until results from the soil borings east of the site become available.

In the interim, I would like to tour the facility to obtain a firsthand overview of the problems associated with the site. Since the Agency is located in Springfield, I will be able to visit the site at your earliest convenience.

Please submit three (3) copies of reports regarding any future remedial activities at the above referenced site.  If you have any questions, please feel free to call me at the above number.

Sincerely,

Stanley F. Komperda, Project Manager
Pre-Notice Sub-Unit, State Sites Unit
Remedial Project Management Section
Division of Remediation Management
Bureau of Land

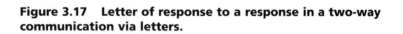

**Figure 3.17   Letter of response to a response in a two-way communication via letters.**

### *Figures 3.18 to 3.20*

The writers for all three letters in Figures 3.18 to 3.20 are attempting to persuade their readers to do business with their companies. The letter in Figure 3.18 is a sales proposal. It was initiated by the writer to inform the reader of his company's services. The writer does not know his reader but, in an attempt to create a rapport with him, he addresses him by his first name.

The letter in Figure 3.19 is a response to the reader's request for a proposal (the letter responds to the PSR [Project Specification Report] in Chapter 1 in Figures 1.2, 1.5b, and 1.5c. The writer attempts to persuade the reader to accept his company's bid. However, the writer estimates the cost at $75,500, although the memo in Figure 1.5 estimates the cost at $10,000. Because the writer does not make any claims to counter the reader's original estimate, the reader does not accept the bid. As a result, the work does not get done while the city seeks other bids. During this time the leak becomes a flood and the cost to the city eventually reaches well over ten times that estimated by the writer's company.

The letter in Figure 3.20 serves as a cover letter to transmit information requested by the reader. However, in the second paragraph, the writer attempts to persuade the reader to continue to work with his company.

## WORLD DISTRIBUTION SERVICES, INC.

872 CAMBRIDGE DR. • ELK GROVE VILLAGE, IL 60007 • (708) 640-7776

December 14, 1989

Mr. Tim Shiveley
Archer Daniels Midland Co.
4666 Faires Parkway
Decatur, IL 62525

Dear Tim:

Please allow me to introduce our services to you and your company.

World Distribution Services is a privately owned company that has been providing international postal services to corporations and associations for over three years. By using a specially designed distribution network, we are able to virtually guarantee your organization a more expedient delivery of its printed materials throughout the world.

**The following will highlight a few features of our service:**

1.  Faster delivery of your Air Mail, Air Printed Matter and Surface Mail, worldwide.
2.  Non-deliverable Air Mail is returned to your company within 30 to 45 days at no charge.
3.  Elimination of immediate expenditure of funds for postage. We invoice your company for postage on most services.

World Distribution Services offers discounts on a wide range of international mail services. Depending upon the volume and frequency of mail, your discounts can vary from 10% to 50% off the USPS rates.

**Levels of service offered:**

Air Mail                            Air Printed Matter
Air Parcel Post                     Air Surface
Surface Mail (First & Second Class) Canadian Bulk Mail

Tim, I sincerely appreciate the time you have taken to review our company's services. I will call you in the future to review your comments. In the interim, if you have any questions please call me at (800) 356-4467.

Sincerely,

WORLD DISTRIBUTION SERVICES, INC.


Robert J. Bruce
Vice President

RJB/jr

**Figure 3.18    A sales proposal.** Source: World Distribution Services, Inc. With permission.

PASCHEN CONTRACTORS, INC.
GENERAL CONTRACTORS
2739 ELSTON AVENUE
CHICAGO, ILLINOIS 60647
(312) 278-4700
TWX 910-2216037

April 8, 1992

City of Chicago
Department of Transportation
Bureau of Bridge Operations,
Repair, and Maintenance
Mr. Robert Serpe, Director
535 West Grand Avenue
Chicago, Illinois 60610

Dear Mr. Serpe:

In response to your request, we offer this proposal and method for bulkheading the damaged service tunnel under the Chicago River.

Due to the potential hazards involved with installing bulkheads at the damaged area locally, we propose to install the bulkheads at the two access shafts.

The work to include:
Removing muck and debris in the pour area
Installation of wooden bulkheads ( to remain);
Provide two matts of rebar (dowel into floor);
Provide 4" pipe and values (each shaft);
4000 psi concrete;
Fill tunnel with river water after cure;
Encase valves with concrete after the tunnel is filled, approx. 2'0" encasement.
See the enclosed sketches for detail.

The remaining void area of the shafts would be filled with stone by others. This work to be completed in a period of three (3) weeks for the sum of Seventy-five thousand five hundred dollars ($75,500.00).

It is also understood that adequate area for machinery, concrete trucks , etc. is reasonably available ( by land) at each access shaft location. This proposal is based upon the information which you have provided to us, and it does not include a reserve for unanticipated and unforeseeable conditions.

This proposal is subject to changes imposed by your engineering staff once it has had an opportunity to analyze and approve the validity and soundness of the proposed method and plan from an engineering prospective, given all the conditions at and around the site.

If you have any questions pertaining to this proposal, please contact either Mr. Dan Simonides or myself at the above phone number.

Sincerely yours,

Peter Carbonaro
Vice President

PC:mr
ENCLOSURES

**Figure 3.19    A response to a request for a proposal.** Source: City of Chicago.

# BENJAMIN ASSOCIATES

**TELECOMMUNICATIONS PROJECT MANAGERS AND CONSULTANTS**

1039 W. Taylor
Chicago, Illinois 60607
TEL. (312) 421-4638
FAX (312) 421-4648

8 November, 1990                                    TCPM0027.212

Mr. James P. McTigue
Engineer Technician V
Construction Services
Department of Public Works
City of Chicago
320 North Clark Street, Room 500
Chicago, Illinois  60610

          Subject:  Teleport Communications Chicago Project
                    Project Management

              Ref:  Tunnel Personnel List

Dear Jim:

Attached, please find the Tunnel Personnel List for the personnel
of Teleport Communications Group, Benjamin Associates and U.S.
Electric.  The list gives all personnel authorized to either enter
for inspection purposes or work within the Chicago Freight Tunnels.

Benjamin Associates thanks you and the Department of Public Works
for your continuing assistance with the Projects within the Chicago
Freight Tunnels.  We also look forward to continuing our good
working relationship with you and Construction Services.  Should
you have any questions regarding the personnel assigned to the
Project, please contact this office for our immediate attention.

Very Truly Yours,

*Robert Barnfield*

Robert Barnfield
Benjamin Associates

RAB/nt

Attachment

**Figure 3.20    A cover letter.** Source: City of Chicago.

*Figure 3.21*

The purpose of the letter in Figure 3.21 is to persuade the reader (1) that the writer's company is not responsible for the severed cable discussed in Figure 3.8 and (2) to meet with the reader to resolve the issue. The writer

Great Lakes
Dredge & Dock
Company

2122 YORK ROAD
OAKBROOK 60521-1930
708 574-3000
FAX 708 574-2980/2981
TELEX 25441GRATLAKCGO

Launching Our Second Century

*Is this in file?*
*Is this in Rem... in 91 -*
*No*
*05*
*11-29*

October 18, 1991

Mr. R.M. Yedavalli
Chief of Construction Engineer
Department of Public Works
Bureau of Construction Management
320 N. Clark Street
Chicago, IL 60610

*c.c. Yedavalli.*

Re:  Various Drawbridges, New Pile Clusters, E-0-469

Dear Mr. Yedavalli:

In response to your letter dated September 10, 1991, regarding the severed power cable near the southwest corner of the Chicago Avenue Bridge, we have reviewed our activities and work procedures at this location and find that we were neither negligent nor improperly positioned at any time as suggested. Our normal precautions were exercised here as those used at the other locations. The barge was positioned perpendicular to the bridge which keeps the holding spuds the maximum distance away from the cables. Also, the cables are visually watched above water for any movement while working near them, and no movement was noticed.

Please refer to the interoffice memo of our Project Superintendent, Mr. Pete McClain. He notes that when the divers pulled the cable end from below the mudline, the armor was unwound, frayed, and rusted in the vicinity of the broken end. He also mentions the large amount of steel scrap and debris on the slopes over the cables near the cluster. What caused this cable to break is only speculation, but we feel Mr. McClain's on-site opinions are very close to what actually happened.

I suggest that we sit down and discuss this matter further and try to resolve this issue.

Sincerely,

Wayne S. Valley
Projects Manager

TO *Crepko/Kext*
FOR PCM OR COMMENT
FROM *10/24/91*

**Figure 3.21   A persuasive letter.** Source: City of Chicago.

has prepared the letter based on the information in the Figure 3.8 memo. Notice that although the letter is written to a single reader, Mr. Yedavalli, the letter is written for additional readers, whose names are penciled onto the letter. In fact, several of these additional readers (Ron and J.S.) have used the letter to correspond back and forth to each other.

As you read, determine the claims the writer makes in presenting his argument and the evidence he uses to support each claim.

---

### Suggestions for Discussion and Writing

1. If you received the letter in Figure 3.13, would you be persuaded to follow the procedures outlined in it? What claims does the writer make? Do you think these reflect readers' attitudes about the topic? What evidence does the writer use to support the claims? Why do you think readers would accept these as valid or reject them as invalid?

2. The letter in Figure 3.22 is a response to the request for a bid discussed in Figure 1.5 on page 23 in Chapter 1. Compare Figure 3.22 with the letter in Figure 3.19. If you were the reader, Mr. Serpe, which bid would you accept? Why?

3. Study the letter in Figure 3.21. What claims does the writer make? What evidence does the writer use to support his claims? The writer of the letter in Figure 3.21 refers to the memo in Figure 3.8. What information from the memo does the writer of the Figure 3.21 letter use to support his argument that his company is not responsible for the broken cable?

4. The letters in Figures 3.18, 3.19, and 3.22 are all proposals. Compare and contrast them in terms of content, organizational pattern, sequence of information, point of view, and style.

5. Contrast the tone of the letters in Figures 3.20 and 3.21. What do you think is the reason for the difference in tones?

6. During the eleventh century, the information, organization, and style of letters were determined by a set of rules known as the *ars dictaminis,* which were reader-based. Correspondence was divided into five parts: (1) *salutio* (salutation), (2) *captatio benevolentiae* (persuasive introduction), (3) *narratio* (background), (4) *petitio* (request) or *dispositio* (de-

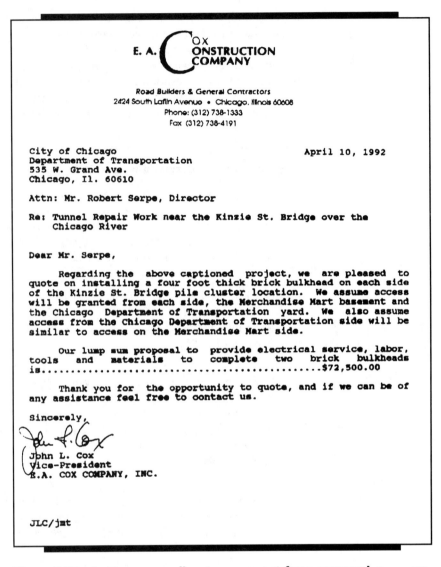

**Figure 3.22 Letter responding to a request for a proposal.** Source: City of Chicago.

mand), and (5) *conclusio* (conclusion). Discuss the relationship between these five parts and the parts of today's business letters.

7. *On Style* is the earliest extant rhetorical work that treats letter writing.
   Attributed to Demetrius of Phalerum, it probably dates from the first
   century B.C. In the section on "Epistolary Style," Demetrius provides
   the following suggestions for writing effectively:

   - Diction must consist of current and usual words.
   - Connectives should be used to integrate ideas.
   - Lucidity often requires that words should be repeated.
   - Lucid writing should avoid ambiguities.
   - The natural order of words should be followed, . . . the subject
     comes first, then the description, then the rest follows.
   - Avoid long clauses.
   - Persuasiveness requires two things: clarity and the avoidance of
     the unusual, for people are not persuaded by what is obscure or un-
     familiar to them. We should therefore avoid uncommon or swollen
     language when we are trying to be convincing.

   Demetrius makes the following recommendations specifically in relation to writing letters:

   - "The style of letter-writing . . . requires the simple manner."
   - "A letter should be written rather more carefully than dialogue,
     though not obviously so. Dialogue imitates impromptu conversa-
     tion, whereas a letter is a piece of writing and is sent to someone as
     a kind of gift."
   - "Disjointed sentences frequently occur (in conversation) but they
     are out of place in letters. In written work disjointed clauses are ob-
     scure; nor is the imitation of conversation appropriate, as it is in an
     actual debate."
   - "The letter, like the dialogue, should be very much in character. You
     might say that everyone draws, in his letters, an image of his per-
     sonality. A writer's character may be seen in all his works, but
     nowhere so clearly as in his letters."
   - "The beauty of a letter lies in the expression of affection and cour-
     tesy."

   Source: Grube, G. M. A. 1961. *A Greek critic, Demetrius on style.* Reprinted by
   permission of University of Toronto Press.

   Evaluate the letters in Figures 3.13, 3.19, and 3.21, using Demetrius'
   criteria for effective letters.

## CHAPTER SUGGESTIONS FOR DISCUSSION AND WRITING

1. According to Demetrius, people draw an image of their personality in their letters: "A writer's character may be seen in his works." Does Demetrius' statement apply to today's technical correspondence? Discuss your answer in terms of some of the memos and letters in this chapter.

2. The expansion of an airport can have a significant effect on the environment, especially in terms of noise. The Federal Aviation Administration requires that any airport authority planning to expand its runway or facilities must submit a noise compatibility plan. Prior to submitting the plan, the airport authority needs to conduct a study to determine present and future noise levels. The results of the study are made public, and citizens are invited to comment on them. These comments are then incorporated into the final draft of the study. When the Greater Peoria, Illinois, airport conducted a noise compatibility study, it arrived at the following results.

➤ Aircraft noise currently impacts 1,911 people at levels above Ldn [day/night noise level] 65, the level which is considered under the Federal [Aviation Regulations] Part 150 guidelines to impact residential land [negatively]. Approximately 652 people are exposed to noise in excess of 70 LDN. . . .

The total LWP [level weighted population] value for 1988 conditions is 1,365 for people exposed to noise exceeding 65 LDN. These are the noise impact conditions against which the effectiveness of noise abatement alternatives will be measured. . . . Population impacts in the years 1993 and 2008 are greater than the current impacts because of anticipated increase in air traffic at the Airport, especially in air cargo traffic. The LWP value of the number of people affected by 65 and greater LDN noise levels in the 1993 and 2008 scenarios are 2,275 and 3,465 respectively. The actual number of people estimated to be affected by noise levels greater than 70 LDN are 1,399 and 2,090 for the 1993/2008 scenarios respectively.

Source: Noise Compatibility Study by Crawford, Murphy, and Tilly, Inc., for the Greater Peoria Regional Airport, 1990. ➤

The letter in Figure 3.23 from two residents affected by airport noise is a response to the study.

a. What is the tone of the letter? Is it personal or impersonal? Social and friendly or businesslike? Formal or informal? What are the dif-

February 15, 1990

RECEIVED
FEB 16 '90
C M & T, INC.

Mr. Bruce Jacobson
Chief of Airport Planning
Crawford, Murphy, & Tilly, Inc.
2750 W. Washington
Springfield, IL 62702

Dear Mr. Jacobson:

This is written in response to your request for input to the Noise Study
for Greater Peoria Regional Airport.  Besides volunteering for the PAC
Committee, our residence was one selected for a 24-hour outdoor monitor -- we
are Site #8 on the grid locations.  Our comments reflect our feelings about
how the noise affects us, etc.  However, we also know, through conversations
with various neighbors over time, they tend to have many of these same opinions,
as well.  This is one reason we are interested in this committee in that perhaps we can help our neighborhood.

Items of concern for us:  (not in ranked order)

1. <u>TELEPHONE</u> - When commercial flights are at the noisest times and peaks,
   phone conversations have to be delayed until the noise is gone.  (And if
   we are talking with someone in Bartonville, we have to wait again whenever the noise is there because they can't hear either!)  However, the
   Air Guard jets (when doing their practicing or whatever they do) are the
   worst.  It is very hard to nearly impossible to have phone conversations
   at that time.

2. <u>RUN-UP TIME</u> - A new term we learned at the meeting on February 7th and we
   assume this is what we've heard for years and have called it "warming up".
   Sometimes it isn't done for long periods, but when it is, it seems to be
   done <u>over and over and over</u>.

3. <u>VIBRATION</u> --The lady who spoke at the meeting of February 7th saying she
   was gone for 3 weeks and came home to find broken what-nots on the floor is
   easy for me to understand.  We have a curio cabinet that needs to be straightened always weekly and sometimes twice weekly due to vibration moving the
   objects around.  Pictures have to be straightened periodically on the wall
   and vibrations also cause considerable window rattling.  We have noticed
   cracks in the living room plaster (though we cannot be sure they are from
   the planes, but we suspect this might be the case).

4. <u>SUMMER OUTDOOR ACTIVITIES</u> - This has been <u>quite annoying</u>  to have backyard
   picnics or gatherings planned, and sure enough, sometimes has turned out to
   be one of the days when there is all sorts of activity going on -- both on
   the ground and in the air.  Conversation is nearly impossible.  Now, this
   hasn't happened every time, but it is unpredictable.  (I don't have any
   specific dates to give you as example as we have never logged this.)  This
   hasn't, however, been just the last week, last month, or last year -- just
   "over the past years".

(continues)

**Figure 3.23   A persuasive response letter.** Source: Federal Aviation
Administration.

Mr. Bruce Jacobson                    -2-           February 15, 1990
RE: Greater Peoria Regional
    Airport Noise Study

5. FUMES AND
   SIDING "GUNK" - This is not noise, and we fully realize this, but is
   something we aren't particularly happy about. We do not know
   exactly where the fumes come from, but we can say they are worse on
   damp days, as well as windy days, probably depending on direction of
   wind. The spots on the siding of our house are gray and will come off
   fairly well, but with extremely hard scrubbing, which also takes off
   siding color.

6. PREDICTED CONTOURS - Since our "measured" day-night average noise was
   less than predicted, will this mean ways to lessen the noise for us and
   our neighborhool will not have to be undertaken or considered as seriously
   since it was not the predicted amount?

Hopefully, this is the sort of thing you are asking for and will find helpful. Please let us know if we can be of further assistance.

Sincerely,

*Walter and Wilma Bruninga*

Walter and Wilma Bruninga
2617 S. Avion Drive
Peoria, IL 61607
(309) 697-1642

**Figure 3.23** *(continued)*

ferences in tone between the letter and the excerpt from the study? What strategies do the authors of the excerpt from the study and the authors of the letters use to achieve their respective tones?

b. What claims do the writers of the letter make? Examine the evidence the writers provide for each of their claims. Does the evidence support their claims?

c. The writers of the letter use visual text effectively. What visual cues do you see that help you read the letter more quickly and understand it more easily?

d. Although the letter writers' explicit purpose is to inform the reader about the effect of the noise on their residence, they have an implicit purpose—to persuade the reader to improve the conditions, which involve not only noise but also "fumes and 'gunk.' "

They never specifically state this purpose. Do you think they should? Why or why not?

    e. Consider the headings for the list of items. Do you think these are effective? Are they related to the purpose(s) of the letter? Do they provide the reader with a good cue for predicting what will be discussed in each of the following paragraphs? If you have answered no to any of these questions, revise the headings so that they are related to the purpose of the letter and help the reader predict what he will read.

3. The next time you plan to be absent from class, write a memo to one of your classmates, requesting that he or she take notes for you, pick up an extra set of handouts for you, and provide you with information on assignments.

4. Compare the tone of the Reichssache memo on pages 91–92 to the tone of the excerpt from the noise compatibility study exercises, page 119. Compare the tone of the response letter written by the residents involved in the airport study on pages 121–122 with the tone of the statement made by the Chelmno survivor concerning the gas vans, exercise 6 (page 94).

    Both the survivor's account and the residents' letter emphasize the personal consequences of the actions discussed in the Reichssache memo and the noise study, respectively. How do the survivor's account and the residents' letter point out the ethical responsibilities of the writers of the memo and the noise study? Do you think the writers of the noise compatibility study dehumanize their topic? Should they? Why or why not? Could they have written the study in a more humane fashion? How?

## Notes

Anderson, Paul V. 1985. "What survey research tells us about writing at work. In *Writing in nonacademic settings,* edited by Lee Odell and Dixie Goswami. New York: Guilford Press.

Barabas, Christine. 1990. *Technical writing in a corporate culture.* Norwood, N.J.: Ablex.

Roth, Lorie. 1993. Education makes a difference: Results of a survey of writing on the job. *Technical Communication Quarterly* 2 (Spring): 177–84.

Sims, Brenda. 1991. Electronic mail and writing in the workplace. In *Studies in technical communication,* edited by Brenda R. Sims. Denton, Tex.: University of North Texas, pp. 137–156.

**CHAPTER**

**4**

# Instructions

## INTRODUCTION

**W**ith the advent of high-technology equipment and do-it-yourself kits into the workplace and the home, manufacturers have found an increasing need to provide instructions for their employees and customers. Most people cannot set a digital clock, prepare meals with a food processor, hook up a VCR, or use a software program on their personal computer without a set of instructions. Instructions are also necessary for office workers because new equipment such as computers, fax machines, copiers, and voice mail has replaced pencils and paper, interoffice mail, secretaries, receptionists, and telephone operators. Employees have also become dependent on instructions for operating the new, complicated robots and other innovative machines that are being installed in their plants. They need to understand what these new mechanisms can do and how to use them as well as what safety precautions to take.

Some instructions are easy to follow; others may create monsters out of normally patient people. Improper instructions may block a user from performing a task correctly or cause the user to perform a task incorrectly. The consequences can range from a consumer's inability to assemble a child's toy, to the return of computer programs users can't operate, to a nuclear accident. Because of poor instructions, an operator at Caterpillar Inc. spent twelve hours doing a task that should have taken only seven, costing the company hundreds of dollars in unnecessary labor hours.

Instructions assume several forms, including directions for doing something, procedures to be followed, and documentation for software programs. Instructions can involve such simple processes as putting together a child's mini bike or setting a digital clock. They can also involve much more complex tasks, such as repairing a 125-horsepower turbocharged combine or operating a nuclear reactor. Although instructions for simple mechanisms can often be printed on a single page, those for operating complicated machinery may fill hundreds of pages.

People have been providing instructions since prehistoric times. At some point in time, a member of a Cro-Magnon clan explained to a member of another clan how to design a wheel or start a fire. We know that by the time of the Roman empire, people were writing complex sets of instructions. *Ten Books of Architecture,* an elaborate set of instructions for tasks ranging from the building of the walls of Rome to the construction of the Roman aqueducts, was written in 27 B.C. by Marcos Vitruvius Pollio to ensure uniformity in the building of Rome.

The "how to" technical book had its beginnings in the sixteenth century. Instructional "how-to" books on everything from gardening to midwifery to cooking were as popular then as they are now (Tebeaux and Lay 1992). The following is an excerpt from a book on midwifery.

> As two twins may change to offer themselves with their heads, so likewise sometimes they may happen to come with their feet foremost. When this chanceth, the Surgeon ought to observe, whether the twins be separated, (as we said before), whether they be unnaturall, as having four legs, one or more bodies, and likewise one or two heads. Now the best way to find it is, that the Surgeon having his hand annointed (as before) flip it up gently wide open, as high as he can, and finding that the twins are not joined together, but divided and distinguished, he shall bring down the hand between one of their thighs. . . .(Guillemeau 1612).

As technology developed over the centuries, instructions assumed the form of compendious textbooks. Voluminous sets of instructions and procedures continue to be published today in the form of manuals. However, companies are beginning to put their manuals on computers using hypertext, a special type of software program that allows users to search through layers of information quickly and easily. Instead of opening a manual, users navigate through a series of menus to locate the particular instructions they need.

This section includes both historic and contemporary documents so that you can see how instructional texts have evolved. The contemporary documents exemplify efficient instructions that readers can use quickly and easily. Study the strategies that the writers use so that you can employ similar strategies in writing your own sets of instructions and procedures.

# CONTEXT

## Situation

Most instructions have multiple readers who use instructions under a wide variety of situations for a wide range of purposes. These readers may need to perform a task or to repair a product. They may work on equipment all of the time, or they may do it only occasionally. They may have plenty of time, or they may need to take quick and effective action to prevent a catastrophe.

### *The Writer's Situation*

Technical and adjunct specialists often write procedures for those in their organizational community, and they usually write about a product or process related to their work. As new machinery is introduced into the workplace, technical specialists are increasingly finding a need to write instructions for the assembly, operation, and repair of the equipment. Often instructions are not included with the machinery, or instructions are written in a foreign language. Even if they are written in the English language, they may have been written by someone who is not a native English speaker and contain nonstandard sentence structures and inappropriate or unfamiliar vocabulary, which can cause users to have difficulty in understanding them. Because the writers often write from their own viewpoints rather than from the users' point of view, the instructions may be too complex and detailed to follow or may be missing necessary basic data or steps.

Technical writers may be asked to write instructions for a myriad of processes or products with which they have little or no knowledge. One of the major tasks technical writers face is gathering the information for writing these instructions. These writers may need to interview people who have developed or worked on the equipment, or they may need to observe someone using the machinery or engaging in the process, or they may want to work on the equipment or engage in the process themselves.

### *The Reader's Situation*

Readers read instructions under a wide variety of situations. They may be in the privacy of their home, reading a manual to understand how a software program works, or they may be in their yard, trying to put up a swing for their four-year-old child. They may be at their desk in the middle of an open area at their office, trying to learn how to record a message on their voice mail; or they may be in an assembly line, trying to fix a jammed piece of equipment; or they may be on a shop floor, trying to install a new assembly line. In some cases, they may be able to hold the instructions in one hand and perform the task with the other. In other cases, they may need to put the instructions down and go to another area to perform the task. They may be working alone, or they may be working with others. They may have all the materials and tools in the immediate vicinity, or they may need to obtain them.

### Usability Tests

Because writers are often unaware of the situations in which people use a set of instructions or exactly how readers interpret them, they need to consider all contingencies. Writers should therefore evaluate their instructions before publishing them by having end users test them. These evaluations are called "usability tests." The object of these tests is to determine whether readers can follow the instructions and do what they are supposed to do.

### Legal Implications

Improper or inadequate instructions can result in injury to users as well as to those affected by the processes or products. Product liability laws hold a company responsible for injuries caused by a company's failure to provide clear instructions or to warn users of potential dangers. Writers of instructions have an ethical responsibility to strive to provide users with whatever warnings and advice are necessary to ensure their health and safety. For these reasons, warnings and cautionary notes are often placed at the beginning of a set of instructions and again at the specific step to which they relate. Visual text in the form of boldface type, large lettering, or all capital letters is often used to emphasize these precautions, and graphic symbols may be used as well.

## Purpose

Writers write instructions so readers can *do* something. They write procedures so readers will follow rules and regulations. Both instructions and procedures require writers to be persuasive. Writers need to persuade readers to follow their instructions and to do as the procedures require.

All readers read instructions to do something. Some read instructions to perform a task; others read to learn what they can do. Some read instructions to do something for the first time, and others read instructions to remind themselves how to do something they've done previously. Still others read instructions to find out how to correct something they've done.

Instructions need to be written to provide for each of these purposes. For example, readers who want to do something for the first time need step-by-step instructions that focus on the project they want to do. If users need to leave their instructions to perform their work and then return to the instructions to learn what to do next, they need to be able to find their place quickly.

## Audience

Readers of instructions invariably assume the role of users and novices; if they were experts, they wouldn't need instructions.

### Readers' Reading Behaviors

Readers have different reading styles. Some users skim through all of the instructions before they begin a task, others read only as much as they need to perform a step, and still others begin to work on a task without ever looking at the instructions and read them only when they can no longer figure out what to do or when they have done something obviously incorrect.

Because readers read instructions in so many different ways, instructions must allow users not only to do a task from the very beginning, but also to enter into it in the middle. Instructions must help users not only *perform* a task but also *solve* a problem or *undo* a step. Instructions need to help users locate specific information relevant to their needs as well as help them perform a task.

# DISCOURSE STRATEGIES

## Navigation and Location Aids

Manuals may contain hundreds of sets of instructions, but readers need to read only those sections related to their specific purpose. Therefore, instructions need to provide readers with aids to help them locate the specific information they need. These aids should help readers navigate and locate information at both the global and local levels. Global cues include tables of contents and indices or menus in on-line documentation. On the local level, visual cues, such as titles, headings, and subheadings within a text, help readers discover information they want as they skim, scan, or search a document.

## Ancillary Documents

Because so many different readers use a manual for so many diverse purposes, manuals should contain a variety of ancillary documents to facilitate readers' fluency, reading processes, and behaviors. These documents include title pages, tables of contents, and lists of figures and tables. They also include glossaries for readers who are not familiar with the terminology and appendices for readers who want additional data that is not essential for carrying out a task.

## Organizational Patterns

Manuals need to be organized both systemically and functionally. Readers need to understand a mechanism's system, how it operates, and how the various parts relate to one another and to the mechanism as a whole; they also need to know how to put it together, operate it, diagnose problems with it, and repair it.

Chapters and sections usually follow an analytic organizational pattern. A chapter or section usually begins with an overview to help readers understand how the details in the chapter or section fit into the general picture as well as how they relate to each other. The chapter or section then presents the details, which are categorized either systemically or functionally. The major categories of an equipment manual usually include assembly and disassembly, operation, and maintenance. These categories are usually subdivided further. The information within each category may then be sequenced from most to least important; from top to bottom, from outside to inside a mechanism, or from the beginning step to the final step of a process.

## Point of View and Focus

Instructions need to be written from a user's point of view. A reader who is using a word-processing program to write a paper wants to know how to create outlines, not how to use the ruler (the area in which margins are controlled in certain software programs). The instructions should include a functional section on creating outlines as well as a systemic section on the ruler.

## Graphics

Graphics—diagrams and photographs—can help readers understand instructions. They allow readers to *see* a specific part and its location in a mechanism. Graphics are very helpful to users who do not speak English as their primary language.

## Conventional Formats

The following conventions govern instructions, procedures, and documentation.

- Steps in instructions or items in procedures are listed.
- Steps are listed in sequence. Procedures are listed in sequence and/or order of importance.
- Related steps are chunked together in sections and subsections.
- Headings and subheadings introduce sections/subsections.
- Numbers identify each step in instructions. Bullets identify each item in procedures.
- Lists are indented.
- Hanging indentation is used for steps requiring more than a single line.
- Double spacing is used between steps; single spacing is used between lines of a single step.
- Appropriate text grammar is used:
  — active voice.
  — imperative mood.
  — begins with a verb.
  — simple sentences.

# THE DOCUMENTS

We will begin by looking at several short sets of instructions. Then we'll see how these sets are collated to create a complete manual. We will also look at several historical sets of instructions to learn how they have influenced the manuals of today.

## Single Sets of Instructions

Instruction manuals are composed of many sets of instructions. Figures 4.1 and 4.2 are examples of individual sets of instructions contained within a larger manual. Each set is concerned with a single aspect of a complex piece of equipment.

Figure 4.1 contains instructions for calibrating a new type of dipstick. The instructions are included in an operation and maintenance manual for industrial and EPG generator set diesel engines. Users will need to move back and forth between the instructions and the equipment. Each operation is placed in a separate step. The instructions are listed in sequential order. They are numbered, and the numbers are set in boldface type so users can easily locate the step on which they have been working. Notice that step six includes a special note, which is printed in all capital letters to catch readers' attention.

A graphic is placed at the top of the instructions so readers can *see* the aspect of the piece of equipment being discussed. The artist's arrow further directs the readers' focus. Notice that the text in the graphic is easily legible.

**Page number.**
**Header**

**Title**

**Graphic: Diagram**

**Visual "Note" to catch readers' attention.**

**Caption**

**Heading**
**Introduction**

**List of steps.**

**Items are numerated.**

**Numbers are bold-faced.**

---

## Dipsticks

**New Dipsticks With Full Range Readings**

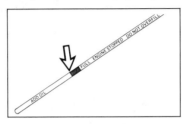

The crankcase oil level gauges are used with dipsticks that have the traditional FULL mark or FULL RANGE dipsticks with a knurled FULL zone. They can be used for applications with various installations.

### Calibration

Use the following procedure to calibrate the dipstick. A convenient time to calibrate the dipstick is at the first oil change.

Refer to the 250 Hour or 500 Hour interval of the Maintenance Management Schedule of this publication for the proper oil change interval.

To verify the ADD mark and establish the actual FULL mark in the FULL RANGE zone, use the following procedure:

**1.** Operate the engine until it reaches normal operating temperature.

**2.** Stop the engine. Drain the oil and change the oil filter.

**3.** Fill the crankcase with 23.5 L (25 U.S. qt) of oil for 3306B engines or 15 L (16 U. S. qt) for 3304B engines. Allow the oil to drain back to the sump for a minimum of five minutes.

**4.** Remove the dipstick. The oil level should be at the ADD mark. If it is not, mark the actual level on the dipstick. This is now the correct ADD mark.

**5.** Add an additional 3.8 L (4 U.S. qts) of oil to the sump. Allow enough time for the oil to drain into the sump. Again, check the level on the dipstick.

**6.** This is the correct FULL mark in the FULL RANGE zone on the dipstick. If it is not, mark the new FULL level on the dipstick.

NOTE: Remote mounted or auxiliary filters require additional oil also. For all information pertaining to auxiliary oil filters, refer to the OEM or manufacturer's instructions.

**7.** Start the engine and operate until it reaches normal operating temperature. Stop the engine.

**8.** Allow the oil to drain back to the sump for a minimum of 10 minutes.

**9.** Remove the dipstick. The oil level should be near the FULL mark. If it is not, add oil until the level reaches the new calibrated FULL mark on the dipstick.

This procedure is correct for use with either the non-spacer plate oil pan or the spacer plate oil pan. Sump capacities and oil levels are the same for each oil pan.

Any dipstick for 3300 Series engines can be calibrated using this procedure.

## CATERPILLAR®

---

**Figure 4.1   A simple set of functional directions.**   Source: *Operation and Maintenance Manual: 3304B and 3306B Industrial and EPG Generator Set Diesel Engines.* Reprinted with permission of Caterpillar, Inc.

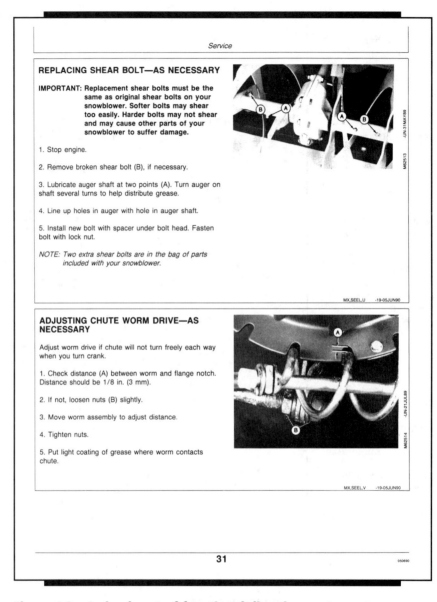

**Figure 4.2   A simple set of functional directions.**   Source: *Operator's Manual: TRS24, TRX24, TRS26, and TRX26 Walk-Behind Snowblowers.* Reprinted by permission of Deere & Company, © 1990 Deere & Company. All rights reserved.

An introductory statement at the beginning of the directions provides readers with a frame within which to read the steps that follow. Thus, the reader who reads one step at a time will have a general idea of what all the steps involve.

Figure 4.2 is a page from an operator's manual for a snowblower. This page comes from the section on servicing the equipment. Users may not be accustomed to operating mechanical equipment, or they may operate equipment but may never have operated equipment similar to a snowblower. The people who operate the equipment may not be the same ones who service it. Those who service it are probably familiar with mechanical equipment in general and with the snowblower specifically.

Like the previous set of instructions, each step contains only a single operation and each step is numbered. These instructions also contain a special note and an "important" message introducing the set of instructions. Graphics are also used in this set of instructions so readers can locate the different parts of the engine. Photographs are used rather than a diagram because the parts are large enough to see and are accessible to a photographer's camera. Call-out letters identify the various parts discussed in the text.

As you study these instructions, consider whether a novice as well as someone familiar with snowblowers could follow the instructions.

---

### Suggestions for Discussion and Writing

1. Discuss how each set of instructions in Figures 4.1 and 4.2 adheres to the conventions for writing instructions discussed on page 130.

2. Compare the format, visual text, and graphics in Figures 4.1 and 4.2. Which do you prefer? Why? Why do you think Caterpillar uses a diagram instead of a photograph in Figure 4.1? Why do you think Deere uses photographs instead of diagrams in Figure 4.2?

## A Complex Set of Instructions: *Operation and Maintenance Manual for a Hydraulic Hammer*

The following operation and maintenance manual provides owners with information for using a hydraulic hammer. As you examine it, consider the following aspects.

### Facilitates Navigation and Location of Information

The manual includes a cover page so readers can recognize they have located the appropriate manual. It also provides a table of contents to help users locate the specific information they need. The information is chunked into four main categories: introduction, operation, maintenance, and disassembly and assembly. These categories are then further subdivided and listed in the table of contents.

At the local level, in the upper outside corners of each page, headers indicate the chapter title and page number to help users skimming through the document to locate the appropriate section. Readers can also use the headings, subheadings, and numerical designations to locate the information they need. Because these use various typographical markers (e.g., all capital letters, boldface type, centering), users can see them easily as they skim the manual or scan a page.

### Meets Multiple Readers' Needs with Ancillary Documents

Appendices are included for readers who want such additional information as the technical specifications for the hammer and tools or a spare parts list. This information is not essential for carrying out the task and is therefore not included in the main text.

### Fulfills Ethical and Legal Responsibilities

Safety is obviously a concern. The manual begins with a safety warning immediately after the cover page, and the first four pages of the introduction are devoted to safety measures. These safety discussions are emphasized visually. By using reverse printing, the document's designer boxes the warning and caution sections, making it almost impossible for users to miss them.

### Facilitates Readers' Reading Processes and Behaviors

The introduction, which provides readers with a discussion on the purpose of the manual, a description of how to use it, and an overall description of the hammer and its parts, provides readers with a frame in which to understand the details that follow. The introduction also ensures that readers' predictions will be aligned with the content they read. Each chapter provides a summary of the information in that chapter and a forecast sequencing the categories of information to be discussed. The chapters are sequenced from general information to specific.

By dividing the page into three columns, the writer makes it easy for readers to scan an instruction item. By breaking large categories of instructions into several subcategories and by including a lot of white space around the items, the designer gives the document an airy look so readers don't feel inundated by the information.

The white space also makes it easy for users to locate their place as they move between the instructions and their task. Numerating the items and placing them in lists using hanging indentation also helps readers locate their next step.

As you study the manual, compare the document designer's treatment of the individual sets of directions with the individual sets in Figure 4.1 and 4.2.

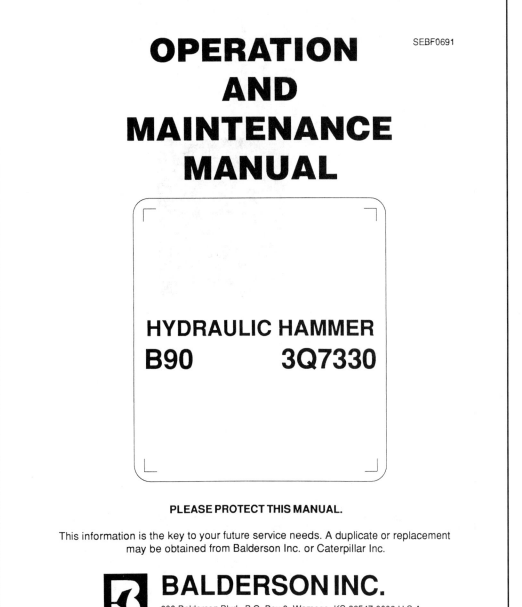

SEBF0691

# OPERATION AND MAINTENANCE MANUAL

## HYDRAULIC HAMMER
## B90          3Q7330

**PLEASE PROTECT THIS MANUAL.**

This information is the key to your future service needs. A duplicate or replacement may be obtained from Balderson Inc. or Caterpillar Inc.

# BALDERSON INC.

600 Balderson Blvd., P.O. Box 6, Wamego, KS 66547-0006 U.S.A.
Phone: 913-456-2224   Fax: 913-456-2027   TWX: 910-749-6524

(continues)

> ⚠ **WARNING**
>
> Study this manual before installing, operating or maintaining this equipment. You must understand and follow the instructions in this manual. You must observe all relevant laws and regulations. Otherwise You and/or others can be seriously injured.

Publications:  B90 Operation and Maintenance Manual,   SEBF0691
B90 Service Manual,   SEBF0690

Date 1991-03-01
Printed in U.S.A.

Specifications and design presented in this manual are subject to change without notice.

# C O N T E N T S

(continues)

# 1. INTRODUCTION

### General Information

This manual instructs you on your Balderson hammer and its safe operation and maintenance. Study this manual before installing, operating or maintaining this equipment. The hammer is a powerful tool. Used without proper care, it can cause damage. Use it properly and use it well. Chapter 2 *Operation* is arranged to guide you how to operate the hammer safely.

Pay particular attention to all safety messages. They are there to warn you of possible hazards.

### ⚠ WARNING

**Denotes a hazard exists. If proper precautions are not taken you/others could be seriously injured.**

### CAUTION!

**Denotes a reminder of safety practices. Failure to follow these safety practices could result in injury to the operator/others and possible damage to the equipment.**

To use the hammer correctly, you must also be a competent operator of the carrier machine. Do not use or install the hammer if you can not use the carrier machine.

Do not rush the job of learning. Take your time and most important, take it safely.

If there is anything you do not understand, ask your Cat dealer. He will be pleased to advise you.

REMEMBER

BE CAREFUL
BE ALERT
BE SAFE

**Hammer model and serial number**

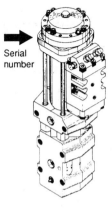

Serial number

This manual deals with the Balderson B90. The equipment serial number is stamped on the rear body, near low pressure accumulator.

It is important to make correct reference to the serial number of the hammer when making repairs or ordering spare parts. Identification of serial number is the only proper means of maintaining and identifying parts for specific hammer.

### Using the manual

This manual is arranged to give you a good understanding of the equipment and its safe operation. It also contains maintenance information and technical specifications. Read this manual from front to back before using or maintaining for the first time. If you do not understand something or you are unsure, ask your Cat dealer. Do not guess. Read all safety statements carefully. Be safe and be careful.

Page numbering system in this manual is not continuous. There is a gap of few pages between sections. This allows for the insertion of new pages in later issues.

In this manual, the units of measurement are metric. For example, weights are given in kilograms (kg). In some cases the other unit follows in parenthesis (). For example 28 liters (7.4 USgal).

## Introduction
A-2

### 1.1 SAFETY - YOURS AND OTHER PEOPLE'S

All mechanical equipment can be hazardous if operated without care or correct maintenance.

In this manual you will find warning messages. Read them and understand them. They tell you of hazards and how to avoid them. If you do not understand the messages, ask your employer or your Cat dealer.

Safety is not just a matter of responding to the warnings. All the time you are working with your Balderson hammer you must be thinking what hazards there might be and how to avoid them.
Do not work with the hammer until you are sure that you control it.

Do not start any job until you are sure that you and those around you will be safe.

If you are unsure of anything, about your Balderson hammer or the job, ask someone who knows. Do not assume anything - check it out.

### SAFETY CHECK LIST
#### General Safety

### ⚠ WARNING

Read carefully following warning messages. They tell you of different hazards and how to avoid them. If proper precautions are not taken you/others could be seriously injured.

#### Manuals
Study this manual before installing, operating or maintaining the hammer. If there is anything you don't understand, ask your employer or your Cat dealer to explain it. Keep this manual clean and in good condition.

Study also the operating and maintenance manual of your carrier before operating Balderson hammers.

#### Clothing
You can be injured if you do not wear proper clothing. Loose clothing can get caught in the machinery. Wear protective clothing to suit the job.

Examples are: a safety helmet, safety shoes, safety glasses, well-fitting overalls, ear-protectors and industrial gloves. Keep cuffs fastened.

Do not wear a necktie or scarf.

Keep long hair restrained.

#### Care and alertness
All the time you are working with the hammer, take care and stay alert. Always be careful. Always be alert for hazards.

#### Lifting equipment
You can be injured if you use faulty lifting equipment.

Make sure that lifting equipment is in good condition. Make sure that lifting tackle complies with all local regulations and is suitable for the job. Make sure that lifting equipment is strong enough for the job and you know how to use it.

#### Tools
Do not use Balderson hammers or hammer tools for lifting. Contact your Cat dealer to find out how to lift with your carrier.

### CAUTION!

Read carefully following safety messages. Failure to follow these safety practices could result in injury to the operator/others and possible damage to the equipment.

#### Regulations and laws
Obey all laws, worksite and local regulations which affect you and your equipment....

(continues)

### 1.2 BALDERSON B90

The purpose of this part of the manual is to introduce you to the Balderson B90. The main parts of the hammer will be named and identified. What they do and how they do it will be briefly explained.

Your Cat dealer will gladly give you more information if you want it.

**General description**

Key

1  Hammer
2  Sideplate
3  Bolt
4  Lock washer
5  Hose connections
6  Grease fitting
7  Threaded rod

The hammer is a hydraulically operated breaker. It can be used on any carrier which meets the necessary hydraulic and mechanical installation requirements (See *Technical specification*).

However, you should pay attention where and how you use your hammer.

**B90 is designed to be used in breaking concrete, asphalt and frozen ground. It can be used also in breaking small and soft boulders.**

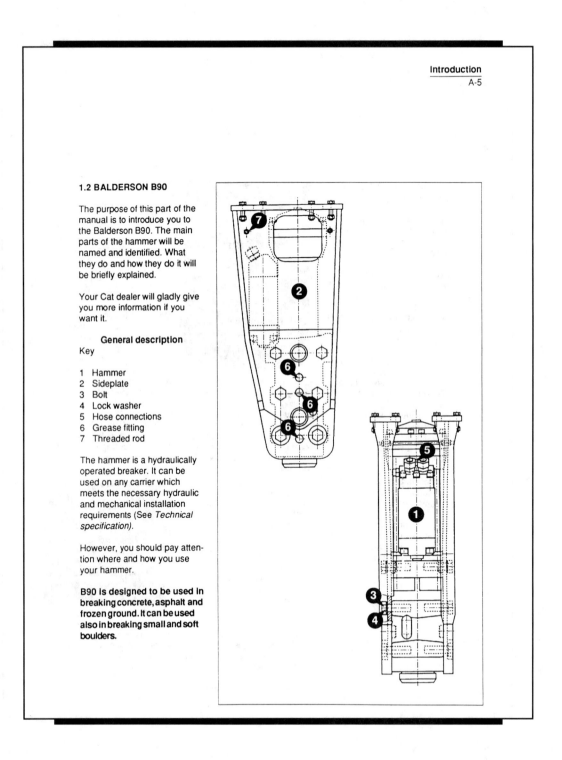

Introduction
A-6

The hammer is located between two sideplates which protect the hammer's mechanism during operation.

Basically, the unit functions by repeatedly raising a steel piston and driving it down onto the head of a removable breaking tool.

The impact energy of the hammer is constant and independent of the carrier's hydraulic system.

The operating principles of the hammer are described in the end of the Chapter 1 *Introduction.*

(continues)

**142**     The Documents

## 1.2 PRINCIPLES OF OPERATION

1   Piston
2   Distributor
3   High pressure accumulator
4   Check valve
5   Pressure adjusting valve
6   Low pressure accumulator

A   Piston head
B   Upper ring surface area
C   Ring surface area

X, Y, Z Distributor ring surfaces

High pressure

Low pressure

Tank pressure

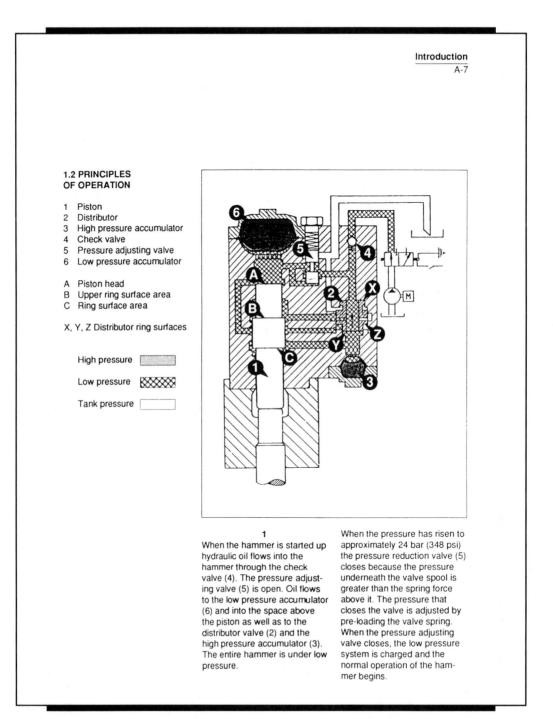

**1**

When the hammer is started up hydraulic oil flows into the hammer through the check valve (4). The pressure adjusting valve (5) is open. Oil flows to the low pressure accumulator (6) and into the space above the piston as well as to the distributor valve (2) and the high pressure accumulator (3). The entire hammer is under low pressure.

When the pressure has risen to approximately 24 bar (348 psi) the pressure reduction valve (5) closes because the pressure underneath the valve spool is greater than the spring force above it. The pressure that closes the valve is adjusted by pre-loading the valve spring. When the pressure adjusting valve closes, the low pressure system is charged and the normal operation of the hammer begins.

**Introduction**
A-8

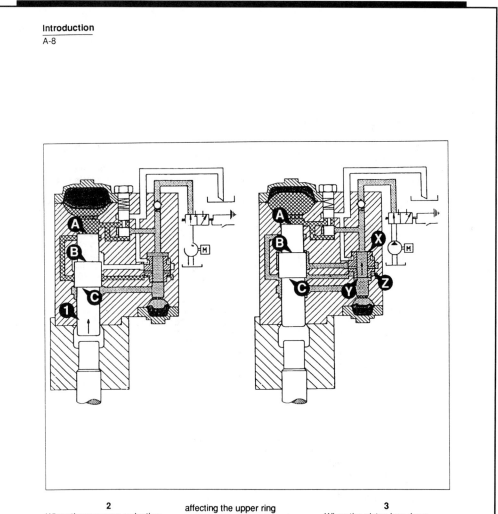

**2**

When the pressure reduction valve closes, the pressure in the high pressure system rises. The ring surface area (C) of the piston is affected by high pressure. The upper ring surface area (B) is affected by tank pressure and the piston head (A) is affected by low pressure. The force affecting the ring surface area (C) is greater than the sum of the forces affecting the upper ring surface area (B) and the piston head (A). Thus the piston (1) begins to travel upwards and the oil above the piston is pressed into the low pressure accumulator in which the pressure now rises to the pre-adjusted level for opening the pressure reduction valve (about 38 bar/551 psi).

**3**

When the piston has risen almost to its upper limit the piston shoulder (C) connects the high pressure to pilot pressure channel. Now the distributor ring surfaces X, Y and Z are affected by high pressure. Because the surface area Y + Z is greater than the surface area X, the distributor begins to move upwards. The borehole in the distributor connects the high. . .

(continues)

# 2. OPERATION

This chapter describes the correct working methods and how to choose the correct tool for the job. To increase the hammer's working life, pay particular attention to correct working methods.

The hammer is a powerful machine and lots of damage can be done if you do not know how to use the hammer well and safely.

Read this chapter before operating Balderson hammers.

### Choosing tools

Balderson can offer a selection of standard and special tools to suit each application. The correct type of tool must be selected to get the best possible working results and longest life time for tool.

1. Blunt
• For igneous (e.g. granite) and tough metamorphic rock (e.g. gneiss) into which tool doesn't penetrate
• Concrete
• Breaking boulders

2. Chisel and Moil
• For sedimentary (e.g. sandstone) and weak metamorphic rock into which tool penetrates
• Concrete
• Trenching and benching

3. Spade tool
• Frozen or compact ground
• Asphalt

4. Compacting plate
• Ground compacting

### Principles of breaking

There are basically two ways of breaking with a hydraulic hammer:

a) Penetrative (or cutting) breaking

In this form of breaking moil point or chisel tool is forced inside the material. This method is most effective in soft, layered or plastic, low abrasive material. Using of chisel in hard material will cause the sharp edge to wear very quickly. The high impact rate of the B90 makes it very suitable for penetrative breaking.

b) Impact breaking

In impact breaking , material is broken by transferring very strong mechanical stress from the tool into material to be broken. Best possible energy transfer between tool and object is achieved with a blunt tool. Impact breaking is most effective in hard, brittle and very abrasive materials

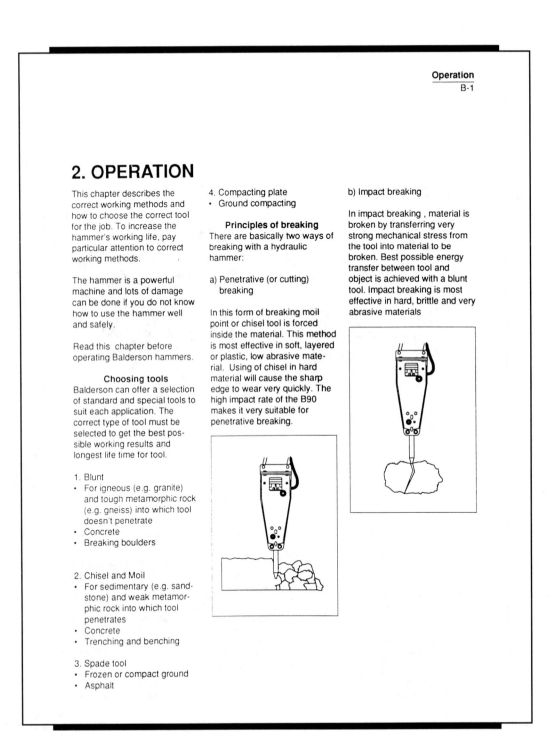

Operation
B-2

## 2.1 CORRECT WORKING METHODS

1. Prepare the carrier for normal excavation work.

a) Move the carrier to required position.
b) Engage the parking brake.
c) Set the drive to neutral.
d) Disengage the boom lock (if fitted).

2. Set the engine speed to the recommended engine RPM.

3. Place the tool against the object at 90 degrees angle. Avoid small irregularities on the object which will break easily and cause either idle strokes or incorrect working angle.

4. Use the excavator boom to press the hammer firmly against or the object. Do not pry the hammer with the boom. Do not press too much or too little with the boom.

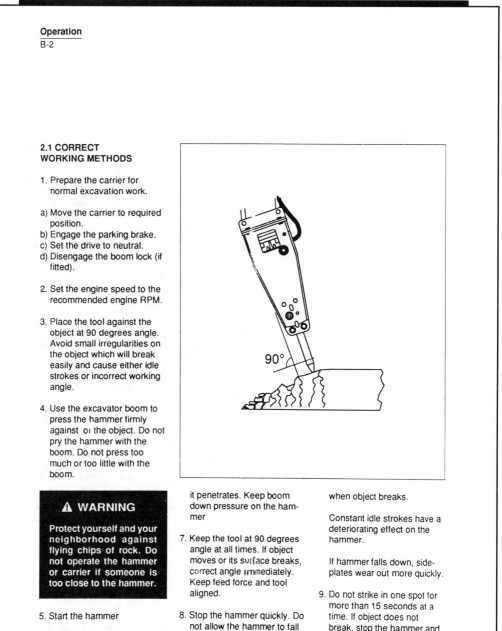

> ⚠ **WARNING**
>
> **Protect yourself and your neighborhood against flying chips of rock. Do not operate the hammer or carrier if someone is too close to the hammer.**

5. Start the hammer

6. Do not let the tool move outwards from hammer when it penetrates. Keep boom down pressure on the hammer

7. Keep the tool at 90 degrees angle at all times. If object moves or its surface breaks, correct angle immediately. Keep feed force and tool aligned.

8. Stop the hammer quickly. Do not allow the hammer to fall down and make idle strokes when object breaks.

Constant idle strokes have a deteriorating effect on the hammer.

If hammer falls down, side-plates wear out more quickly.

9. Do not strike in one spot for more than 15 seconds at a time. If object does not break, stop the hammer and change position of the tool. . . .

(continues)

# 3. MAINTENANCE

This section of the manual describes how to maintain your hammer.

**General Information**
Whenever maintenance work is carried out, keep following basic rules in mind:

1. The hydraulic hammer is a precision made hydraulic machine. Absolute cleanliness and great care are basic and essential matters in the handling of any hydraulic components. Dirt is the worst enemy in hydraulic systems.

2. Handle hammer parts carefully and remember to cover cleaned and dried parts with clean lint free cloth.

3. Do not use any other than purpose designed materials for cleaning hydraulic parts. Never use water, paint thinners or carbon tetrachloride.

4. Components, gaskets and seals in the hydraulic system should be oiled with very clean hydraulic oil before assembly.

### 3.1 MAINTENANCE INTERVALS

*Note:* Times given refer to hours of hammer operation.

**Every two hours or daily**
- Grease the tool shank and the tool bushings. See page C-4.
- Observe hydraulic oil temperature, all lines and connections as well as impact efficiency and evenness of operation.
- Tighten loose connections.

**Every 10 hours or weekly**
- Remove the retaining pin and the tool and check their condition. Grind the burrs away if necessary.
- Check that the tool has received sufficient greasing. Grease more frequently, if necessary.

**Every 50 hours or monthly**
- Check the tool shank and tool bushings for wear. See page D-4.
- Check the hydraulic hoses. Replace if necessary. Do not let dirt get into the hammer or hoses.

### 3.2 INITIAL 50-HOUR INSPECTION

It is recommended to have the initial 50-hour inspection done by your Cat dealer after **50–100 operating hours.**

**Contact your Cat dealer for more information about the initial 50-hour inspection.**

- Check all hydraulic connections.
- Check that the hydraulic hoses do not rub against anything in any boom position.
- Replace the hydraulic oil filters of the carrier.

### 3.3 EVERY 600 HOURS OR YEARLY

Yearly service is recommended to be done by your Cat dealer after **600 operating hours.**

Your cat dealer will reseal the hammer, replace accumulator membrane(s) and replace all safety decals.

**Contact your Cat dealer for more information about yearly service.**

- Check all hydraulic connections.
- Check that the hydraulic hoses do not rub against anything in any boom position.
- Replace the hydraulic oil filters of the carrier.

Maintenance
C-2

### 3.4 HYDRAULIC OIL

Generally speaking the hydraulic oil originally intended for the carrier can be used in the Balderson hydraulic hammer. However, since working with the hydraulic hammer will heat the oil much more than the usual excavation work, the viscosity of the oil must be checked.

When the hammer is used continuously, the temperature of the hydraulic oil normalizes at a certain level depending on conditions and on the carrier. At this temperature, the viscosity of the hydraulic oil should be 20–40 cSt (2.90–5.35 °E).

The Balderson hydraulic hammer must not be started if the viscosity of the hydraulic oil is above 1000 cSt (131 °E) or operated when the viscosity of the hydraulic oil is below 15 cSt (2.35 °E).

Table 1 shows hydraulic oils recommended for hammer use. The most suitable oil is selected in such way that the temperature of the hydraulic oil in continuous use is in the ideal area on the chart and the hydraulic system is used to best advantage.

Failures due to incorrect hydraulic oil in hammer:

Oil too thick

- Difficult start up
- Stiff operation
- Hammer strikes irregularly and slowly
- Danger of cavitation in the pumps and hydraulic hammer
- Sticky valves
- Filter bypass, impurities in oil not removed

Oil too thin

- Efficiency losses (internal leaks)
- Damage to gaskets and seals, leaks
- Accelerated wearing of parts, because of decreased lubrication efficiency

*Note:* Balderson strongly recommends different hydraulic oils for use in summer and winter if there is an average temperature difference of more than 35 °C (95 °F). The correct hydraulic oil viscosity would thus be ensured.

#### Special oils
In some cases special oils (e.g. biological oils and non-inflammable oils) can be used with Balderson hydraulic hammers. Observe following aspects when considering the use of special oils:

- The viscosity range in the special oil must be in the given range (15–1000 cSt)
- The lubrication properties

must be good enough
*Note:* although special oil could be used in carrier, always check suitability with hammer due to the high piston speed in hammer.
- The corrosion resistance properties must be good enough

Contact oil manufacturer or Balderson for more information about special oils.

#### Hydraulic oil purity
No separate filter is required when the Balderson hammer is installed in the hydraulic circuit. The hydraulic oil filter of the carrier will clean the oil flowing through the hammer.
The purpose of the oil filter is to remove impurities from the hydraulic oil since they cause accelerated component wear, blockages and even seizure. Impurities also cause the oil to heat and deteriorate. Air and water are also impurities in oil. Not all impurities can be seen with the naked eye.

Impurities enter the hydraulic system:

- During hydraulic oil changes and refilling
- When components are repaired or serviced
- When the hammer is being installed on the carrier
- Because of component wear

#### Oil filter
In hydraulic hammer work, the carrier oil filter must fulfill the following specifications: . . .

(continues)

# 3.8 TROUBLE SHOOTING

| TROUBLE | PROBABLE CAUSE | REMEDY |
|---|---|---|
| 1. The hammer does not start. | 1. The piston is in its lower hydraulic brake. | Keep the hammer control valve open and push the tool firmly against an object. The tool head lifts the piston out of the brake. |
| | 2. The hammer control valve does not open. | When operating the hammer control valve, check that the pressure line will pulsate (that means the hammer control valve opens). If the valve does not operate, check the operating means: mechanical connections, pilot pressure or electrical control. |
| | 3. The relief valve in hydraulic circuit opens at too low pressure. The hammer operating pressure is not reached. | Contact your Cat dealer. |
| | 4. Leakage from pressure to return in excavator hydraulic circuit. | Check the installation. Check the pump and other hydraulic components. |
| | 5. Too much back pressure. | Check the installation. |
| | 6. Failure in hammer valve or distributor operation. | Contact your Cat dealer. |
| | 7. Piston failure | Contact your Cat dealer. |
| 2. The hammer operates irregularly but the blow has full power. | 1. The relief valve in hydraulic circuit opens at too low pressure. The hammer operating pressure is not reached. | Contact your Cat dealer. |
| | 2. Failure in hammer valve or distributor operation. | Contact your Cat dealer. |
| | 3. Not enough feed force from the carrier. | Refer to correct working methods in Chapter 2 *Operation.* |

Maintenance
C-8

| TROUBLE | PROBABLE CAUSE | REMEDY |
|---|---|---|
| 3. The hammer operates poorly and the blow has no power. | 1. The relief valve in hydraulic circuit opens at too low pressure. The hammer operating pressure is not reached. | Contact your Cat dealer. |
| | 2. There is no pressure in the pressure accumulator. | Contact your Cat dealer. |
| | 3. The working method is not correct. | Refer to correct working methods in Chapter 2 *Operation*. |
| | 4. Failure in hammer valve operation. | Contact your Cat dealer. |
| 4. Impact rate slows down. | 1. Oil has overheated (over + 80°C/176°F) | Check for fault in oil cooling system or internal leak in hammer. Check hydraulic circuit of the carrier. Install extra oil cooler. |
| | 2. Too much back pressure. | Contact your Cat dealer. |
| | 3. The relief valve in hydraulic circuit opens at too low pressure. The hammer operating pressure is not reached. | Contact your Cat dealer. |
| | 4. Leakage from pressure to return in excavator hydraulic circuit. | Contact your Cat dealer. Check the pump and other hydraulic components. |
| | 5. Failure in hammer valve or distributor operation. | Contact your Cat dealer. |
| | 6. There is no pressure in the pressure accumulator. | Contact your Cat dealer. |
| | 7. Hydraulic oil viscosity is too low. | Check hydraulic oil viscosity.... |

(continues)

# 4. DISASSEMBLING AND ASSEMBLING OF B90

This part of the manual contains detailed information how to disassemble and assemble your Balderson B90.

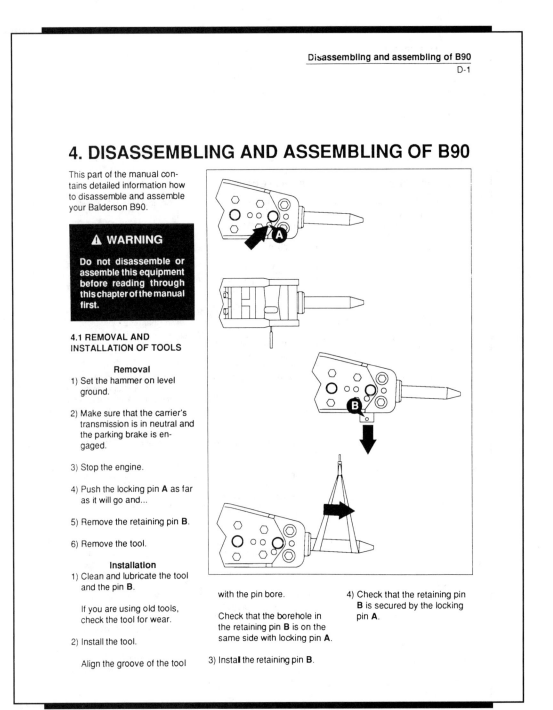

> ⚠ **WARNING**
>
> **Do not disassemble or assemble this equipment before reading through this chapter of the manual first.**

### 4.1 REMOVAL AND INSTALLATION OF TOOLS

#### Removal

1) Set the hammer on level ground.

2) Make sure that the carrier's transmission is in neutral and the parking brake is engaged.

3) Stop the engine.

4) Push the locking pin **A** as far as it will go and...

5) Remove the retaining pin **B**.

6) Remove the tool.

#### Installation

1) Clean and lubricate the tool and the pin **B**.

   If you are using old tools, check the tool for wear.

2) Install the tool.

   Align the groove of the tool with the pin bore.

   Check that the borehole in the retaining pin **B** is on the same side with locking pin **A**.

3) Install the retaining pin **B**.

4) Check that the retaining pin **B** is secured by the locking pin **A**.

**Disassembling and assembling of B90**
D-2

**4.2 CHANGING LOWER
TOOL BUSHING**

**Removal**

1) Set the hammer on level
   ground.

2) Make sure that the carrier's
   transmission is in neutral and
   the parking brake is engaged. Stop the engine.

3) Remove the tool

4) Push pins **C** (2 pcs) through
   the holes in sideplates.

5) Remove the O-ring **E** from
   the pin **C** (2 pcs).

6) Remove the lower tool
   bushing **F** from the front
   head.

7) Remove seal **G** and O-ring **H**
   from the lower tool bushing.

   Check tool and lower tool
   bushing for wear.

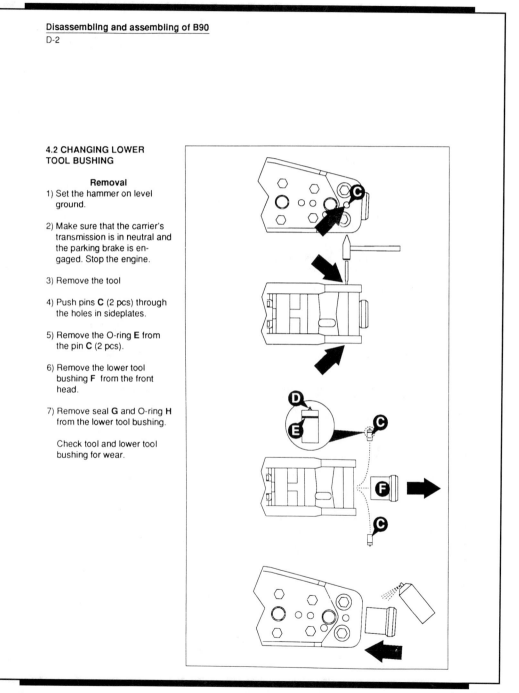

**152** The Documents

### Installation

1) Clean all parts

2) Install new seal **G** and O-ring **H**.

3) Check the condition of grease fitting **D**.

4) Apply MoS$_2$ spray to the contact surfaces of lower tool bushing and front head.

4) Install lower tool bushing **F**.

Align the two holes in the lower tool bushing with the holes in front head.

5) Install pins **C** (2 pcs).

6) Install the tool  . . .

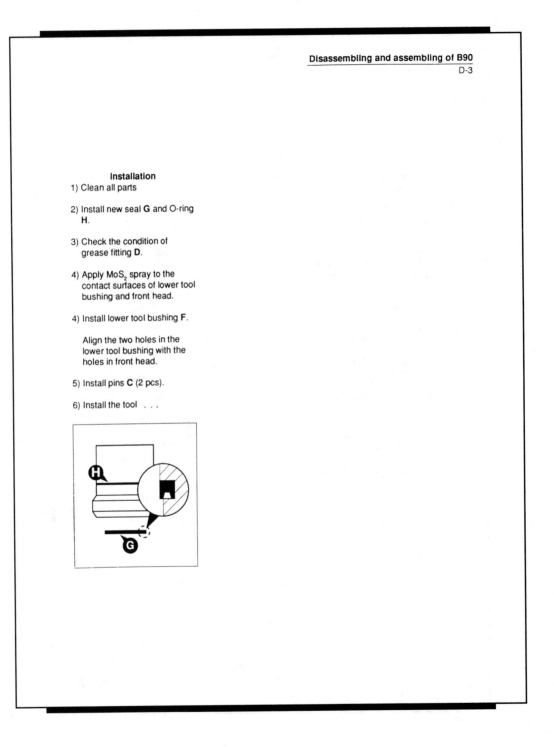

# BALDERSON B90
# TECHNICAL SPECIFICATION

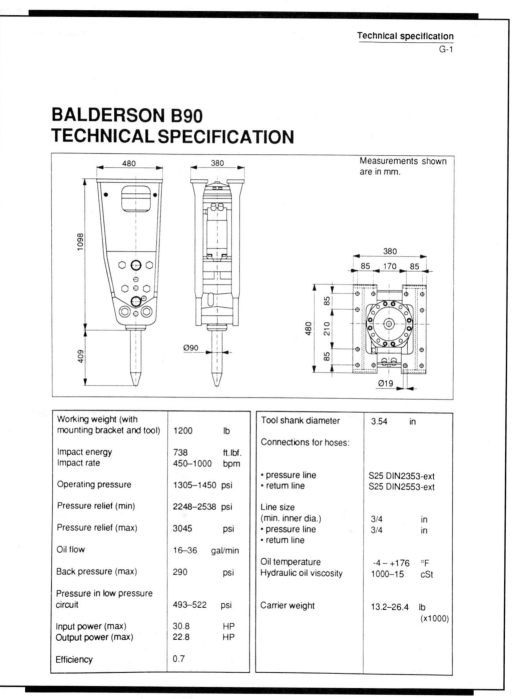

Measurements shown are in mm.

| | | | | Tool shank diameter | 3.54 | in |
|---|---|---|---|---|---|---|
| Working weight (with mounting bracket and tool) | 1200 | lb | | Connections for hoses: | | |
| Impact energy | 738 | ft.lbf. | | • pressure line | S25 DIN2353-ext | |
| Impact rate | 450–1000 | bpm | | • return line | S25 DIN2553-ext | |
| Operating pressure | 1305–1450 | psi | | Line size (min. inner dia.) | | |
| Pressure relief (min) | 2248–2538 | psi | | • pressure line | 3/4 | in |
| Pressure relief (max) | 3045 | psi | | • return line | 3/4 | in |
| Oil flow | 16–36 | gal/min | | Oil temperature | -4 – +176 | °F |
| Back pressure (max) | 290 | psi | | Hydraulic oil viscosity | 1000–15 | cSt |
| Pressure in low pressure circuit | 493–522 | psi | | Carrier weight | 13.2–26.4 | lb (x1000) |
| Input power (max) | 30.8 | HP | | | | |
| Output power (max) | 22.8 | HP | | | | |
| Efficiency | 0.7 | | | | | |

(continues)

# BALDERSON B90
# TOOLS

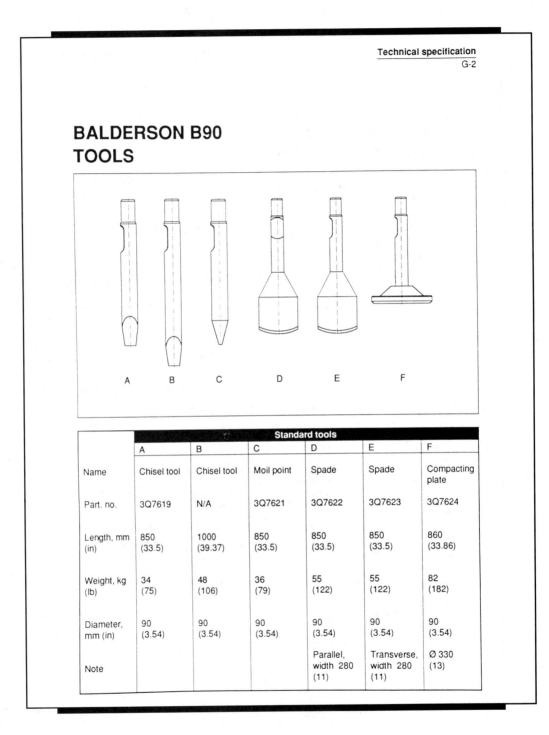

| | Standard tools | | | | | |
|---|---|---|---|---|---|---|
| | A | B | C | D | E | F |
| Name | Chisel tool | Chisel tool | Moil point | Spade | Spade | Compacting plate |
| Part. no. | 3Q7619 | N/A | 3Q7621 | 3Q7622 | 3Q7623 | 3Q7624 |
| Length, mm (in) | 850 (33.5) | 1000 (39.37) | 850 (33.5) | 850 (33.5) | 850 (33.5) | 860 (33.86) |
| Weight, kg (lb) | 34 (75) | 48 (106) | 36 (79) | 55 (122) | 55 (122) | 82 (182) |
| Diameter, mm (in) | 90 (3.54) | 90 (3.54) | 90 (3.54) | 90 (3.54) | 90 (3.54) | 90 (3.54) |
| Note | | | | Parallel, width 280 (11) | Transverse, width 280 (11) | Ø 330 (13) |

SPARE PART LIST

BALDERSON B90

19.03.1991/TI

Group no. B90B01XXX/B01

| Ref.no. | Name | Cat part no. | Qty |
|---|---|---|---|
|  | HYDRAULIC HAMMER | 3Q7330 | 1 |
| 402 | HOUSING PLATES | 3Q7302 | 1 |
| 403 | HEX. SCREW | 3Q7403 | 12 |
| 404 | LOCK WASHER | 3E6502 | 24 |
| 405 | THREADED ROD | 3Q7694 | 2 |
| 406 | HEX. NUT | 8T4244 | 8 |
| 407 | HEX. SCREW | 8T0357 | 12 |
| 408 | LOCK WASHER | 3E5824 | 48 |
| 409 | HEX. NUT | 985747 | 12 |
| 506 | PIN | 3Q7449 | 4 |
|  | ACCUMULATOR WARNING DECAL | 3E5776 |  |
|  | ACCUMULATOR CHARGING SYSTEM | 3Q7751 |  |
|  | NOISE DECAL | 3E5753 |  |
|  | DEPRIS DECAL | 3E5774 |  |

(continues)

# B9O

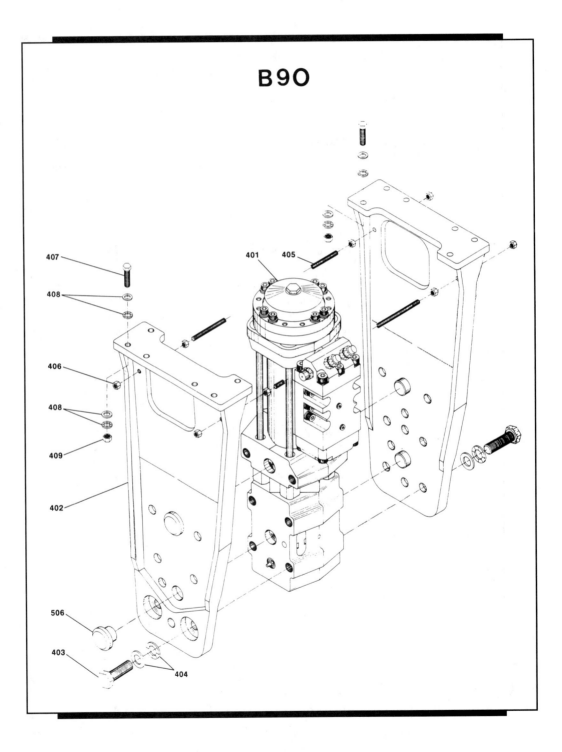

407

401 405

408

406

408

409

402

506

403

404

# B 90

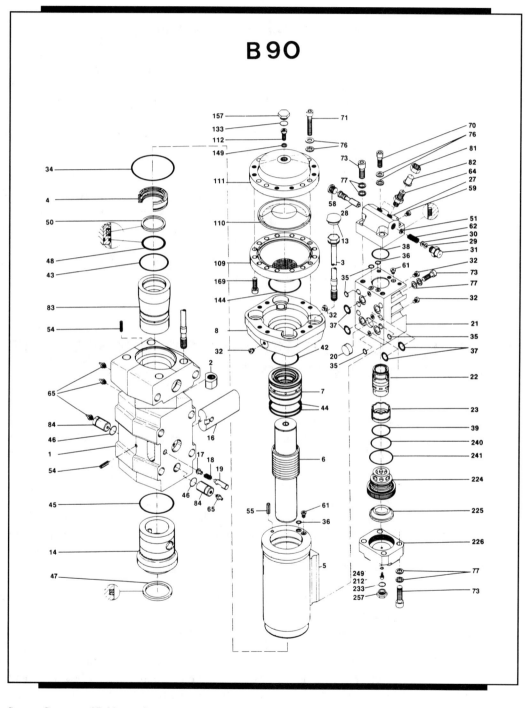

Source: Courtesy of Balderson Inc.

---

### Suggestions for Discussion and Writing

1. How does the writer attempt to protect the reader from possible injury and the company from a potential suit? Can you find anything discussed in the manual that might cause users to become injured for which the writer has not issued a warning or a caution notice?

2. Why is the table for the trouble-shooting subsection a good device for helping readers locate the specific information they need? How does it help users?

3. There should be a close relationship between graphics and a text. Circle the places in the text that reference a graphic. How do readers identify the parts of the graphic that are being discussed in the text?

4. Is the manual organized systemically or functionally? Is there any part of the manual you would revise to improve it?

---

# *DE ARCHITECTURA* [THE TEN BOOKS OF ARCHITECTURE]
# MARCUS VITRUVIUS POLLIO, 27 B.C.

Some of the most majestic architecture in our country today is based on instructions written by an architect 2,000 years ago. The Roman-style bridges spanning Philadelphia's Schuylkill River and the capital building rising over Washington, D.C., are both based on the Roman architectural style described by Marcus Vitruvius Pollio in *De Architectura*. A theoretical as well as practical guide for planning and building civil and military structures, *De Architectura* served as a textbook to guide Roman architects throughout the Roman period and well into the Renaissance, influencing such sixteenth century architects as Palladio. Our knowledge today of the procedures for building the Roman aqueducts, for constructing the double-tier arches in the Roman walls, and for tunneling through steep hills comes from Vitruvius's manual.

Vitruvius's purpose for writing the instructions was to preserve the classical tradition in the design of Rome's temples and public buildings. He dedicated the book to Augustus Caesar, attempting in his introduction to Book I to persuade Caesar to accept his "manual." Although his primary audience was Augustus Caesar, he wrote for a secondary audience of architects who, he hoped, would adhere to his principles. His arguments were sufficiently persuasive to influence architects for two centuries.

Unlike today's impersonal instructional manuals, *De Architectura* focuses not only on the "how" of doing something but also on the "why" of doing something. The style includes features common to philosophical treatises and historical narratives as well as those related to instructions as the writer attempts to persuade readers to accept his ideas for the way in which something should be done. His instructions often serve as claims for his arguments, which he supports with evidence as he tries to persuade readers to follow his advice. For example, before readers decide on a site to develop housing, he instructs them to examine the liver of the cattle grazing on the land, claiming that if the livers of the slaughtered cattle are unhealthy, the liver of residents living on that site will become unhealthy and the site should not be developed. As evidence, he cites an incident in Crete in which the size of the spleen found in cattle depended on which side of a river they had fed.

Vitruvius also uses metaphorical language to help readers accept his instructions. For example, in his discussion of the proper location of a city, he compares the influence of the site of a town on its residents to the location of a granary on the grain.

As you read the document, consider other examples of arguments and metaphorical language that the writer uses to persuade the reader to follow his instructions.

# VITRUVIUS
## THE TEN BOOKS ON ARCHITECTURE

TRANSLATED BY

MORRIS HICKY MORGAN, PH.D., LL.D.

WITH ILLUSTRATIONS AND ORIGINAL DESIGNS

PREPARED UNDER THE DIRECTION OF

HERBERT LANGFORD WARREN, A.M.

(continues)

## BOOK I

### PREFACE

**Personal (uses first person).**

**Writer attempts to establish a positive attitude on the part of the reader toward himself.**

1. WHILE your divine intelligence and will, Imperator Caesar, were engaged in acquiring the right to command the world, and while your fellow citizens, when all their enemies had been laid low by your invincible valour, were glorying in your triumph and victory,—while all foreign nations were in subjection awaiting your beck and call, and the Roman people and senate, released from their alarm, were beginning to be guided by your most noble conceptions and policies, I hardly dared, in view of your serious employments, to publish my writings and long considered ideas on architecture, for fear of subjecting myself to your displeasure by an unseasonable interruption.

**Notes reader's attitude toward society is similar to his own.**

2. But when I saw that you were giving your attention not only to the welfare of society in general and to the establishment of public order, but also to the providing of public buildings intended for utilitarian purposes, so that not only should the State have been enriched with provinces by your means, but that the greatness of its power might likewise be attended with distinguished authority in its public buildings, I thought that I ought to take the first opportunity to lay before you my writings on this theme. For in the first place it was this subject which made me known to your father, to whom I was devoted on account of his great qualities. After the council of heaven gave him a place in the dwellings of immortal life and transferred your father's power to your hands, my devotion continuing unchanged as I remembered him inclined me to support you. And so with Marcus Aurelius, Publius Minidius, and Gnaeus Cornelius, I was ready to supply and repair ballistae, scorpiones, and other artillery, and I have received rewards for good service with them. After your first bestowal of these upon me, you continued to renew them on the recommendation of your sister.

**Reminds reader of his own expertness.**

3. Owing to this favour I need have no fear of want to the end of my life, and being thus laid under obligation I began to write this work for you, because I saw that you have built and are now building extensively, and that in future also you will take care that our public and private buildings shall be worthy to go down to posterity by the side of your other splendid achievements. I have drawn up definite rules to enable you, by observing them, to have personal knowledge of the quality both of existing buildings and of those which are yet to be constructed. For in the following books I have disclosed all the principles of the art. . . .

**Purpose of document.**

**Forecast, so reader knows what the document is about.**

### CHAPTER IV

#### THE SITE OF A CITY

1. FOR fortified towns the following general principles are to be observed. First comes the choice of a very healthy site. Such a site will be high, neither misty nor

frosty, and in a climate neither hot nor cold, but temperate; further, without marshes in the neighbourhood. For when the morning breezes blow toward the town at sunrise, if they bring with them mists from marshes and, mingled with the mist, the poisonous breath of the creatures of the marshes to be wafted into the bodies of the inhabitants, they will make the site unhealthy. Again, if the town is on the coast with a southern or western exposure, it will not be healthy, because in summer the southern sky grows hot at sunrise and is fiery at noon, while a western exposure grows warm after sunrise, is hot at noon, and at evening all aglow. . . .

2. These variations in heat and the subsequent cooling off are harmful to the people living on such sites. The same conclusion may be reached in the case of inanimate things. For instance, nobody draws the light for covered wine rooms from the south or west, but rather from the north, since that quarter is never subject to change but is always constant and unshifting. So it is with granaries: grain exposed to the sun's course soon loses its good quality, and provisions and fruit, unless stored in a place unexposed to the sun's course, do not keep long.

3. For heat is a universal solvent, melting out of things their power of resistance, and sucking away and removing their natural strength with its fiery exhalations so that they grow soft, and hence weak, under its glow. We see this in the case of iron which, however hard it may naturally be, yet when heated thoroughly in a furnace fire can be easily worked into any kind of shape, and still, if cooled while it is soft and white hot, it hardens again with a mere dip into cold water and takes on its former quality. . . .

7. If one wishes a more accurate understanding of all this, he need only consider and observe the natures of birds, fishes, and land animals, and he will thus come to reflect upon distinctions of temperament. One form of mixture is proper to birds, another to fishes, and a far different form to land animals. Winged creatures have less of the earthy, less moisture, heat in moderation, air in large amount. Being made up, therefore, of lighter elements, they can more readily soar away into the air. Fish, with their aquatic nature, being moderately supplied with heat and made up in great part of air and the earthy, with as little moisture as possible, can more easily exist in moisture for the very reason that they have less of it than of the other elements in their bodies; and so, when they are drawn to land, they leave life and water at the same moment. Similarly, the land animals, being moderately supplied with the elements of air and heat, and having less of the earthy and a great deal of moisture, cannot long continue alive in the water, because their portion of moisture is already abundant.

8. Therefore, if all this is as we have explained, our reason showing us that the bodies of animals are made up of the elements, and these bodies, as we believe, giving way and breaking up as a result of excess or deficiency in this or that element, we cannot but believe that we must take great care to select a very temperate climate for the site of our city, since healthfulness is, as we have said, the first requisite.

9. I cannot too strongly insist upon the need of a return to the method of old times. Our ancestors, when about to build a town or an army post, sacrificed some

(continues)

of the cattle that were wont to feed on the site proposed and examined their livers. If the livers of the first victims were dark-coloured or abnormal, they sacrificed others, to see whether the fault was due to disease or their food. They never began to build defensive works in a place until after they had made many such trials and satisfied themselves that good water and food had made the liver sound and firm. If they continued to find it abnormal, they argued from this that the food and water supply found in such a place would be just as unhealthy for man, and so they moved away and changed to another neighbourhood, healthfulness being their chief object.

10.  That pasturage and food may indicate the healthful qualities of a site is a fact which can be observed and investigated in the case of certain pastures in Crete, on each side of the river Pothereus, which separates the two Cretan states of Gnosus and Gortyna. There are cattle at pasture on the right and left banks of that river, but while the cattle that feed near Gnosus have the usual spleen, those on the other side near Gortyna have no perceptible spleen. On investigating the subject, physicians discovered on this side a kind of herb which the cattle chew and thus make their spleen small. The herb is therefore gathered and used as a medicine for the cure of splenetic people. The Cretans call it ασπληνον. From food and water, then, we may learn whether sites are naturally unhealthy or unhealthy. . . .

## CHAPTER V

### THE CITY WALLS

1.  AFTER insuring on these principles the healthfulness of the future city, and selecting a neighbourhood that can supply plenty of food stuffs to maintain the community, with good roads or else convenient rivers or seaports affording easy means of transport to the city, the next thing to do is to lay the foundations for the towers and walls. Dig down to solid bottom, if it can be found, and lay them therein, going as deep as the magnitude of the proposed work seems to require. They should be much thicker than the part of the walls that will appear above ground, and their structure should be as solid as it can possibly be laid.

2.  The towers must be projected beyond the line of wall, so that an enemy wishing to approach the wall to carry it by assault may be exposed to the fire of missiles on his open flank from the towers on his right and left. Special pains should be taken that there be no easy avenue by which to storm the wall. The roads should be encompassed at steep points, and planned so as to approach the gates, not in a straight line, but from the right to the left; for as a result of this, the right hand side of the assailants, unprotected by their shields, will be next the wall. Towns should be laid out not as an exact square nor with salient angles, but in circular form, to give a view of the enemy from many points. Defence [sic] is difficult where there are salient angles, because the angle protects the enemy rather than the inhabitants.

3.  The thickness of the wall should, in my opinion, be such that armed men meeting on top of it may pass one another without interference. In the thickness

there should be set a very close succession of ties made of charred olive wood, binding the two faces of the wall together like pins, to give it lasting endurance. For that is a material which neither decay, nor the weather, nor time can harm, but even though buried in the earth or set in the water it keeps sound and useful forever. And so not only city walls but substructures in general and all walls that require a thickness like that of a city wall, will be long in falling to decay if tied in this manner.

4. The towers should be set at intervals of not more than a bowshot apart, so that in case of an assault upon any one of them, the enemy may be repulsed with scorpiones and other means of hurling missiles from the tower to the right and left. Opposite the inner side of every tower the wall should be interrupted for a space the width of the tower, and have only a wooden flooring across, leading to the interior of the tower but not firmly nailed. This is to be cut away by the defenders in case the enemy gets possession of any portion of the wall; and if the work is quickly done, the enemy will not be able to make his way to the other towers and the rest of the wall unless he is ready to face a fall.

5. The towers themselves must be either round or polygonal. Square towers are sooner shattered by military engines, for the battering rams pound their angles to

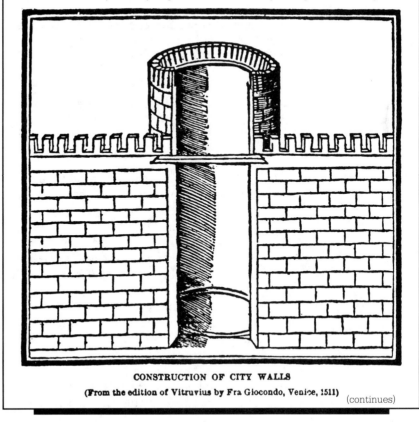

**CONSTRUCTION OF CITY WALLS**

**(From the edition of Vitruvius by Fra Giocondo, Venice, 1511)**

(continues)

pieces; but in the case of round towers they can do no harm, being engaged, as it were, in driving wedges to their centre. The system of fortification by wall and towers may be made safest by the addition of earthen ramparts, for neither rams, nor mining, nor other engineering devices can do them any harm.

6. The rampart form of defence, however, is not required in all places, but only where outside the wall is high ground from which an assault on the fortifications may be made over a level space lying between. In places of this kind we must first make very wide, deep ditches; next sink foundations for a wall in the bed of the ditch and build them thick enough to support an earthwork with ease. . . .

8. With regard to the material of which the actual wall should be constructed or finished, there can be no definite prescription, because we cannot obtain in all places the supplies that we desire. Dimension stone, flint, rubble, burnt or unburnt brick,—use them as you find them. For it is not every neighbourhood or particular locality that can have a wall built of burnt brick like that at Babylon, where there was plenty of asphalt to take the place of lime and sand, and yet possibly each may be provided with materials of equal usefulness so that out of them a faultless wall may be built to last forever.

**Source:** Vitruvius. *Ten Books of Architecture,* translated by Morris Hicky Morgan, 1960. Reprinted by permission of Dover Publications, Inc.

## Suggestions for Discussion and Writing

1. Vitruvius' preface seems more like a cover letter than part of the book. Explain why.

2. Readers often have to be persuaded to use procedures. How does Vitruvius attempt to enlist the attention and sympathy of his audience in his preface? How does he try to persuade them to use his procedures? What primary claim does he make?

3. Does the figure on page 163 help the reader understand the writer's discussion? If not, why not? If it helps, how does it do so?

4. Why does Vitruvius place considerable importance on the characteristics of the natural environment in planning a city? What primary claim does he make in the section on the site of a city? What evidence does he use to support his claims?

5. Vitruvius appeals to the past as one method of persuading his audience to adopt his instructions. He comments, "I cannot too strongly

insist upon the need to return to the methods of old times." Locate a set of instructions for operating equipment of the past (e.g., a manual for an electric typewriter, a teletypesetter, a telephone prior to 1950, a copy machine prior to 1960, a computer prior to 1970). Compare those instructions with a set of instructions for a wordprocessor, facsimile machine, contemporary telephone, copier, or personal computer. What differences do you notice in the layout, visual text, graphics? Do contemporary instructions contain more or fewer details and warnings? Why do you think these differences exist?

## *DE RE METALLICA,* GEORGIUS AGRICOLA, 1556

*De Re Metallica* was written to provide information and instructions in the art of metalworking. The book includes methods of prospecting, surveying, smelting, refining, and assaying as well as discussions on ownership laws and management practices. Though published 1500 years after *De Architectura,* it resembles the early text in many ways. The preface, like the introduction to Book I of *De Architectura,* attempts to persuade the readers to whom the book is dedicated of the importance of the information. Also like *De Architectura,* the book includes a vast number of subjects that will appeal to a wide variety of readers. Agricola is careful to use language familiar to both citizens and engineers and to discuss topics so that all readers understand. In fact, Agricola is especially conscientious in ensuring that readers follow his instructions easily and accurately. He includes numerous illustrations throughout the manual and provides readers with a framework within which to understand the information that follows. The preface includes an overview of the topics to be considered in the book, and each chapter begins with a forecast for that chapter.

The style is personal; Agricola uses first person and often interjects his own opinion, as in his discussion of the mining of veins coming from the south (p. 171). Like Vitruvius, he is interested in providing the "why" as well as the "how." Thus, in Book V, he explains why a dwelling house should be built in addition to a shaft house (p. 169). He is very specific in providing directions so that readers will know exactly what to do and will not have to guess. For example, he provides the exact measurements for a shaft. To help readers comprehend an overall object, he uses illustrations, but he seldom uses metaphorical language as Vitruvius does.

*De Re Metallica* bears some resemblance to today's functional manuals, with chapters devoted to digging, assaying, and smelting.

# TO THE MOST ILLUSTRIOUS AND MOST MIGHTY DUKES OF

Saxony, Landgraves of Thuringia, Margraves of Meissen,
Imperial Overlords of Saxony, Burgraves of Altenberg
and Magdeburg, Counts of Brena, Lords of
Pleissnerland, To MAURICE Grand Marshall
and Elector of the Holy Roman Empire
and to his brother AUGUSTUS,
GEORGE AGRICOLA S. D.

Metaphor.

Purpose.

Reviews previous literature on the subject. Attempts to persuade readers of subject's importance by relating to reader's attitudes and values toward subject.

MOST illustrious Princes, often have I considered the metallic arts as a whole, as Moderatus Columella considered the agricultural arts, just as if I had been considering the whole of the human body; and when I had perceived the various parts of the subject, like so many members of the body, I became afraid that I might die before I should understand its full extent, much less before I could immortalise it in writing. This book itself indicates the length and breadth of the subject, and the number and importance of the sciences of which at least some little knowledge is necessary to miners. Indeed, the subject of mining is a very extensive one, and one very difficult to explain; no part of it is fully dealt with by the Greek and Latin authors whose works survive; and since the art is one of the most ancient, the most necessary and the most profitable to mankind, I considered that I ought not to neglect it. Without doubt, none of the arts is older than agriculture, but that of the metals is not less ancient; in fact they are at least equal and coeval, for no mortal man ever tilled a field without implements. In truth, in all the works of agriculture, as in the other arts, implements are used which are made from metals, or which could not be made without the use of metals; for this reason the metals are of the greatest necessity to man. When an art is so poor that it lacks metals, it is not of much importance, for nothing is made without tools. Besides, of all ways whereby great wealth is acquired by good and honest means, none is more advantageous than mining; for although from fields which are well tilled (not to mention other things) we derive rich yields, yet we obtain richer products from mines; in fact, one mine is often much more beneficial to us than many fields. For this reason we learn from the history of nearly all ages that very many men have been made rich by the mines, and the fortunes of many kings have been much amplified thereby. . . .

Seeing that there have been so few who have written on the subject of the metals, it appears to me all the more wonderful that so many alchemists have arisen who would compound metals artificially, and who would change one into another. Hermolaus Barbarus, a man of high rank and station, and distinguished in all kinds of learning, has mentioned the names of many in his writings; and I will proffer more, but only famous ones, for I will limit myself to a few. Thus Osthanes has written on χνμεντικα; . . .

**Review of market.**

There are many other books on this subject, but all are difficult to follow, because the writers upon these things use strange names, which do not properly belong to the metals, and because some of them employ now one name and now another, invented by themselves, though the thing itself changes not. These masters teach their disciples that the base metals, when smelted, are broken up; also they teach the methods by which they reduce them to the primary parts and remove whatever is superfluous in them, and by supplying what is wanted make out of them the precious metals— that is, gold and silver,—all of which they carry out in a crucible. Whether they can do these things or not I cannot decide; but, seeing that so many writers assure us with all earnestness that they have reached that goal for which they aimed, it would seem that faith might be placed in them; yet also seeing that we do not read of any of them ever having become rich by this art, nor do we now see them growing rich, although so many nations everywhere have produced, and are producing, alchemists, and all of them are straining every nerve night and day to the end that they may heap a great quantity of gold and silver, I should say the matter is dubious. But although it may be due to the carelessness of the writers that they have not transmitted to us the names of the masters who acquired great wealth through this occupation, certainly it is clear that their disciples either do not understand their precepts or, if they do understand them, do not follow them; for if they do comprehend them, seeing that these disciples have been and are so numerous, they would have by to-day filled whole towns with gold and silver. . . .

**Forecast of entire book so readers know what they will read.**

Since no authors have written of this art in its entirety, and since foreign nations and races do not understand our tongue, and, if they did understand it, would be able to learn only a small part of the art through the works of those authors whom we do possess, I have written these twelve books *De Re Metallica.* Of these, the first book contains the arguments which may be used against this art, and against metals and the mines, and what can be said in their favour. The second book describes the miner, and branches into a discourse on the finding of veins. The third book deals with veins and stringers, and seams in the rocks. . . . The fifth book describes digging of ore and the surveyor's art. . . .

Although I have not fulfilled the task which I have undertaken, on account of the great magnitude of the subject, I have, at all events, endeavoured to fulfil it, for I have devoted much labour and care, and have even gone to some expense upon it; for with regard to the veins, tools, vessels, sluices, machines, and furnaces, I have not only described them, but have also hired illustrators to delineate their forms, lest descriptions which are conveyed by words should either not be understood by men of our own times, or should cause difficulty to posterity, in the same way as to us difficulty is often caused by many names which the Ancients (because such words were familiar to all of them) have handed down to us without any explanation.

**Statement similar to that made by Thucydides in writing about the plague (see p. 300).**

I have omitted all those things which I have not myself seen, or have not read or heard of from persons upon whom I can rely. That which I have neither seen, nor carefully considered after reading or hearing of, I have not written about. The same rule must be understood with regard to all my instruction, whether I enjoin things

(continues)

which ought to be done, or describe things which are usual, or condemn things which are done. Since the art of mining does not lend itself to elegant language, these books of mine are correspondingly lacking in refinement of style. The things dealt with in this art of metals sometimes lack names, either because they are new, or because, even if they are old, the record of the names by which they were formerly known has been lost. For this reason I have been forced by a necessity, for which I must be pardoned, to describe some of them by a number of words combined, and to distinguish others by new names,—to which latter class belong *Ingestor, Discretor, Lotor,* and *Excoctor.* Other things, again, I have alluded to by old names, such as the *Cisium;* for when Nonius Marcellus wrote, this was the name of a two-wheeled vehicle, but I have adopted it for a small vehicle which has only one wheel; and if anyone does not approve of these names, let him either find more appropriate ones for these things, or discover the words used in the writing of the Ancients.

**Attempts to persuade readers to accept his document by relating to reader's values.**

These books, most illustrious Princes, are dedicated to you for many reasons, and, above all others, because metals have proved of the greatest value to you; for though your ancestors drew rich profits from the revenues of their vast and wealthy territories, and likewise from the taxes which were paid by the foreigners by way of toll and by the natives by way of tithes, yet they drew far richer profits from the mines. Because of the mines not a few towns have risen into eminence, such as Freiberg, Annaberg, Marienberg, Schneeberg, Geyer, and Altenberg, not to mention others. Nay, if I understand anything, greater wealth now lies hidden beneath the ground in the mountainous parts of your territory than is visible and apparent above ground. Farewell.

*Chemnitz, Saxony,*
  *December First,* 1550. . . .

# BOOK V.

**Provides coherence between books, forecast for Book V.**

IN the last book I have explained the methods of delimiting the meers along each kind of vein, and the duties of mine officials. In this book I will in like manner explain the principles of underground mining and the art of surveying. First then, I will proceed to deal with those matters which pertain to the former heading, since both the subject and methodical arrangement require it. And so I will describe first of all the digging of shafts, tunnels, and drifts on *venae profundae;* next I will discuss the good indications shown by *canales,* by the materials which are dug out, and by the rocks; then I will speak of the tools by which veins and rocks are broken down and excavated; the method by which fire shatters the hard veins; and further, of the machines with which water is drawn from the shafts and air is forced into deep shafts and long tunnels, for digging is impeded by the in-rush of the former or the failure of the latter; next I will deal with the two kinds of shafts, and with the making of them and of tunnels; and finally, I will describe the method of mining *venae dilatatae, venae cumulatae,* and stringers.

**Provides safety precautions.**

Now when a miner discovers a *vena profunda* he begins sinking a shaft and above it sets up a windlass, and builds a shed over the shaft to prevent the rain from falling in, lest the men who turn the windlass be numbed by the cold or trou-bled by the rain. The windlass men also place their barrows in it, and the miners store their iron tools and other implements therein. Next to the shaft-house an-other house is built, where the mine foreman and the other workmen dwell, and in which are stored the ore and other things which are dug out. Although some persons build only one house, yet because sometimes boys and other living things fall into the shafts, most miners deliberately place one house apart from the other, or at least separate them by a wall.

Now a shaft is dug, usually two fathoms long, two-thirds of a fathom wide, and thirteen fathoms deep; but for the purpose of connecting with a tunnel which has already been driven in a hill, a shaft may be sunk to a depth of only eight fath-oms, at other times to fourteen, more or less. A shaft may be made vertical or in-clined, according as the vein which the miners follow in the course of digging is vertical or inclined. A tunnel is a subterranean ditch driven lengthwise, and is nearly twice as high as it is broad, and wide enough that workmen and others may be able to pass and carry their loads. It is usually one and a quarter fathoms high, while its width is about three and three-quarters feet. Usually two workmen are required to drive it, one of whom digs out the upper and the other the lower part, and the one goes forward, while the other follows closely after. Each sits upon small boards fixed securely from the footwall to the hangingwall, or if the vein is a soft one, sometimes on a wedge-shaped plank fixed on to the vein itself. Miners sink more inclined shafts than vertical, and some of each kind do not reach to tun-nels, while some connect with them. But as for some shafts, though they have al-ready been sunk to the required depth, the tunnel which is to pierce the mountain may not yet have been driven far enough to connect with them.

It is advantageous if a shaft connects with a tunnel, for then the miners and other workmen carry on more easily the work they have undertaken; but if the shaft is not so deep, it is usual to drift from one or both sides of it. From these openings the owner or foreman becomes acquainted with the veins and stringers (thin offshoot vein) that unite with the principal vein, or cut across it, or divide it obliquely; however my discourse is now concerned mainly with *vena profunda,* but most of all with the metallic material which it contains.

(continues)

THREE VERTICAL SHAFTS, OF WHICH THE FIRST, A, DOES NOT REACH THE TUNNEL; THE SECOND, B, REACHES THE TUNNEL; TO THE THIRD, C, THE TUNNEL HAS NOT YET BEEN DRIVEN. D—TUNNEL.

Excavations of this kind were called by the Greeks ρνπτας for extending along after the manner of a tunnel, they are entirely hidden within the ground. This kind of an opening, however, differs from a tunnel in that it is dark throughout its length, whereas a tunnel has a mouth open to daylight. . . .

I have spoken of shafts, tunnels, and drifts. I will now speak of the indications given by the *canales,* by the materials which are dug out, and by the rocks. These indications, as also many others which I will explain, are to a great extent identical in *venae dilatatae* and *venae cumulatae* with *venae prefundae.* . . .

The common miners look favourably upon the stringers which come from the north and join the main vein; on the other hand, they look unfavourably upon those which come from the south, and say that these do much harm to the main vein, while the former improve it. But I think that miners should not neglect either of them: as I showed in Book III, experience does not confirm those who hold this opinion about veins, so now again I could furnish examples of each kind of stringers rejected by the common miners which have proved good, but I know this could be of little or no benefit to posterity. . . .

Let us now consider the metallic material which is found in the *canales* of *venae profundae, venae dilatatae,* and *venae cumulatae,* being in all these either cohesive and continuous, or scattered and dispersed among them, or swelling out in bellying shapes, or found in veins or stringers which originate from the main vein and ramify like branches; but these latter veins and stringers are very short, for after a little space they do not appear again. . . .

Gold, silver, copper, and quicksilver are often found native; less often iron and bismuth; almost never tin and lead. Nevertheless tin-stone is not far removed from the pure white tin which is melted out of them, and galena, from which lead is obtained, differs little from that metal itself. . . .

*Margin notes:*

**Introduces new category. (Refers to indications that a payload of ore exists.)**

**Inserts personal opinion. Uses first person. Tries to persuade readers of his opinion by relating to experience which readers value.**

**Introduces new category.**

Source: Agricola. *De Re Metallica,* translated by Herbert Clark Hoover and Lou Henry Hoover, 1950. Reprinted by permission of Dover Publications, Inc.

## Suggestions for Discussion and Writing

1. What claims does Agricola make in persuading the readers to whom he has dedicated his book to accept it? How do these claims relate to the economic, political, and social context in which the book is written?

2. Agricola discusses the strategies he has used to write the book on pages 167 to 168 in his preface. How do these strategies compare with the strategies for writing instructions today?

3. Agricola is aware of the ethical responsibilities of a technical writer. Select a passage in Book V that demonstrates the author's concern for the safety of the miners as well as his concern that procedures be executed properly.

4. Instead of headings, Agricola uses paragraphs to serve as transitions between subtopics, thus depriving readers of cues for locating information. One of these passages is included in the excerpt from Book V. Bracket it and write a heading to replace it.

5. Examine the illustration. Is it helpful in understanding the information? Why or why not? Compare it with iillustrations in today's manuals. What similarities and differences do you find?

## CHAPTER SUGGESTIONS FOR DISCUSSION AND WRITING

1. Compare the instructions for using a dipstick, servicing a snowblower, and operating a hydraulic hammer. To which writing conventions do the writers of each adhere? Which of the layouts is easiest to read? Why? How do the graphics in each facilitate readers' ability to comprehend the instructions? The instructions for the snowblower use photographs, and the instructions for the hydraulic hammer use diagrams. Why do you think the documents' designers use different types of graphics? Are both types successful in helping the reader comprehend the instructions, or do you think one type is more effective? Which one and why? How do the writers indicate to the readers that information is important or that something may be dangerous? How do writers include additional information that may not fit neatly into a step?

2. Compare Vitruvius's handbook with the manual on the hydraulic hammer. What are the similarities and differences in conventions, format, visual text, and graphics?

3. Rewrite the chapter on the city walls in *De Architectura* to follow today's conventions for writing instructions.

## References

J. Guillemeau. 1612. *Child-birth or, the happy deliverie of women.* London: A. Hatfield. Cited in Tebeaux, Elizabeth, and Mary Lay. 1992. Images of women in technical books from the English renaissance. *IEEE Transactions on Professional Communication* 35 (December): 196–207.

Tebeaux, Elizabeth, and Mary Lay. 1992. Images of women in technical books from the English renaissance. *IEEE Transactions on Professional Communication* 35 (December): 196–207.

# Proposals

## INTRODUCTION

Proposals make the technical world "go 'round." New designs for cars and aircraft, improved processes for the safe disposal of toxic wastes, and the initiation of services to provide mobile health care to the elderly all start with a proposal. Proposals may provide the life blood of a company, especially a company that engages in consulting or that relies on subcontracts from the federal government for a large percentage of its business. Because these companies must submit proposals to obtain contracts for most of their projects, they are often willing to spend large amounts of time and manpower writing them. In developing a 225-page proposal to respond to a $3 million request for a proposal (RFP) from the Air Force, one company involved the full-time efforts of eight engineers, five technical managers, and two editors as well as the part-time assistance of six text processors, seven artists, and two photographers.

But proposals are not limited to large projects. Today, with industry emphasizing employee responsibility and encouraging innovation that cuts costs and improves quality, personnel have the opportunity to suggest improvements in the form of proposals. Whether a company agrees to fund ideas for a new or altered product or a change in the manufacturing process or delivery system depends on the proposal. Employees are increasingly becoming involved with the production process of their companies and writing these proposals. You will have an opportunity to read one proposal in which a mechanic and his supervisor propose a change in a mechanism to simplify repairs (see page 232).

## Definition

A proposal is a special form of request in which a writer proposes an idea, project, service, or product. The writer tries to persuade readers to approve the ideas, provide financial or other support for the projects, or accept the service or product proposed. To persuade their audience, writers demonstrate that their proposal solves a problem. For example, a writer proposing a new dunking process for painting engines might argue that the process solves the problem of toxic waste disposal by reusing excess paint.

## Types of Proposals

Proposals are written to make a change, to obtain support for a project, or to provide services or products.

- **To make a change.** These proposals are usually internal and relate to the writer's organization. The writer seeks to persuade readers to approve the change. The proposal for adding a valve to reduce the time spent repairing a circuit breaker (page 232) exemplifies this type of proposal.
- **To obtain support for a project.** These proposals are often related to the research and development or design phase of a project. The writer seeks to persuade readers to approve the project and to provide financial and other types of support for it. The policy proposal for legislative action on lead paint (p. 184) and the proposal for funding to protect murres in Alaska (p. 229) fall under this category.
- **To provide services or products.** These proposals are usually related to marketing a company, product, or service. The writer seeks to persuade readers to purchase the organization's services or products. The proposal to provide services to clean up a leaking storage tank, included in this chapter on page 222, is an example of this type of proposal.

# CONTEXT

Proposals exist in a competitive environment. Sometimes a company's profits or an individual's reputation rests on whether a proposal is accepted.

Individuals who submit research proposals are often competing with other individuals, just as companies that submit proposals are often competing with other companies. Regardless of whether the proposers are individuals or companies, they must demonstrate that their solution is as good, if not better, than their competitors' solutions, that they are at least as qualified to do the research or perform the service, and that they can do the work at least as cost effectively as their competitors.

## Solicited and Unsolicited Proposals

Proposals may be solicited by readers who are searching for organizations or persons to perform a specific task or they may be unsolicited, the result of a writer's need to do something or obtain something. Companies solicit proposals when they have work for an individual or another company to do. These companies publish an RFP (Request for Proposal), RFS (Request for Specifications), or PSR (Project Specification Report, see p. 15) in which they describe the work they want done. Organizations such as the National Science Foundation often solicit individuals and institutions to submit proposals for research and development projects. Readers of solicited proposals have already decided to approve a proposal; they must simply determine whether a proposal meets their specifications and, if there are several proposals, which one is the best. Therefore, writers of solicited proposals concentrate on the solution section of a document.

Unsolicited proposals are initiated by writers who have ideas, products, or services that they would like implemented or supported. These proposals are usually more difficult to write than solicited ones because they may be written for readers who are unaware of the problem for which a solution is being proposed, who see the problem as irrelevant to them, who have no interest in solving the problem, or who disagree with the writer's particular solution.

## Situation

Economic and political factors are usually involved in writing proposals. Proposals related to environmental projects received little financial support while George Bush was president because Bush did not set much money aside for such projects and because the government did not place a priority on improving the environment. Such projects have received better support under the Clinton administration because of the support of Vice President Albert Gore.

Regardless of whether the purpose of a proposal is to create a change, obtain support for a project, or provide a service, economic factors are extremely important in readers' decisions to approve a proposal. If the proposed change, project, or service will cost money, readers need to be persuaded that it is worth the cost. Moreover, they need to be persuaded that they should spend the money for the proposed solution rather than for something else. Therefore, most proposals include a category in the solution section that demonstrates the solution is cost effective, will save money in the short or long run, or will provide a satisfactory return for the investment. The latter may be measured in terms of self-satisfaction, the good of society, or monetary value.

Political factors also play a deciding role in a proposal's acceptance. For example, a proposal from a technician to a supervisor suggesting a more cost-effective method for painting engines may not be approved because the supervisor is afraid that if something goes wrong with the new method, he won't receive a promotion. On the other hand, if the supervisor knows that management is looking for new methods, then he may be more likely to approve the suggestion.

## Purpose

The writer's purpose is to persuade readers to approve a proposal, and the reader's purpose is to decide whether a proposal should be approved or supported. When readers are faced with a number of proposals, their purpose is also to decide which of the many proposals meets their needs best.

## Audience

All readers assume the role of decision makers who must decide whether to agree to a proposal.

Although proposals are occasionally read by a single reader, they are more often read by multiple readers. Readers may be experts in the field to which a topic is related, or they may know very little about a topic. Readers may read an entire proposal or only those parts of it that are related to their specific specialty. For example, in helping the government determine whether to approve the proposal for the construction of a federal courthouse in Foley Square, Manhattan, New York (discussed in Part III), architects may have been asked to evaluate the part of the proposal related to the design of the building, contractors may have been asked to evaluate the part related to construction specifications, and budget analysts may have been asked to evaluate the budget section. Staff members at the General Services Administration, who are generalists in the field, may only have read the executive summary and the experts' evaluation of that proposal before deciding whether to approve the project.

Readers may be faced with a large number of documents to read and evaluate if proposals have been solicited. The Trustee Council, which is responsible for allocating funds from Exxon for the Valdez oil spill, received nearly 450 proposals for cleanup projects. (One of these is reprinted on page 229.) These proposals are only three pages in length, and did not overwhelm readers. However, in response to a PON (program opportunity notice) for clean coal technology pilot projects, the Department of Energy (DOE) received 33 proposals that were each several hundred pages. Long documents such as these are often read by assistants, who evaluate the proposals to determine whether they meet the specifications. Qualifying proposals are then sent to those higher up in the hierarchy for a final selection.

If a proposal is not solicited, then it may be one of many documents that a reader receives in the course of a work day. If a reader is busy, and if a proposal does not appear to be closely related to ongoing projects, the reader may place the proposal in the "to do," box where it may remain unexamined for months.

Regardless of whether a document is solicited or unsolicited, readers' attitudes toward a proposal are usually skeptical because they are trying to determine whether a project is worthwhile, whether a writer or a writer's organization is qualified to do a project, and whether a writer or a writer's organization is more qualified than others who have submitted proposals to do a project.

Readers usually read a proposal quickly, scanning to determine if an idea, product, or project meets specifications or appears relevant to their needs and purposes. They also check whether the proposal meets their basic values and criteria. Once readers have determined that a proposal is worth considering, they evaluate it to determine whether to approve it.

# DISCOURSE STRATEGIES

## Persuasion

All proposals are persuasive. The writer must persuade the reader that the change, project, or service being proposed solves a problem. Thus, the writer must develop two arguments. First the writer must persuade the reader that a problem exists. Then the writer must persuade the reader that the most appropriate solution is the one being proposed. To persuade a university to establish a desktop publishing laboratory, students needed to convince university officials that the lack of such a lab was a problem. They based their argument on the claim that graduates in business, professional writing, and engineering were at a disadvantage in competing for jobs against students from other colleges who had had opportunities to acquire desktop publishing skills. Once officials accepted this claim, students then proposed an open lab in the university library as a solution to the problem.

Readers judge a proposal according to standards established by their communities. For readers to be persuaded to accept a proposal, they must agree with the values implied in the problem section of a proposal, and they must accept the evidence. The values implied by the student's proposal for the desktop publishing lab coincided with the values held by the university's officials, who believed their graduates should be competitive with graduates of other colleges.

Readers also judge the merits of the solution section of a proposal according to criteria developed by their communities. Many scientists will not accept the results of a research study unless a control group has been used as a comparison and an ANOVA (a complex statistical formula) has been used to analyze the data. Because many universities across the country had already approved computer laboratories as a valid expenditure for improving students' skills, the proposed computer lab for students to learn the specialized skills of desktop publishing met the criteria of the university officials.

## Navigation and Location Aids

Because long proposals may be read by multiple readers who read only those sections that relate to their fields, proposals need to help readers locate the information in which they are specifically interested. Tables of contents, lists of tables and figures, and indices help readers locate the specific data they want to read. Headers, pagination, chapter titles, and headings and subheadings help readers recognize the information when they skim a document.

Even when documents are short, they still should include visual cues so readers can scan a proposal for information they consider crucial in determining whether to approve a proposed idea, project, or change. Readers may want to locate specific information concerning a budget or a method to be used. Subheads, marginal notes, and typographic markers such as boldface type and capital lettering help these readers.

## Ancillary Documents

Because proposals are often too complicated and too long for a single reader to read and understand, ancillary documents are important. Cover letters or letters of transmittal provide readers with a quick way of determining when they want to read a document and whether they want to read it themselves or whether it should be submitted to another person. The letters also provide readers with a quick means of identifying a specific proposal if they wish to locate it in a pile of similar documents.

Executive summaries or abstracts are usually included to permit multiple readers to see how a specific section fits into an entire proposal. A summary or abstract also provides readers with an overview of the problem and solution so they can get a quick grasp of the issues without having to read the entire document.

Appendices are often used to include copies of surveys, breakdowns of statistical data, and additional documents and data that support the argument in the main text.

A glossary may be included for readers who are generalists or novices.

## Organizational Patterns and Sequence

Proposals can be preliminary, brief descriptions of a proposed plan, or they can be final, detailed versions. Information contained in a preliminary proposal is general; the writer simply tries to catch the readers' attention. However, final proposals are detailed and specific.

Guidelines for the content and organizational pattern and sequence of a solicited final proposal are usually provided by the organization issuing an RFP or a PON. Writers need to be especially careful to cover all of the areas delineated in these guidelines.

Most final proposals, regardless of whether they are solicited or unsolicited, follow a problem/solution pattern. They usually include three major sections: (1) a discussion of the problem, (2) a listing of long- and short-term objectives, and (3) a discussion of the solution. The main organizing idea for the problem section usually occurs at the *conclusion* of that section, and the main organizing idea for the solution section usually occurs at the *beginning* of that section. Both short-term and long-term objectives are often included. If there are multiple problems, objectives, or solutions, they are sequenced from most to least important.

The solution section often consists of six parts: (1) an explanation of how the solution can solve the problem, (2) the methods to be used and the procedures to be followed, (3) the personnel involved, (4) the material, equipment, and facilities to be used, (5) the amount of time it will take, and (6) the cost.

## Graphics

Graphics are an important aspect of proposals. Diagrams and photographs can help readers perceive a proposed mechanism or process with which they are unfamiliar. Tables, charts, and graphs help readers quickly *see* numerical data that serve as evidence to support claims in the problem section.

## Conventional Format

Proposals can be as short as a single page or as long as several volumes with hundreds of pages. They can be incorporated into a memo or letter or they can stand alone. They are usually impersonal and formal because writers rarely know their readers. In addition, proposals can serve as legal documents in cases where intent is questioned.

A heading should introduce each of the three major categories: the problem, the objectives, and the solution. The heading helps readers quickly locate the respective sections. Headings also facilitate readers' comprehension by helping them accurately predict what they will read.

Objectives follow specific text grammars. To write an objective, use the following conventions:

- Truncate the sentence.
- Begin with an infinitive phrase (*to* + verb + phrase).
- Use an action verb.
- Limit the objective to a single sentence.

For example,

Objective: To set up a desktop publishing laboratory in the university
library.

Sequence multiple problems, objectives, or solutions from most to least important.

# THE DOCUMENTS

The chart in Figure 5.1 indexes the various sample documents according to their respective categories.

We will examine the first proposal closely to see how it adheres to the conventions. Marginal notes will help you study the various aspects of the next document. You will then have an opportunity to apply what you have learned to the remaining documents.

As you study these proposals, notice that writers adhere to the conventions, regardless of whether they are writing a solicited or unsolicited proposal, a short or long proposal, or a preliminary or final proposal. Also notice that wide variation exists within these conventions.

### Policy Proposal: *Legacy of Lead: America's Continuing Epidemic of Childhood Lead Poisoning*

Because of industry's and government's emphasis on improving the workplace environment and our lifestyles in general, and because of industry's new openness toward employee suggestions, you may occasionally have an opportunity to write a proposal that significantly affects people's lives. Such was the case for the writers of this unsolicited proposal, sponsored by the Environmental Defense Fund (EDF). EDF was established in 1967 to find constructive solutions to the world's environmental problems. One of its goals has been to eliminate childhood lead poisoning, which can result in

| | Lead Paint | Leaking Storage Tanks | Oil Spill | Valve Change |
|---|---|---|---|---|
| Preliminary | | | | |
| Final | XXX | XXX | XXX | XXX |
| Solicited | | XXX | XXX | XXX |
| Unsolicited | XXX | | | |
| Make Changes/ Improvements | XXX | | | XXX |
| Obtain Support/ Resources | | | XXX | |
| Provide Services/ Resources | | XXX | | |
| For Experts | | | | XXX |
| For Generalists | XXX | XXX | XXX | XXX |
| For Novices | XXX | | XXX | |

**Figure 5.1   Types of proposals.**

mental retardation. This proposal was written to persuade the U.S. Congress to eradicate the lead poisoning of children that occurs in poor urban areas.

The proposal received a great deal of attention from both Congress and the media. A bill based on the proposal was drawn up for the 102nd Congress but never made it to the floor. However, as of the writing of this book, the bill has been revised and will be resubmitted.

The document was written collaboratively by an attorney knowledgeable in legislative law, a toxicologist who is an expert on lead poisoning, and a technical writer. Other people provided data and information. Still others reviewed the information and offered suggestions for improving it (see p. 185).

The audience for the proposal was members of Congress as well as interested citizens and organizations who would be apt to lobby their Congressional representatives to pass the necessary legislation. Peripheral readers such as the news media were also expected to read the document and publish a synopsis of it to inform all citizens of the proposed legislation.

The writers were aware that their readers had little knowledge about the topic and that some of the knowledge they had was incorrect. Most people believed that lead poisoning was no longer a problem because legislation to outlaw lead in paint had been passed years earlier.

In addition, the writers recognized that economic and political factors would play a substantial role in determining whether their idea was accepted. The proposal was written during the administration of President George Bush. At that time, relatively little money was available for environmental or health issues. Furthermore, most Congressional representatives, in agreement with the president, were unwilling to pass any legislation that could be construed as a tax, and most citizens were unwilling to be taxed. However, there was a powerful movement among citizens to buy American products, and the Congress was split over whether to pass protectionist legislation against foreign trade that hurt American businesses. The arguments in Chapter 8 of the proposal are based on claims that recognize these factors.

**Presents Effective Arguments**   Writers must persuade their readers that the problem of lead poisoning among children exists and that the solution is viable. To help readers recognize and understand the problem as well as accept their claims, the writers use the following strategies to persuade their readers to accept their proposal.

- Present background information indicating that the problem of lead poisoning persists despite the outlawing of lead paint.
- Reflect readers' attitudes toward protecting children, who are the nation's future resource.
- Cite the results of recent research studies that reflect data readers consider valid and the ideas of people that readers consider as authorities.
- Describe the effects of lead poisoning on the brain so readers understand how lead poisoning affects children.
- Discuss the various sources of lead so readers understand how children become exposed to it.
- Claim that the problem affects all society to counter lack of interest on the part of readers who consider the problem limited to poor children living in urban areas.
- Discuss the types and numbers of individuals affected to support their claim that lead poisoning among children constitutes an epidemic.
- Discuss evidence indicating that lead poisoning is incurable and the resulting neurological damage cannot be reversed to support their argument that lead poisoning needs to be prevented rather than treated.

Once the writers complete their presentation of the problem, they present their solution, which is the establishment of a trust fund to finance the elimination of lead paint. In Chapter 7 of the proposal, the writers present their claims to support their argument that the solution can solve the problem. In Chapter 8 they present claims to support their argument that the solution is good for the American people.

**Facilitates Navigation and Location of Information**   The table of contents includes three levels of headings so readers can look up specific areas of concern. Pages are numbered and chapter titles and headings and subheadings are in boldface type so readers skimming the document can easily locate the information they seek. Furthermore, margin notes call readers' attention to important information.

**Meets Multiple Readers' Needs with Ancillary Documents**   The proposal includes a cover sheet. It also includes an executive summary providing a general overview and the major recommendations of the proposal. Most legislators will not read beyond this. If they are interested, they will probably ask their aides to read the entire report, give them an analysis of it, and recommend whether they should sponsor the proposed legislation.

Supplementary information is included in the back of the proposal. The bibliography provides interested readers with additional references to study. It also provides authoritative support for the evidence introduced in the text.

**Facilitates Readers' Reading Processes**   A brief summary placed at the beginning of each chapter helps readers accurately predict what they will read. Headings and subheadings within the chapters also provide readers with accurate cues for the information that follows.

The layout helps readers quickly scan the text. By printing on only three-quarters of the page, the designer shortens the line of text sufficiently that readers can scan straight down the page without having to scan across the page. The addition of white space at the bottom of various pages provides some airiness to the text.

The few graphics that are included help readers comprehend the information. These graphs emphasize the major findings and summarize and synthesize the statistics the authors use to support their claims.

**Follows the Conventional Format**   The proposal follows a problem/solution organizational format. The problem is discussed in Part I, and the solution is discussed in Part II of the proposal. Chapter 7 serves as a bridge between the two parts and provides a set of short-term and long-term objectives.

As you read, consider how the writers develop their argument for legislative action.

# LEGACY OF LEAD:
# AMERICA'S CONTINUING EPIDEMIC
# OF
# CHILDHOOD LEAD POISONING

A Report
and Proposal
for Legislative Action

Written by

Karen L. Florini
Senior Attorney
Environmental Defense Fund

George D. Krumbhaar, Jr.
Consulting Author

Ellen K. Silbergeld
Senior Toxicologist
Environmental Defense Fund

March 1990
Environmental Defense Fund
Washington, DC

© Copyright 1990, Environmental Defense Fund
3rd Printing.  Printed on recycled paper.

Acknowledgements

The authors gratefully acknowledge the assistance of numerous individuals in preparing this report.  Special thanks are due to Annemarie Crocetti, Ph.D., who prepared the regional data analysis in Appendix I; Stephanie Pollack, Esq., of the Conservation Law Foundation for her extensive comments on an earlier draft; and to Dr. Lawrence Goulder, Department of Economics, Stanford University, and Dr. Susan Cohen, Columbia University, for their assistance on economic issues.

Additional extremely helpful comments were received from Elizabeth Feuer, MD; Cheryl Burke, Esq., of Aiken, Gump; Herbert L. Needleman, MD, of the University of Pittsburg Department of  Psychiatry; Keith Winston; Mary Lynn Sferrazza; and Jerry McLaughlin.

All errors are entirely the responsibility of the authors.

Environmental Defense Fund
1616 P st. NW
Washington, DC  20036
(202) 387-3500


Document layout by Keith Winston.

(continues)

# TABLE OF CONTENTS

*(continues)*

## Introduction

As a tragic legacy of the decades-long use of leaded products on a vast scale, lead today pervades America's environment. The result is a nation-wide epidemic of low-level lead poisoning, an epidemic that is causing permanent neurologic damage to millions of American children. Recent studies demonstrate that the long-term consequences of this disease are profound: children who had moderately elevated lead levels in early child-hood later exhibited seven-fold increases in school dropout rates, six-fold in-creases in reading disabilities, and lower final high school class standing.[1] These effects occurred even though the initial exposures caused no overt symptoms.

Although no precise national measurements have been collected, the federal government estimates that well over three million pre-school chil-dren -- more than 1 in every 6 -- have dangerously elevated lead levels. Poor and minority children are disproportionally affected, but the problem cuts across all socioeconomic lines.

The consequences of low-level lead poisoning are devastating not only for the affected children and their families, but also for society as a whole. As the Secretary of Education observed earlier this year, reading and writing skills of the nation's children remain "dreadfully inadequate" despite a decade of educational reform. The new data suggest that lead is partly to blame. By the same token, until children's lead exposures are substantially curtailed, the nation will continue to fall short of its educational goals.

The severity of the nation's lead-poisoning crisis has gone generally unrecognized for decades largely because the great majority of cases have never been diagnosed. The effects of low-level lead poisoning, through severe, are not unique or obvious. Unlike the readily observable signs of chicken pox, for example, the impairment of intellectual ability caused by low-level lead poisoning is hard to pinpoint in individual children. Even when identified, such symptoms overlap with those of a variety of other biological and socioeconomic factors. Only recently, with the completion of sophisticated long-term studies, was the compelling association between childhood lead poisoning and significant neurologic impairment recognized.

## The Toxicity of Lead

In the human body, lead is a potent poison that can affect individuals in any age group. Children and fetuses are particularly vulnerable, because their rapidly developing nervous systems are sensitive to lead's potency as a neurotoxin. Moreover, children generally are exposed to more lead than are

**The federal government estimates that well over three million pre-school children -- more than 1 in every 6 -- have dangerously elevated lead levels.**

---

[1] See HL Needleman, A Schell, D Bellinger, A Leviton, and EN Allred (1990), "The Long-Term Effects of Exposure to Low Doses of Lead in Childhood," New England Journal of Medicine, Vol. 322, pp. 83-88.

adults, and their absorption rates are substantially higher.

Lead's specific neurotoxic effects include impairments to IQ level, short-term memory, and reaction time; it also impairs the ability to concentrate. In adults, low-level lead exposure has been associated with hypertension in men and pregnancy complications in women, including minor birth defects.

Once absorbed, lead is stored primarily in bone. To a lesser degree, storage also occurs in the kidneys and the brain, while a small portion remains in circulation in the blood. Lead's persistence in the body is unequalled by virtually any other toxin. Its "half life" in bone -- the time it takes half of a given dose to be removed -- exceeds twenty years. As a result, even small amounts of lead accumulate in the body, and can cause effects that endure long after exposure ends. Further, because stored lead can be released during pregnancy and readily transferred to the fetus, lead poisoning is, in effect, a heritable disease.

**Many public health experts now believe that lead presents a "continuum of toxicity," in which the slightest exposure contributes to an adverse result somewhere in the body**

In the early and middle decades of this century, lead was generally thought to be harmful only at high doses. Subsequent research, however, has uncovered a variety of effects at lower and lower levels. This trend has accelerated within the last few years, as increasingly sensitive analytic techniques allow investigators to document consequences that persist for years after initial exposure. Many public health experts now believe that lead presents a "continuum of toxicity," in which the slightest exposure contributes to an adverse result somewhere in the body.

Because lead causes neurologic damage even at doses that do not cause overt toxicity, levels of lead in blood are generally used in identifying lead exposures of concern. The federal government's Centers for Disease Control (part of the Public Health Service) is currently reviewing its definition of "lead toxicity," which is now set at 25 micrograms of lead per deciliter of blood (ug/dl); CDC is expected to adopt a new definition of between 10 and 15 ug/dl within the year. The U.S. Environmental Protection Agency, along with many public health experts, has already recognized that blood-lead levels of 10 to 15 ug/dl cause neurotoxic effects in children. This report uses the term "low-level lead poisoning" to denote these levels and the associated health effects.

Even at the 25 ug/dl level, the very limited lead-screening programs now in place uncover over 10,000 previously unreported cases of poisoning each year. Indeed, though little recognized by the general public, the scale of this insidious epidemic makes it among the most common diseases of childhood.[2] It is also nationwide in scope, as an analysis of the estimated numbers of affected children throughout the country reveals. Exposures are

---

[2] Some common childhood illnesses and their reported 1988 incidence rates include:

| | |
|---|---|
| Lead Poisoning (25+ ug/dl) | 11,793 |
| Viral meningitis | 6,927 |
| Mumps | 4,730 |
| Whooping cough | 3,008 |
| Measles | 2,933 |

Source: Centers for Disease Control. The Centers' records include only cases that were identified through screening programs and reported by health officials; therefore, these must be considered minimum figures.

2

(continues)

endemic in some urban regions, with **over 50%** of children under 6 estimated to have blood lead levels over 10 ug/dl.[3]

As a practical matter, prevention is the only realistic "cure" for lead poisoning. Available treatments are expensive and painful, do not completely remove lead from the body, and are powerless to undo neurologic damage. But little has been done to prevent childrens' exposures to lead already dispersed into existing environmental reservoirs.

## Exposure Levels and the Environmental Reservoir of Lead

Most children are exposed to lead as a result of its presence in paint, plumbing, gasoline, solder, and other products. Over many decades, these uses have dispersed millions of tons of lead throughout the environment. And that reservoir continues to grow each year, as the United States uses another million-plus tons of lead in products such as automotive batteries, construction materials, gasoline, and other items. Because lead is an element, no force save a nuclear reaction can transform it into a more innocuous material; once excavated from the earth and distributed in commerce, lead can exert its inherent toxicity on the biosphere almost indefinitely.

Of all the sources that make up the existing reservoir of environmental lead, one is responsible for especially intense exposures for many children: the three million tons of leaded paint remaining on the walls and woodwork of American homes. Though banned for most uses in 1977, leaded paint applied during the preceding decades continues to present a hazard. An estimated 1.2 million children under 6 absorb enough lead from deteriorated paint to elevate their blood-lead levels beyond 15 ug/dl, with a significant chance of subsequent neurologic impairment.

Although the pronounced long-term consequences of childhood lead poisoning have only recently been identified, its more obvious manifestations have been a focus of concern for decades. As early as 1904, reports of childhood lead intoxication appeared in the medical literature. But due to the limited diagnostic capabilities of the time, only the most obvious cases were identified -- those involving high doses of lead resulting in readily observable effects such as convulsions, coma, and even death. With the advent of blood-lead determinations in the 1940s and '50s, however, it became increasingly apparent that the problem was far greater in scope than had been recognized previously.

Citing the "epidemic proportions" of childhood lead poisoning, Congress first took action in 1970 to eliminate a primary source of children's exposures. The Lead-Based Paint Poisoning Prevention Act of 1971[4] authorized a wide range of actions designed to identify and treat those already harmed, to remove lead-based paint from federally-assisted homes, and to prohibit its use in areas thought to be accessible to children. Unfortunately, implementation of key provisions faltered badly almost from the start. As a result, two decades later the epidemic persists.

The reservoir continues to grow each year, as the United States uses another million-plus tons of lead in products such as automotive batteries, construction materials, gasoline, and other items.

---

[3] Geographical distributions of lead-affected children are described in Appendix 1.

[4] 42 U.S.C. sections 4801-4846.

The role of paint is even clearer today than it was twenty years ago, for many other major sources of lead have been at least partially controlled in the interim. For the American populace as a whole, the most significant reductions in lead exposure have resulted from the phase-down in use of leaded gasoline over the last fifteen years. But while this step has provided important benefits in reducing lead exposures for many people, it has done little to aid those children whose primary source of lead is from paint. And these children -- many of them poor and/or minorities -- are precisely the same individuals who are most disadvantaged by a myriad of other social and economic factors.

## Toward a Solution:
## A Proposal for Legislative Action

The massive amounts of information on lead's toxicity -- bolstered by recent findings on low exposure level effects -- as well as indications of children's current exposure levels, reveal an urgent need for an aggressive federal program to control America's continuing epidemic of lead poisoning. To be effective, such a program must provide a mechanism not only to *stop adding* lead to children's environments, but also to *remove* it from the areas where they are most heavily exposed: their homes. And, to be politically feasible, it must respond to current budgetary realities the nation now faces.

The Environmental Defense Fund proposes creation of a National Lead Paint Abatement Trust Fund, to be financed by placement of a substantial excise fee on the production and importation of lead. Proceeds from the fund initially would be devoted to the removal of deteriorating lead-based paint from the group of highest-risk homes. In addition, a portion of the monies could be made available for research to develop more effective lead-removal methods.

The program would be implemented jointly by the Environmental Protection Agency and the Department of Health and Human Services. It would contain provisions to enable it to reflect market conditions and, where possible, accomplish secondary goals of improving housing and creating employment opportunities by hiring and training workers for abatement programs. In addition, by avoiding the slow and resource-intensive process of developing a regulatory approach to control continuing uses of lead in products, it would yield results far more quickly than would more traditional approaches.

While the proposed program would not alleviate every aspect of the nation's current lead poisoning epidemic, it would constitute a pragmatic and timely next step. Lead poisoning already burdens America with millions of dollars of costs each year -- both the direct costs of medical treatments, and the indirect social costs of special education, lost income, and a less productive citizenry. It also imposes grave handicaps on individual children, their families, and their communities. For them and for the nation as a whole, these handicaps will only intensify as the transition to the twenty-first century's "information age" continues. By creating a nationwide paint-abatement program funded by a lead excise tax, America can permanently reduce lead exposures and bring about a significant improvement in the health and abilities of the nation's children -- now and for generations to come.

> **"Ideally, in keeping with the precepts of primary prevention, lead should have been prohibited from ever having been dispersed in the modern environment."**
> **American Academy of Pediatrics**

4

(continues)

## 1. THE TOXICITY OF LEAD[1]

*Lead's primary effect of concern is neurotoxic damage to fetuses and preschool children, for this effect occurs at levels of exposure that are commonplace in contemporary society. Low levels of lead exposure can also cause kidney damage and high blood pressure in adults.*

Upon entering the body, lead makes its way into the blood stream; into soft body tissue, including the brain and kidneys; and into the "hard tissues," such as bone and teeth.[2] Blood-lead content is generally considered to be the most accurate measure of short-term lead exposure. The estimated half life of blood lead (i.e., the time required for one half of the lead to disappear) is 35 days.[3] While about 50 to 60 percent of the lead entering a person's body is eliminated fairly rapidly,[4] most of the remainder is stored in bone, where it stays for far longer periods. In fact, lead in bone has an estimated half life of about 20 years.[5]

Long thought to be inert, bone-based lead is now looked on as a double threat to the body. Bone is a living tissue that is itself sensitive to toxic assaults.[6] Many conditions, moreover, can rapidly release bone-based lead back into the blood stream. For example, pregnancy and osteoporosis, both

---

[1] The documentation of lead's toxic effects is immense. Key sources include: Agency for Toxic Substances and Disease Registry (1988), The Nature and Extent of Lead Poisoning in Children in the United States: A Report to Congress (Atlanta: U.S. Dep't of Health and Human Services/Public Health Service), Doc. No. 99-2966, especially Chapters III and IV; U.S. Environmental Protection Agency (1986a), Air Quality Criteria Document for Lead, Vols. I through IV; Centers for Disease Control (1985), Preventing Lead Poisoning in Young Children, (Atlanta: Dept of Health and Human Services/ U.S. Public Health Service). Excellent review articles include HL Needleman (1988a), "Why We Should Worry About Lead Poisoning," Contemporary Pediatrics, pp. 34 - 56; JM Davis and DJ Svensgaard (1987), "Lead and Child Development," Nature, Vol. 329, pp. 299-300; HL Needleman (1988b), "The Persistent Threat of Lead: Medical and Sociological Issues," Current Problems in Pediatrics, Vol. XVIII, pp. 703-76; EK Silbergeld (1985), "Neurotoxicology of Lead," in K Blum and L Manzo (eds.), Neurotoxicology (Amsterdam: Dekker).

[2] MB Rabinowitz, GW Wetherill, and JD Kopple (1976), "Kinetic Analysis of Lead Metabolism in Healthy Humans," Journal of Clinical Investigation, Vol. 58, p. 260.

[3] Ibid.

[4] Agency for Toxic Substances and Disease Registry (1988), p. III-7.

[5] Rabinowitz et al. (1976).

[6] Agency for Toxic Substances and Disease Registry (1988), p. III-7.

---

## PART I: UNDERSTANDING THE PROBLEM

**Lead makes its way into the blood stream; into soft body tissue, including the brain and kidneys; and into the "hard tissues," such as bone and teeth.**

of which cause demineralization of bone, have been associated with sharp rises in blood lead levels.[7] Indeed, lead moves from bone to other parts of the body readily enough that it may well be an "insidious source" of long term lead poisoning.[8]

An important aspect of lead's menace, therefore, is its *cumulative effect*. Even seemingly trivial exposures, if often repeated, can add up to doses that exert toxic effects.[9]

And virtually no part of the body is immune from lead. As one recent analysis put it, "At a sufficient level of lead exposure, virtually all body systems will be injured or have a high risk of injury."[10] While researchers have not yet discerned the exact biological mechanisms of lead toxicity, they have extensively documented its effects on a number of organ systems at the cellular level.

The most important effects of lead involve disruption of energy metabolism at the cellular level and interference with neural cell function in the brain. Specifically, lead interferes with the formation of heme, the molecule that carries oxygen in all cells.[11] In the nervous system, lead has a unique ability to inhibit communication and slow motor nerve conduction velocity[12] -- the speed at which nerves process signals.[13]

**Neurotoxic Effects:** Lead's neurotoxic effects at relatively low exposure levels include decreased intelligence, short-term memory loss, reading and spelling under-achievement, impairment of visual-motor functioning, poor perceptual integration, poor classroom behavior, and impaired reaction time. Children and fetuses are especially susceptible to these effects, because their neurologic systems are rapidly developing.[14] Growing

> "At a sufficient level of lead exposure, virtually all body systems will be injured or have a high risk of injury."

---

[7] EK Silbergeld, J Schwartz, and KR Mahaffey (1988), "Lead and Osteoporosis: Mobilization of Lead from Bone in Menopausal Women," Environmental Research, Vol. 47, p. 79.

[8] Environmental Protection Agency (1986a), Vol. IV, p. 13-16.

[9] Centers for Disease Control (1985), p. 3.

[10] Agency for Toxic Substances and Disease Registry (1988), p. IV-3.

[11] Silbergeld (1985).

[12] Ibid.; see also, PJ Landrigan (1989), "Toxicity of Lead at Low Dose," British Journal of Industrial Medicine, Vol. 46, pp. 593-4.

[13] In addition to these effects of low-level lead exposure, effects of high-level exposures are also varied, and include anemia, brain damage, muscle palsy, kidney failure, headache and vomiting, convulsions, and death. These high-dose effects have been known for centuries. The first known clinical account of lead poisoning comes from the first century B.C., while Hippocrates offered unconfirmed descriptions two centuries earlier. See HA Waldron (1973), "Lead Poisoning in the Ancient World," Medical History, Vol. 17, pp. 391-99. In eighteenth century Massachusetts, lawmakers enacted one of the country's first public health statutes after recognizing the health effects of drinking "rum and other strong liquors" from leaded containers. See CP McCord (1953), "Lead and Lead Poisoning in Early America: Benjamin Franklin and Lead Poisoning," Industrial Medicine and Surgery, Vol. 22, p. 397.

[14] Centers for Disease Control (1985), p.1.

6

(continues)

evidence indicates that the effects of lead poisoning occur before any overt symptoms appear and often constitute a serious health problem even in the absence of obvious symptoms.

Several key epidemiologic studies in recent years have compellingly demonstrated the range of lead's effects on a variety of populations.[16] Most of these studies are retrospective, meaning that researchers identify a group of children, determine their lead levels, and evaluate their current health status in an attempt to ascertain the effects of prior lead exposure. Although lead poisoning is often viewed as primarily a disease of the poor, wealth and social status confer no immunity. Indeed, a recent government study concluded that children living *above* the poverty level comprise the largest category of people in danger of undue exposure.[15]

A series of landmark studies on lead neurotoxicity have been conducted by Dr. Herbert Needleman of the University of Pittsburgh and his colleagues. The researchers collected baby teeth -- which, like bone, serve as long-term storage sites for lead -- from over 2300 first and second graders in two suburban Boston school districts. They then categorized the children according to dentine (tooth) lead levels [17], and identified two groups for further study: a low-lead group of 100 children who had extremely low levels and a high-lead group 58 children who had relatively high levels but who had no symptoms of overt lead poisoning.[18] Those 158 children were then evaluated using an array of standardized and some nonstandardized neuropsychological tests.

When the results were controlled for 39 other factors (such as socioeconomic status, family size, and mother's IQ), children in the high-lead group had a median IQ deficit of six points compared to their low-lead classmates, as well as shorter attention spans and imparied language skills. Even more striking was the effect on the overall distribution on IQ scores: the children in the higher-lead group were almost four times as likely to have an IQ below 80, while none of them scored above 125.[19]

Also striking were the results of evaluations by the children's teachers (who did not know their pupils' lead status). Using an 11-item scale that examined classroom behavior, attention, and overall functioning, teachers concluded that children with elevated lead levels scored significantly worse than the low-lead group.

Five years later, the researchers re-examined these two groups of

**Although lead poisoning is often viewed as primarily a disease of the poor, wealth and social status confer no immunity. Indeed, a recent government study concluded that children living *above* the poverty level comprise the largest category of people in danger of undue exposure.**

---

[15] Agency for Toxic Substances and Disease Registry (1988), p. I-48.

[16] For an outline of epidemiological considerations, see sidebar below and Needleman (1990b), p. 677.

[17] The high lead group had dentine levels above 24 parts per million (ppm); the low-lead group had dentine levels below 6 ppm.

[18] These 158 children were a subset of the 270 children with levels above 24 or below 6. Others were excluded to avoid possible confounding factors such as head injuries, acute lead poisoning, and variable lead levels in different teeth.

[19] Needleman (1988a).

[20] Bellinger, D, HL Needleman, R Bromfield, and M Montz (1984), "A Follow-up Study of the Academic Attainment and Classroom Behavior of

children.[20] The high-lead group had lower IQ scores, needed more special academic services, and had a higher rate of school failure. Eleven years after the initial study, a second follow-up was conducted, to determine whether lead's effects persist into young adulthood.[22] The findings were dramatic: compared to the lower-lead classmates, the higher-lead group showed a 7.4 increase in school dropout rates, and a 5.8 increase in reading disabilities (defined by scoring two or more grade levels below that expected for the highest grade completed). The higher-lead group also exhibited lower class rank and higher absenteeism.

Other researchers have also found effects in epidemiological studies on lead-exposed children, though some have conducted similar studies and reported no effects. All studies published since 1972 were recently evaluated using meta-analysis, a technique that allows investigators to pool data across studies and to draw conclusions as to the statistical reliability of the data taken collectively.[23]

After eliminating studies that failed to meet key criteria such as adequate sample size, exclusion of acutely poisoned children, and controls for socioeconomic factors, data from the remaining twelve studies were pooled. The outcome strongly supports a linkage between low-dose lead exposure and intellectual deficits in children.

Further evidence of lead's neurotoxicity comes from a series of prospective studies, in which investigators measure variables over an extended period of time into the future. Recent studies have found notable effects from prenatal lead exposures at very low levels.[24] In fact, one study found effects from prenatal exposures as low as 6 to 7 ug/dl.[25]

For example, in a study of several hundred children whose prenatal lead exposure had been determined from umbilical cord blood samples at the time of birth, investigators found that even moderate lead levels affected the

---

**THE IMPORTANCE OF SMALL NUMBERS**

Figure 1 shows the frequency distribution of IQ scores between the "low lead" and "high lead" children, and indicates that high blood lead levels are associated with a left-ward shift in the overall IQ distribution curve. In addition to showing that the median IQ deficit is 6 points, these data also illustrate two other key points: (1) High lead children in this case were almost *four times as likely to have IQs of less than 80*; and (2) *five percent of the low lead group had IQs of more than 125, while none of the high lead group did.*[21]

In other words, lead's effect on a population *as a whole* is more dramatic than its effects on individuals, by affecting the frequency of high and low scores. The disadvantaged are further harmed, while the truly gifted are deprived of their potential.

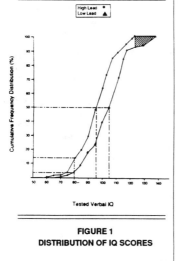

**FIGURE 1**
**DISTRIBUTION OF IQ SCORES**

---

Children with Elevated Dentine Lead Levels," <u>Biological Trace Elements Research</u>, Vol. 6, pp. 207-223.

[21] Diagram adapted from HL Needleman (1988a).

[22] HL Needleman, et al. (1990a), "The Long-Term Effects of Exposure to Low Doses of Lead in Childhood," <u>New England Journal of Medicine</u> Vol. 322, pp. 83-88. Researchers were able to trace and evaluate about half of the original participants. The others could not be located or refused to participate. The group that was retested tended to have lower dentine levels, higher IQs, and better school behavior reports. As a result, it seems likely that the 11-year follow-up may underestimate lead's long-term effects, since a higher percentage of the most severely affected individuals did not participate.

[23] HL Needleman, CA Gatsonis (1990b), "Low Level Lead Exposure and the IQ of Children," <u>Journal of the American Medical Association,</u> Vol. 263, pp. 673-678.

[24] For detailed discussion of these studies, see Agency for Toxic Substances and Disease Registry (1988), pp. IV-8 to IV-13.

[25] Bellinger, D, A Levitan, C Waternaux, HL Needleman, and M Rabinowitz (1989),"Low-level Lead Exposure, Social Class, and Infant Development," <u>Neurotoxicology and Teratology,</u> Vol. 10, pp 497-503.

(continues)

children's performance on mental-development tests up to two years later.[26] Similar outcomes have been found in studies in Port Pirie, Australia and Cincinnati, Ohio.[27] In the words of the American Academy of Pediatrics, the available data have "shown conclusively" that reduction in intelligence and alteration in behavior occur in children with elevated blood lead levels.[28]

In addition to these extremely disturbing findings on the consequences of lead exposure in children and fetuses, a growing body of research is showing that low levels of lead also exert toxic effects on adults, including cancer, reproductive effects, and high blood pressure.

**Cancer:** The U.S. Environmental Protection Agency has classified lead as a "probable human carcinogen," based on data from animal studies.[29] Recently, EPA's Science Advisory Board, which is comprised of outside experts from industry and academia, formally reviewed EPA's classification and endorsed it.[30] Researchers are currently comparing lead's potency as a carcinogen to its potency as a neurotoxin.[31]

**Reproductive Effects:** Experiments on laboratory animals give ample evidence of lead's toxic effects on the reproductive system (e.g., failure of ovulation, delayed sexual maturity, impotence, sterility, spontaneous abortions).[32] While there are fewer data on the reproductive effects in humans, there are numerous reports of an increase in spontaneous abortions, structurally abnormal sperm, and decreased fertility in lead-poisoned adults.[33]

**Effects on Blood Pressure:** An additional threat to adult males is indicated by evidence showing a link between low-level lead exposure and

[26] D Bellinger, A Leviton, C Watermaux, HL Needleman, and M Rabinowitz (1987), "Longitudinal Analyses of Prenatal and Postnatal Lead Exposure and Early Cognitive Development, New England Journal of Medicine, Vol. 316, pp. 1037, 1039.

[27] AJ McMichael, PA Baghurst, NR Wigg, GV Vimpai, EF Robertson, RJ Roberts (1988), "Port Pirie Cohort Study: Environmental Exposure to Lead and Children's Abilities at the Age of Four Years," New England Journal of Medicine, Vol. 319, pp. 468-75; KN Dietrich, KM Krafft, RL Bornschein (1987), "Low Level Fetal Exposure Effect on Neurobehavioral Development in Early Infancy," Pediatrics, Vol. 5, pp. 721-30.

[28] American Academy of Pediatrics (1987), "Statement on Childhood Lead Poisoning," Pediatrics, Vol. 79, pp. 457.

[29] See 50 Fed. Reg. 46936 (Nov. 13, 1985).

[30] Environmental Protection Agency, Scientific Advisory Board (December 1989), "Report of the Joint Study Group on Lead: Review of LEad Carcinogenicity and EPA Scientific Policy on Lead," (Doc. No. EPA-SAB-EHC-90-001), p. 1.

[31] Research underway at University of Maryland, Program in Toxicology.

[32] HL Needleman and PJ Landrigan (1981), "The Health Effects of Low Level Exposure to Lead," Annual Review of Public Health, Vol. 1981, pp. 277-98.

[33] Ibid.

**DETERMINING CAUSE AND EFFECT THROUGH EPIDEMIOLOGICAL STUDIES**

In epidemiological studies, groups of people are studied in order to determine patterns of disease. Those patterns are then analyzed statistically in order to reveal links between a particular substance and certain health effects.

Scientists attempting to show cause and effect through epidemiological studies use five rules of thumb that, taken together, serve as a rigorous test of causality. There are:

1. Order of precedence. The "cause" must precede the "effect."

2. Consistency. There must be broad consistency among data both internally, and among different studies.

3. Dose-response. Causality can be more strongly inferred when variations in the "cause" are associated with variations in the "effect."

4. Specificity. If the same effect can be produced by other means, the cause-effect relationship under scrutiny is weakened. If the effect can be produced only by the cause, the relationship is strengthened.

5. Biological plausibility. Put simply, this test asks whether, in light of current knowledge of human biology, the cause/effect relationship seem likely.

All five criteria are met for studies on the neurotoxicity of lead.

high blood pressure.[34] Although differences between blood-pressure values were relatively small, the effect nonetheless is of concern from a public health perspective. Like lead's effects on IQ distribution, the consequences of even a small shift in the distribution curve for blood pressures can be severe on a population-wide basis. Given the role of cardiovascular disease as the number one cause of death in America, even "small" increases in average blood pressure are of significant concern.

**Given the role of cardiovascular disease as the number one cause of death in America, even "small" increases in average blood pressure are of significant concern.**

---

[34] In a statistical analysis based on a national health survey of 9,932 persons of all ages, one researcher found a "robust relationship between low-level lead exposure and blood pressure" in adult males.  J Schwartz (1988), "The Relationship Between Blood Lead and Blood Pressure in the NHANES II Survey," Environmental Health Perspectives, Vol. 78, pp. 15-22. A reanalysis of the same data for males between the ages of 12 and 74, using a different and rather conservative statistical technique, also found a significant linear association between blood lead levels and blood pressures. JR Landis and KM Flegal (1988), "A Generalized Mantel-Haenszel Analysis of the Regression of Blood Pressure on Blood Lead Using NHANES II Data," Environmental Health Perspectives, Vol. 78, pp. 35-42. While the actual differences in blood pressure in these and other studies are small, the consistency across studies is strong. W Victery, HA Tyroler, R Volpe, and LD Grant (1988), "Summary of Discussion Sessions: Symposium on Lead-Blood Pressure Relationships," in U.S. Department of Health and Human Services, Environmental Health Perspectives, Vol. 78, pp. 139-155.

10

*(continues)*

## 2. EVALUATING EXPOSURES TO LEAD

*Most public health experts now agree that lead exhibits a "continuum of toxicity," where the smallest exposure can have a consequence somewhere in the body. This marks a radical departure from the approach to the problem only a few years ago.*

Because most cases of lead poisoning have no overt symptoms, screening programs are critically necessary to identify children in need of treatment. Unfortunately, screening programs in many cities were curtailed or eliminated in the early 1980s after the federal government discontinued funding for such programs, and nationwide data-collection efforts were also dropped. As a result, estimates of numbers of affected children must be derived from limited sampling programs, and extrapolated using figures on other variables known to be related to lead poisoning.

In measuring the amount of lead absorbed by an individual and determining whether treatment is needed, doctors generally rely on measurements of the amount of lead in the individual's blood.[1] Though such measurements do not reveal the individual's lifetime history of lead exposure or the amount that is currently stored in bone, blood-lead levels can provide a "snapshot" of recent lead exposures.[2] Results are generally expressed as micrograms of lead per deciliter of blood, or ug/dl.

One practical way to obtain some longer-term data for children involves collecting children's teeth as they are naturally shed (generally between ages five and nine). This approach, however, means that parents must be informed of the need to collect teeth and must agree to participate, and is obviously inapplicable to adults and older children.

Prior to the mid 1960s blood lead levels of 60 micrograms of lead per deciliter of blood (ug/dL) or less were generally considered as not dangerous

**Though such measurements do not reveal the individual's lifetime history of lead exposure or the amount that is currently stored in bone, blood-lead levels can provide a "snapshot" of recent lead exposures.**

---

[1] Because analysis of blood samples takes up to two weeks, screening tests are sometimes used to give a preliminary indication of whether further testing is warranted. The so-called "EP" test, which is a finger-prick test that gives immediate results, was used as a screening tool for many years. The test measures the presence of a naturally occurring protein that is produced at higher levels in response to lead exposure. Unfortunately, the accuracy of the EP test is limited for blood-lead levels below 40 ug/dl. State of California, Department of Health Services (1989), Childhood Lead Poisoning in California: Causes and Prevention, p. 8, 14 (interim report). As a result, it is not useful in screening for exposures at current levels of concern. Researchers at the University of Maryland Department of Toxicology are attempting to develop a substitute test that will serve as a preliminary screening test for blood-lead levels around 10 ug/dl.

[2] While promising, methods of directly measuring the total amount of lead stored in an individual's body are still undergoing development and are not yet widely available. See, for example, JF Rosen et al. (1988), "L-Line X-ray fluorescence of Cortical Bone Lead Compared with the CaNa2EDTA Test in Lead-Toxic Children: Public Health Implications," Proceedings of the National Academy of Sciences (USA), Vol. 86, pp. 685-689.

enough to require monitoring or treatment.[3] This became an official standard in October 1970 when the U.S. Surgeon General issued a report defining 60 ug/dL as a level of "undue lead absorption." But even within the year, further analysis prompted the Public Health Service to circulate a draft lowering that threshold for undue absorption by a third, to 40 ug/dL.[4] Within five years the threshold fell again, to 30 ug/dL, while the threshold for outright lead poisoning was set at 80.[5]

In 1978, the Public Health Service revised its finding, with 30 ug/dL as the threshold for undue lead absorption and 70 ug/dL as the threshold for poisoning.[6] In 1985 the Service's Centers for Disease Control (CDC) issued a statement lowering its thresholds for both excessive lead absorption and lead toxicity to 25 ug/dL.[7] And in late 1989, CDC announced that it was convening an Advisory Committee to update its statement on preventing lead poisoning in young children, to reflect research findings since 1985. Public health experts interpret this action as portending another downward revision of the standard, probably to 15 ug/dl or lower.[8]

The Environmental Protection Agency has also evaluated lead's toxicity in developing regulations under a variety of environmental statutes, including the Clean Air Act, the Clean Water Act, and the Safe Drinking Water Act. A 1986 report prepared as a background document on air regulations cited 10-15 ug/dL as the range associated with "neurological

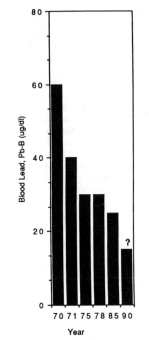

80 —

60 —

Blood Lead, Pb-B (ug/dl)

40 —

20 —

?

0 —

70 71 75 78 85 90

Year

**FIGURE II: RECOMMENDED BLOOD LEAD LEVELS FOR MEDICAL INTERVENTION**

Figure II shows the erosion of the recognized threshold for lead toxicity as new evidence has emerged through increasingly sophisticated scientific studies.

---

[3] Centers for Disease Control/Public Health Service (1985), p. 1.

[4] Public Health Service, Bureau of Community Environmental Management, Control of Lead Poisoning in Children - pre publication draft (Washington, U.S. Department of Health, Education, and Welfare, December 1970), p. 2.

[5] Public Health Service, Increased Lead Absorption and Lead Poisoning in Young Children: A Statement by the Center for Disease Control (Washington, U.S. Department of Health, Education, and Welfare, March 1975), p. 1. That document used lead "poisoning" or "toxicity" to mean a condition showing acute or obvious symptoms; "undue lead absorption" or "elevated blood lead level" means a level warranting medical intervention but where obvious symptoms may not be present.

[6] Public Health Service, Preventing Lead Poisoning in Young Children: A Statement by the Center for Disease Control (Washington, U.S. Department of Health, Education, and Welfare, April 1978), p. 1.

[7] Centers for Disease Control/Public Health Service, (1985), pp. 1-2. Lead toxicity is defined by two factors: a blood lead level of 25 ug/dl together with an erythrocyte protoporphyrin ("EP") of 35 ug/dl. EP is a naturally occurring protein that plays a key role in the manufacture of hemoglobin; an elevated EP level is one of the earliest and most reliable signs of impaired function due to lead. Ibid., p. 3.

[8] 54 Fed. Reg. 48026 (November 20, 1989). The CDC convenes such committees at irregular intervals, when it considers the evidence strong enough to warrant re-examination of the standard.

(continues)

deficits."[9]  In its proposed regulations on lead in drinking water, the EPA again cited 10-15 ug/dL as an "appropriate range of concern for health effects that warrant avoidance."[10]  A December 1989 statement of EPA's Science Advisory Board concludes that "...there is likely to be no threshold for lead neurotoxicity, at least within the contemporary range of blood lead levels (i.e., 1-10 ug/dl)."[11]  And as another EPA advisory group pointed out even more recently, "[t]he value of 10 ug/dl refers to the maximum blood-lead level permissible for all members of these groups, and not mean or median values."[12]

The steady erosion of the accepted threshold for lead's toxic effects, coupled with lead's known biochemical properties, has convinced many public health experts that lead has no threshold.  Rather, the emerging view is that lead presents a "continuum of toxicity" in which traditional symptoms associated with high dosage, such as kidney failure and anemia, have their low dosage counterparts, such as IQ deficits and decreased nerve conduction.[13]

Public health experts have also found that the same level of lead exposure may affect different people unequally.  Among the chief variables appear to be dietary intake of essential trace minerals.  Lead's effects are aggravated in people who lack adequate dietary calcium, iron, zinc, or phosphorous.[14]  This places poor families, where malnutrition may be more common, at greater risk of adverse effects from lead.  . . .

> **"Because the 'baseline' level of lead in blood in the U.S. population is apparently about 10 - 15 ug/dl, it is virtually impossible to demonstrate effects of lead at lower blood levels... the physiological states now defined as "normal" might actually be "abnormal" conditions associated with typical levels of lead in the body. The hypothesis that people would be healthier in subtle ways if the average blood lead level were 1 - 2 ug/dl (or less) deserves sober consideration..."**
>
> National Research Council (1980), p.137

---

[9] Environmental Protection Agency (1986a), Addendum to Vol. 4, p. A-48.

[10] 53 Fed. Reg. 31524 (August 18, 1988).

[11] Environmental Protection Agency, Science Advisory Board (1989), p. 11.

[12] Environmental Protection Agency, Science Advisory Board (1990), p. 1.

[13] Ibid.  See also PJ Landrigan (198_), p. 593.

[14] Centers for Disease Control (1985), p. 3.

## 3. SOURCES AND PATHWAYS OF
## LEAD EXPOSURE

*Lead can enter the environment directly, as from industrial emissions, or indirectly, as when automotive batteries are incinerated or when dust from lead paint forms part of household dust. The most critical source of lead exposure for most children is deteriorating lead paint in dwellings; lead from gasoline and from drinking water are also significant. Additional exposures occur through continuing uses of lead. Each year industry produces, and consumers use and discard, products containing well over a million tons of lead. The lead in each of those products is indestructible.*

Because lead is an element, it cannot break down or decompose into something less toxic. Both pure lead and its compounds are harmful to humans. Once introduced into the biosphere -- that part of the earth's surface and atmosphere where living organisms exist -- lead remains toxic indefinitely.

Lead's widespread usage and its resulting dispersal into the environment have been no accident. Its convenient properties have been recognized from the earliest historical times. It is malleable and easy to work. It insulates well and does not rust. It alloys readily. Lead compounds make excellent pigments in paints that also weather especially well. Egyptians in the time of the pharaohs used lead in ornaments and cosmetics.[1] Chalices made of lead-silver alloys carried wine for the ancient Greeks and lead piping still carries rainwater from the roofs of medieval cathedrals. Indeed, the word "plumbing" is itself derived from the Latin word for lead, "plumbum" (as is its chemical symbol, Pb).

The United States currently consumes well over one million tons of lead per year.[2] A substantial fraction of that amount -- approximately 60% -- comes from secondary refining (recycling). About a third comes from primary refining of lead, while imports (primarily from Canada) slightly exceed exports.[3] Almost 90% of U.S. lead mining occurs in Missouri, with some operations in Alaska, Colorado, Idaho, and Montana, as well as very limited mining in half a dozen additional states.[4] Major industrial sources alone dispose of or release 15,000 tons of lead wastes in the U.S. annually, in forms ranging from placement in landfills to fugitive emissions from facilities.[5] . . .

Figure III. Above, the medieval chemical symbol for lead. Below, the modern equivalent.

---

[1] See HA Waldron (1973), p. 392.

[2] Bureau of Mines (1989c) Mineral Commodity Summaries 1989, pp. 90 - 91 (Washington: U.S. Dept of the Interior) (1988, data converted from metric tons to short tons; one metric ton equals about 2,200 pounds, while a U.S. or "short" ton equals 2,000 pounds).

[3] Ibid.

[4] Ibid.

[5] Environmental Protection Agency, Office of Pesticides and Toxic Substances (1989), The Toxics Release Inventory: A National Perspective,

But other sources also make a notable contribution. Some of these -- such as leaded gasoline -- are products whose use has fallen in recent years, but their legacy of contamination persists. In addition, over a million tons of lead continue to be introduced into commerce in the form of new products each year; the manufacture, use, and disposal of those products adds still more lead to the environment.

### Lead-based paint

During the early and middle decades of this century, lead-based paint was the preferred medium in millions of homes.[9] The lead content of such paint varied, with some -- particularly in earlier years -- containing as much as fifty percent lead by dry weight.[10] Even as late as 1971, the New York City Health Department tested 76 different paints and found eight of them to contain lead concentrations between 3 and 11%.[11] Today an estimated three million tons of lead from paint still remains in dwellings.[12]

At one time, the general consensus was that children were exposed to lead-based paint primarily when they actually ate flakes of the sweet-tasting product or chewed on readily accessible surfaces such as window sills. More recently, however, researchers have realized that the primary exposure route starts with the transformation of lead paint into ordinary household dust. Children absorb lead by playing in the dust that is contaminated with these fine particles of paint. Simply by behaving like children, they get dust particles on their clothes and hands, and into their mouths.[13]

Even a well-maintained home is likely to have some deterioration of paint on window sills, particularly in parts of the nation exposed to freeze-thaw cycles. Many older houses, however, have been poorly maintained. Paint from these dwellings can easily be converted into dust-size particles that pose an extra menace to any active child. Over 40 million houses

**Simply by behaving like children, they get dust particles on their clothes and hands, and into their mouths**

[9] Hearings before the Subcommittee on Housing and Community Development of the House Committee on Banking, Finance and Urban Affairs, 100th Cong., 2d Sess. (1988) (testimony of James Keck, Deputy Commissioner, Baltimore Department of Housing and Community Development). See also R. Rabin (1989), "Warnings Unheeded: A History of Child Lead Poisoning," American Journal of Public Health, Vol. 79, pp. 1668-1674.

[10] HL Needleman and PJ Landrigan, (1981), p. 279.

[11] Rabin (1989). p. 1673.

[12] Agency for Toxic Substances and Disease Registry (1988), p. II-5.

[13] Centers for Disease Control (1985), p. 7.

**Over 40 million houses containing leaded paint are the homes for each successive generation of American children, with 1.97 million of these houses particularly unsound from deteriorating paint.**

containing leaded paint are the homes for each successive generation of American children, with 1.97 million of these houses particularly unsound from deteriorating paint.[14]

Lead based paint was also used extensively on exterior applications, and it too presents a hazard. Particles from exterior paint -- along with air borne lead from gasoline and other sources -- can settle on the ground. These particles then mix into the soil and accumulate with other falling particles year after year. A recent study in Oakland, California found that exterior paint had an even higher average lead content than paint on interior walls. The same study identified a strong correlation between the lead levels in a house's exterior paint and the blood lead levels of children residing in that house.[15]

Lead in soil does not come solely from paint. The same sources that have deposited lead in measurable amounts on Greenland's icecap spread dust in the nation's front and back yards each day. In some regions, total soil-lead levels are alarmingly high. For example, a majority of soil samples from Oakland exceeded 1000 parts per million -- a level that defines materials as hazardous waste under California law.[16] The same study also found that a child's blood lead level increased an average 4-5 ug/dL for every 500 ppm rise in front or back yard soil lead level.[17] In light of the emerging consensus among public health experts that blood-lead levels above 10 or 15 ug/dl are associated with significant toxicity, this exposure source is of obvious concern.

## Gasoline

Leaded gasoline was once a major source of lead releases, and lead from this source undoubtedly still remains in the soils of virtually every urban and suburban area throughout the country. EPA estimates that the lead level of soil alongside roadways can reach 10,000 ppm, or more than eight times the levels associated with elevated blood lead levels in the Oakland study.[18] Areas alongside heavily traveled urban arteries -- such as sidewalks, parking lots, or street-front playgrounds -- may have even higher levels of lead in soil and dust.[19] Because farm vehicles long ran on leaded gasoline, and are still allowed to do so, lead is widely dispersed into agricultural soils as well.

Data gathered during the first phasedown of leaded gasoline provides remarkable evidence of the strong association between blood lead levels and the use of leaded gas. The Environmental Protection Agency noted in a 1986 . . .

---

[14] Agency for Toxic Substances and Disease Registry (1988), pp. VI-13, VI-14.

[15] State of California (1989), p. 19.

[16] Ibid.

[17] Ibid.

[18] Environmental Protection Agency (1986a), Vol. IV, p. 13-5.

[19] Ibid., Vol. IV, p. 13-7.

18

(continues)

## 4. NUMBERS OF INDIVIDUALS AFFECTED

*Millions of Americans are at risk of absorbing enough lead to trigger medically adverse outcomes. Because children and fetuses absorb more lead than adults, theirs is a special danger. But adults are far from immune.*

### Children

Precise measurements of the numbers of U.S. children with elevated blood-lead levels are not available. However, a comprehensive set of estimates was compiled as part of a ground-breaking Report to Congress on Child Lead Poisoning. The report, which was prepared by the Agency for Toxic Substances and Disease Registry, or ATSDR -- a component of the U.S. Public Health Service -- used data from blood samples collected in the late 1970s.[1] Those data were then adjusted for changes in lead exposure in the intervening years, and extrapolated based on key factors known to affect blood-lead levels: age, race, family income, and age of housing.

The ATSDR findings are remarkable. An estimated 3 to 4 million children under six have blood lead levels above 15 ug/dl.[2] Of those, about half -- 1.2 million children -- live in housing with deteriorating surfaces: peeling paint, broken plaster, or holes in walls.[3] Children who live in housing with peeling paint are particularly likely to absorb substantial amounts of lead.

In addition to deteriorating paint, several other sources contribute to elevated blood-lead levels.[4] Lead in drinking water is estimated to account for approximately 240,000 cases.[5] Available data do not allow calculation of directly comparable data for remaining major sources such as gasoline, dust/soil, smelters, and food.[6] For most children, however, those sources are less likely to cause significant increases in lead absorption than exposure to deteriorating paint. In the nation as a whole, exposure to paint-derived dust is the source of greatest concern.

Moreover, many children have blood-lead levels well above 15 ug/dl. Over 200,00 children, about 1.5% of the nation's children, are estimated to have a blood lead level of 25 ug/dl or above.[7] At that level, observable IQ deficiencies, poor attention spans, and slow childhood development can be pronounced. Exposure rates are particularly high for poor, urban black

**An estimated 3 to 4 million children under 6 have blood lead levels above 15 ug/dl.**

---

[1] Agency for Toxic Substances and Disease Registry (1988).

[2] Ibid., p. 4.

[3] Ibid., p. I-19.

[4] Ibid., p. 6-8.

[5] Ibid., p. 8.

[6] Ibid.

[7] Ibid., p. 4. (extrapolating from SMSAs to the entire population).

children under age 6, an estimated 10% of whom have blood lead levels above 25 ug/dl.[8]

Several observations are relevant. First, the data possibly understate the problem. Families move, and a new set of children can become exposed to the paint, dust and other contaminants in and around their new home. A California survey noted that 40 percent of the families in its survey moved every 15 months.[9] The lead in the paint, dust and soil pose a continuing risk to each new resident.

Second, the large numbers illustrate the magnitude of the problem from a public health perspective. Preventing exposure is the only appropriate approach, because available lead-poisoning treatments are fundamentally limited: they can neither remove all of the lead from target organs or long-term storage sites (i.e., bone), nor undo neurological damage. Further, such treatments are both expensive and painful for the patient.[10]

Third, lead poisoning is not just a problem of poor children. While children from poorer families are at greater risk, in part because of other factors such as the greater incidence of malnutrition, millions of more affluent persons live in older housing. Indeed, the majority of children living in the nation's oldest (pre-1950) homes come from families above the poverty level.[11]

The problem not only cuts across socioeconomic classes but also across regional boundaries. Appendix I of this report presents detailed estimates of the prevalence of elevated blood-lead levels in over 300 areas throughout the nation.

Patterns revealed by this analysis are alarming. While the national incidence of children with blood-lead levels over 15 ug/dl is 17%, <u>almost 70%</u> of the urban black children from poor families are estimated to exceed that level, as are over 35% of white children in similar circumstances. While these estimates are necessarily crude, given the limits of the available data, they compellingly indicate the breadth -- in the most literal sense -- of America's lead-poisoning problem, and highlight the gravity of current exposure levels and the urgent need for action.

**While the national incidence of children with blood-lead levels over 15 ug/dl is 17%, <u>almost 70%</u> of the urban black children from poor families are estimated to exceed that level, as are over 35% of white children in similar circumstances**

### Fetuses

The ATSDR Report estimates that at current levels of environmental lead, more than four million individual fetuses will suffer toxic effects of cumulative lead exposure over the next ten years.[12] Lead is especially harmful to the fetus because of the ease with which it passes through the placenta. As noted above, several studies have concluded that the fetus is sensitive to lead even at levels of absorption by the mother previously thought to be harmless.[13] . .

---

[8] <u>Ibid.</u>, (1988), p. V-7.

[9] State of California (1989), p. 25.

[10] See Section 5 of this report.

[11] Agency for Toxic Substances and Disease Registry (1988),.p. I-48.

[12] Agency for Toxic Substances and Disease Registry, (1988), p. I-49.

[13] Needleman (1988b).

(continues)

## 5. TREATMENT AND PREVENTION OPTIONS

*Treatment for lead poisoning is a partial cure at best. It is expensive, does not remove all the lead in the body, and cannot undo neurological damage. Nor does it address the conditions that caused the exposure in the first place. Removing lead from the environment is also expensive. Yet its results are more permanent, and probably cheaper in both the short and long run.*

Lead poisoning is treated through the use of chelation, a process in which a drug binds itself to lead in the body and makes the lead easy to excrete. Candidates for possible chelation therapy generally spend a day in the hospital undergoing screening. If test results indicate that chelation is warranted, the patient spends an additional five days there for treatment and post-treatment testing. Not every child with an elevated blood-lead level requires chelation therapy: such treatment generally is considered warranted only in a relatively small fraction of cases, for chelation is not risk-free.[1]

### Limitations of conventional treatment

While chelation can substantially reduce lead levels, it has a number of significant limitations. The three major limitations of chelation are:

1) Chelation cannot repair neurologic impairment, but rather can only keep further damage to the nervous system from occurring. Children treated for lead poisoning are still likely to require special education and other cognitive or behavior-related therapy long after their initial treatment.[2]

2) Chelation generally cannot reach lead that has found its way to long-term storage sites in the hard body tissues (bone, teeth) or the brain and kidneys. As discussed above, lead can re-enter the soft tissues from bone at high levels. Chelation therapy does not prevent this. In fact, doctors have observed a "rebound" phenomenon in some patients, where blood lead levels rise after the cessation of chelation therapy.[3]

3) Finally, chelation has little effect when the patient, as is often the case, returns to the same lead-contaminated environment in which the

> **Not every child with an elevated blood-lead level requires chelation therapy: such treatment generally is considered warranted only in a relatively small fraction of cases.**

---

[1] Centers for Disease Control (1985), p. 26. Using the 1985 definition for lead toxicity -- a level that, as discussed above, may soon be lowered based on more recent data -- EPA estimated that chelation therapy would be needed for about 5% of the children who have blood-lead levels above 25 ug/dl, in order to reduce their overall lead body burden. Environmental Protection Agency (1986b), <u>Reducing Lead in Drinking Water: A Benefits Analysis</u>, pp. III-53 (Washington: U.S. Environmental Protection Agency), Doc. No. EPA-230-09-86-019.

[2] Bellinger et al. (1984).

[3] Centers for Disease Control (1985), pp. 16-17. In one documented case, the rebound took place a full five years after the initial treatment for lead exposure, with no additional exposure being observed during the interim period. See OJ David, S Katz, CA Arcolo, and J Clark (1987), "Chelation Therapy in Children as Treatment of Sequelae in Severe Lead Toxicity," <u>Archives of Environmental Health</u>, Vol. 40, p. 113.

exposure occurred. Unless the source of exposure can be eliminated -- for example, by removal of accessible lead paint, or by substitution of bottled drinking water for lead-contaminated tap water -- it is likely that the problem will recur. In one reported case, a patient required nineteen chelation treatments over his childhood.[4]

Moreover, the costs of chelation therapy are high. They include hospitalization, physician visits, laboratory tests, and psychological testing and evaluation. Using conservative assumptions, EPA estimates the cost for a single course of chelation treatment at $2,980, in 1988 dollars.[5] This estimate does not include the costs of multiple sessions, which EPA estimates may be needed for half of patients undergoing chelation. Nor does it include the costs of follow-up care such as remedial education and psychological testing.

> "Medical treatment with chelating agents must not be considered a substitute for dedicated preventive efforts to eradicate controllable sources of lead..."
> **Centers for Disease Control**

## The Preventive Approach: Getting the Lead Out

Removing lead paint from homes is a complex task. It typically involves testing surfaces to determine where lead paint is present; replacing, encapsulating, or removing paint from woodwork or wall surfaces; careful cleanup of all dusts generated during the process; and post-removal testing to ensure that cleanup was properly completed. Because some of these operations can exacerbate the problem if not done properly, lead removal usually requires skilled labor and special equipment such as respirators and specialized vacuum cleaners. The Consumer Product Safety Commission warns flatly that "[c]onsumers should not attempt to remove lead-based paint."[6]

Unfortunately, safe and effective removal of lead paint is not cheap. The City of Baltimore, which has an active lead abatement program, estimates the per unit cost of lead removal to run from approximately three thousand dollars, at the low end of the scale, to as high as eight to ten thousand dollars.[7] An urgent need exists for research aimed at developing abatement technologies that are fully effective but less expensive.

---

[4] S. Pollack (1989), "Solving the Lead Dilemma," Technology Review, Oct. 1989, pp. 22-31.

[5] U.S. Environmental Protection Agency (1986b), p. III-53. Figures in the cited source are given in 1985 dollars.

[6] Consumer Product Safety Commission (1989), "CPSC Warns About Hazards of 'Do It Yourself' Removal of Lead-Based Paint," Consumer Product Safety Alert, p.1.

[7] Interview with James Keck, Deputy Commissioner, Baltimore Department of Housing and Community Development. The low number assumes abatement being carried out in tandem with other rehabilitation, and reflects abatement contractor costs on a specific modernization project in Baltimore comprising 328 units; the high numbers assume "worst case" conditions of very deteriorated units where paint abatement alone is being carried out.

(continues)

## 6. GOVERNMENT ACTION -- AND INACTION --
## ON LEAD

*To date, Congress has responded to the lead prob-
lem on a number of fronts, enacting legislation to control
lead in paint, ambient air, drinking water and solid waste.
Initiatives to reduce lead in gasoline and lead-soldered
food cans have made headway in "de-leading" the nation
as a whole. But there have been only sporadic efforts to
control the most stubborn and significant source of
children's exposure to lead: house paint.*

### The Lead-Based Paint Poisoning Prevention Act

The United States historically has been slow to respond to the health
threats posed by lead paint. While most European countries signed a treaty
banning the use of lead-based paint in the interior of buildings in 1921,[1] the
federal government took no action at all until 1971. That year, citing the
"epidemic proportions" of childhood lead poisoning in large cities, Congress enacted the first national lead abatement legislation. The Lead-Based
Paint Poisoning Prevention Act[2] sought to address three distinct aspects of
the lead-paint problem. Specifically, the Act set some limits on the use of lead
paint; created grants for lead-poisoning screening and treatment programs;
and required the submission of a report on abatement methods. The three
components have had notably different histories over the intervening two
decades.

**Limits on the Use of Lead Paint**: Contrary to general belief, the Act
did not ban the production of lead paint or even all of its uses in dwellings.
Rather, it merely prohibited the use of leaded paint on surfaces accessible to
children. The Act also authorized the Secretary of Health, Education, and
Welfare to issue regulations prohibiting the use of lead-based paint in Federal construction or rehabilitation of residential housing.

Recognizing the need to strengthen these provisions, Congress amended
the Act in 1973 to prohibit the use of leaded paint (defined as paint containing
0.5 percent lead by dry weight) in federally funded housing and extended the
prohibition to toys and other articles.[3] Not until 1977 was the use of lead paint
in housing actually banned -- and that ban was imposed by a regulation issued
by the Consumer Product Safety Administration rather than by statute.[4] . . .

**Merely prohibiting the use of addi-
tional lead-based paint in dwellings
has proven to be an inadequate response to the problem.**

---

[1] 1921 Convention Concerning the Use of White Lead in Painting.

[2] Pub. L. No. 91-695, 84 Stat. 2087; current version at 42 U.S.C.
sections 4801-4846.

[3] Pub. L. No. 93-151, 87 Stat. 560 (1973).

[4] 42 Fed. Reg. 44199 (Sept. 1, 1977), codified at 16 C.F.R. Part 1303.
Certain products were exempted, including agricultural and industrial coatings, including building coatings, traffic paints, and artists paints. The
regulations also revised the definition of lead paint to mean paint containing
more than 0.06% lead by dry weight.

**WARNING:** This product contains LEAD, known to cause birth defects or other reproductive harm. Federal law prohibits the use of leaded solders in making up joints and fittings in any private or public potable (drinking) water supply system.

The 1986 Amendments to the Safe Drinking Water Act banned the further use of lead products in new public water systems and in new homes connected to them.[46] The ban encompasses both leaded plumbing fixtures and leaded solder (containing more than 0.2% lead). Nonetheless, leaded solder continues to be sold for a variety of other uses. Unfortunately, it is far from clear that the small warning labels printed on solder packages serve as an effective means for deterring its use in drinking water systems, particularly because unleaded solder is not carried by all hardware stores.

Yet another source of lead in drinking water is found in the lead liners or solder in drinking fountains. Discovery of this problem in 1988 prompted enactment of the 1988 Lead Contamination Control Act, which banned the manufacture and sale of drinking water fountains containing lead that comes into contact with drinking water supplies.[47] The Act also directed EPA to compile lists of lead-containing water coolers, and assist schools in detecting lead in school drinking water.[48]

## Summary of Government Efforts

America has made significant progress in the battle to "de-lead" some aspects of its environment -- most notably air following the phase-down of leaded gasoline -- but has had woefully limited success in addressing other exposure sources. Chief among the latter is lead paint in older homes. This is ironic in view of the fact that the Lead-Based Paint Poisoning Prevention Act was enacted nearly two decades ago expressly to control this very source. But because leaded paint continues to be a highly accessible source of lead as long as it remains on the premises, merely prohibiting the use of additional lead-based paint in dwellings has proven to be an inadequate response to the problem. The intractable features of this issue indicate that leaded paint represents a public health threat demanding extraordinary efforts, different in both degree and kind from those of the past.

---

[45] Ibid., p. II-15.

[46] Pub. L. No. 99-339, section 109, 100th Cong., 2d Sess. (1986), codified at 42 U.S.C. section 300g.

[47] Pub. L. No. 100-572, 102 Stat. 2884, 100th Cong., 2d Sess. (1988), codified to 42 U.S.C. section 300j-21 to 300j-25.

[48] The final water cooler list was issued at 55 Fed. Reg. 1772 (Jan. 18, 1990), while guidance for schools was published at 54 Fed. Reg. 14316 (April 10, 1989).

36

(continues)

## 7. SETTING GOALS AND PRIORITIES
## FOR FUTURE ACTION

Lead's persistence and toxicity mandate efforts to minimize blood-lead levels for all Americans. At the same time, special attention must be directed to those who are most heavily exposed and most vulnerable. A new Federal effort should therefore aim at the two million high risk houses that have been and will be home to millions of children. It should also aim to cut the amount of lead being introduced into commerce.

The vast amount of information available on lead's persistence and toxicity, especially in light of new data on lead's long-term neurologic effects provide a compelling basis for an aggressive Federal program of lead removal. Key factors include the following:

❂ Lead in the body has no known biologic or physiologic value. Its only known effect is that of interfering with essential bodily functions.[1]

❂ Lead is indestructible. Virtually all of the reported cases of lead poisoning today stem from decades-old paint from the walls and woodwork of homes.

❂ Lead's presence in the body is largely cumulative and its effects are largely irreversible. The half-life of lead stored in our bones and teeth is approximately twenty years; minute accumulations of lead in the body over time can produce toxic levels of blood lead, and bring about symptoms that last long after treatment.

❂ New research continues to decrease the levels of human absorption that we deem to be dangerous. The federal government now recognizes adverse effects at levels that are a small fraction of the official standard in place twenty years ago.

❂ While the exact number of persons with unduly high blood lead levels cannot be precisely calculated, the number far exceeds that reported through screening programs. That number includes millions of adults as well as children, and more affluent as well as poorer Americans.

As a practical matter, society cannot immediately remove all the lead that may threaten the public health. In the words of one report, "Lead is toxic wherever it is found, and it is found everywhere."[2]

As a result, priorities must be set. One approach is to (1) identify the group or groups who currently have the highest exposures to lead and susceptibility to its effects, (2) establish a "least cost" method of minimizing those exposures, (3) erect administrative and financial safeguards to ensure the realization of policy goals, and (4) to the degree possible, join any new effort with other important policy goals, including stimulating the use of nonleaded products and environmentally responsible recycling of lead, and

**"Lead is toxic wherever it is found, and it is found everywhere."**

---

[1] Centers for Disease Control (1985), p. 1.

[2] Agency for Toxic Substances and Disease Registry (1985), p. I-1.

increasing the supply of safe and affordable housing.

### Identify Groups Most At Risk

As discussed above, children living in homes with deteriorating lead-based paint are at greatest risk of ingesting undue amounts of lead today. An estimated 1.97 million homes with peeling lead-based paint house well over half a million children today, and pose a continuing threat to additional children in the future. The program could be structured so that the first areas addressed are those in which large numbers of lead-poisoned children have already been identified. By beginning lead abatement programs on this highest-risk segment of the population, resources will be spent where the potential public health benefits are the greatest, with concomitant economic benefits. In addition, millions of American families will avoid the financial and emotional costs of lead-induced school failure and reading disabilities.

**An estimated 1.97 million homes with peeling lead-based paint house well over half a million children.**

### Establish "Least Cost" Methods

The $3-10 thousand per unit cost of de-leading a house assumes current market conditions and current operating technology. The average cost could weigh in closer to the bottom end of this scale given a concerted effort to (1) maximize competition among contractors licensed to perform lead removal, (2) train additional workers in lead removal procedures, (3) encourage local governments and owners of multi-family housing units to couple paint abatement with other rehabilitation, and (4) develop new methods of paint abatement.

### Erect Administrative and Budgetary Safeguards

Congress enacted the 1970 Lead-Based Paint Poisoning Prevention Act with the best of intentions, but the Act's effectiveness has been severely limited by HUD's inaction and by inadequate funding. To avoid a recurrence of this problem, a program to combat the problem of lead-based paint should contain administrative and financial safeguards. These could include (1) guaranteed funding, (2) administrative responsibility vested in health and environmental agencies whose expertise best comports with the goals of the program, and (3) a built-in policy approach that emphasizes prevention in addition to treatment. This third safeguard is particularly desirable given the irreversible effects of lead poisoning, for a policy consisting solely of treatment carries countless social, educational, and medical costs.

### Join New Efforts with Secondary Goals

Any new Federal program takes its place beside others attempting to attain related goals. Where a new program can be constructed so as to reinforce the goals of existing programs, one achieves administrative efficiency and a better return on the taxpayer dollar. Over the years, some of the aims of Federal lead programs have included mandating or encouraging the use of substitutes for lead (e.g., the phase-down on leaded gasoline, the ban on leaded solders in drinking water systems, and FDA-National Food Processors Association program for use of non-leaded solder in food cans).

38

(continues)

## 8. AN OUTLINE OF THE PROPOSAL

*The Environmental Defense Fund (EDF) proposes a trust fund, to be financed by the creation of an excise fee on the production and importation of lead. Proceeds from the fund would be devoted first to the goal of paint removal in the high risk group of homes with peeling, lead-based paint. The program would be administered jointly by the Environmental Protection Agency and the Department of Health and Human Services; these agencies would also monitor the health effects of the lead removal actions. The program would contain extra provisions to enable it to adapt to market conditions and, where possible, accomplish secondary goals.*

### The Trust Fund

EDF believes that earmarking funds in the budget within a specific trust fund is the preferred way to ensure that the goals of this program can be met.[1]

Congress has earmarked funds where it has determined that any inflexibility inherent in the earmarking process is more than offset by the need to accomplish specific policy goals, by the importance of generating secure and long-term funding, and where there is complementarity between the specific funding source, on the one hand, and the policy goals on the other. In the environmental arena, the Hazardous Substance Superfund,[2] the Nuclear Waste Fund,[3] and the Leaking Underground Storage Tank Trust Fund,[4] each operate to achieve specific goals comparable to the one at hand.[5]

---

[1] It may be appropriate to adopt a means test or other eligibility test to ensure that trust fund monies go to housing owners (whether public or private) who are not otherwise able to finance abatement of lead paint within their units. If such a restriction were adopted, it might be desirable to provide tax credits for abatement to homeowners not eligible for grants or loans.

[2] Established by the Comprehensive Environmental Response, Compensation, and Liability Act, Pub. L. No. 96-510, section 221, 94 Stat. 2767, 2801 (1980), codified to 26 U.S.C. section 9631 (commonly referred to as "Superfund").

[3] Established by the Nuclear Waste Policy Act of 1982, Pub. L. No. 97-425, section 302, codified to 42 U.S.C. section 10222. Strictly speaking, this is a "special fund" rather than a trust fund, but it involves the same principles of earmarked funds. See General Accounting Office, Trust Funds and Their Relationship to the Federal Budget, (Washington, GAO, September 1988) GAO Document No. GAO/AFMD-88-55, p. 7.

[4] Established by the Superfund Act Amendments and Reauthorization Act of 1986, Pub. L. No. 99-499, section 522, 100 Stat. 1613, 1780, codified to 26 U.S.C. 9508.

[5] A fourth environmental trust fund, the Oil Spill Liability Trust Fund, was established by the Omnibus Budget Reconciliation Act of 1986 (Pub. L. No. 99-509, 100 Stat. 1874, and awaits authorization.

## The Excise Fee

Each of the trust funds mentioned in the previous paragraph is financed by a narrowly based excise or a series of excises that is closely related to the goal of the fund. Overall, on-budget federal trust fund receipts from excise and other levies in fiscal year 1989 were approximately $250 billion.[6] EDF proposes an excise on the introduction of new lead into commerce, including imported lead, for an initial period of seven years.

The goal of de-leading two million homes, at an estimated average cost of $5,000 per unit, determines the total level of needed receipts as $10 billion.[7] If the program is designed to have a seven year life, and to spend funds at a constant rate, then the fee should be set to yield approximately $1.5 billion annually (subject to adjustment for inflation). In 1988, the volume of new and imported lead entering U.S. commerce was approximately 600,000 tons.[8] These data suggest that the fee initially should be set at approximately $2500 per ton, equivalent to $1.25 per pound of lead. Based on the November, 1989 average price of lead at 41.3 cents per pound,[9] the excise would work out to approximately a four-fold increase in the price of lead.[10]

Imposition of a fee of this magnitude is not a novel concept. For example, the Budget Reconciliation Act of 1989 contains a per-pound fee on ozone depleting CFCs that rises to $4.90 per pound by the end of the decade.[11]

A preliminary analysis suggest that impacts on consumers would generally be moderate. Assuming the fee has a linear effect on prices, a new automobile battery (which contains approximately 18 pounds of lead) would cost about $11 more if the battery contained 50% virgin or imported lead. Batteries range significantly in price, from about $50 to about $100, and typically carry a warranty of 5 years or longer.[12] The $11 initial price increase thus could be viewed as an incremental cost of about $2 per year over the guaranteed lifetime of the battery. Moreover, in practice the increase would probably be considerably smaller, since most batteries already contain more than 50% recycled lead. Price increases (in absolute terms) on other classes of products that use less lead by weight would be correspondingly lower.

---

[6] Office of Management and Budget (1989), Budget of the United States Government, Fiscal year 1990: Special Analysis C (Washington, Government Printing Office), p. C-14.

[7] This brief analysis is not intended to be exhaustive or comprehensive, but rather only to provide a starting point for further discussion.

[8] Figure derived from the Bureau of Mines (1989c) and converted from metric tons to short tons.

[9] Bureau of Mines (1989b), Table 10.

[10] By way of comparison, during the last dozen years, lead prices have fluctuated from $0.202 per pound (1985) to $0.789 (1979) (based on constant 1987 dollars). Bureau of Mines (1989a), Nonferrous Metal Prices in the United States Through 1988, (Washington: U.S. Dep't of the Interior), pp. 55-57.

[11] Budget Reconciliation Act of 1989, Pub. L. No. 101-239, section 7506, 103 Stat. 2106, 2364, 101st Cong., 1st Sess. (1989).

[12] Consumer Reports, p. 103 (Feb. 1987).

(continues)

We propose that the level of the fee, the possibility of additional funding sources, and the duration of the program be carefully monitored in order to ensure the viability of the program as lead production levels fluctuate.

We believe this narrow excise fully meets the test of complementarity between funding source and policy objective. It also takes a policy approach to the issue that we believe maximizes the efficiency of the program, by accomplishing a number of other important policy goals.

*First, it would use market signals, rather than cumbersome regulatory processes, to discourage new lead production.* The proposed excise is intentionally large, so as to help begin to internalize the extraordinary social costs of lead exposure and thus create incentives to adopt safer substitutes. The excise fee approach avoids the current resource-intensive federal approach to toxic substance control embodied in the Toxic Substances Control Act (TSCA). That approach, which involves a lengthy and expensive use-by-use investigation of the health effects of a particular substance, fails the pragmatic test. Since TSCA's enactment in 1976, only a handful of substances have been regulated. The most far-reaching set of regulations — those banning many uses of asbestos — took almost a decade to develop and will not take effect for most asbestos products until 1997, almost two decades after the rulemaking process began.[13]

Moreover, TSCA's regulatory approach as implemented to date ignores economic forces, while this proposal seeks to exploit them. Making new lead more expensive will provide a market incentive to use existing substitutes for lead, and to develop additional ones. This approach bypasses TSCA's inefficient reliance on bans coupled with use-by-use waivers, a process that is inherently cumbersome and that does nothing to encourage development of substitutes for any use that is initially granted a waiver.

*Second, the excise would promote the more responsible use and reuse of lead.* Increases in the price of virgin lead would raise the price that smelters would be willing to pay for scrap, because they in turn will be able to receive higher prices for selling secondary lead. Because the excise would apply to virgin lead alone, there would be no corresponding direct increase in the costs of scrap lead. Thus, construction firms, salvage firms, and even individuals would have a strong economic incentive to sell their lead scrap.

*Third, the excise would direct needed attention to the dangers of lead in our environment.* The failure of the 1970s lead-based paint legislation is attributable in part to the fact that lead poisoning slipped from the forefront of the nations's health policy concerns. As a result, competing budget priorities deprived the program of adequate funding. By placing a substantial fee on lead production and assigning specific responsibility for preventing lead poisoning in the nation's health and environmental agencies, Congress would both create a stable funding source for a sustained response, and send a strong signal that lead poisoning is one of the country's top environmental health priorities.

*Fourth, the program would increase the availability of safe housing for low income families.* The deteriorated housing units targeted by the abatement program frequently house poor families with young children. By creating an external source of funding for abatement, the program will help . . .

---

[13] 54 Fed. Reg. 29460 (July 12, 1989).

## CONCLUSION:
## FUTURE CONCERNS

*At the close of the initial seven-year program, adjustments may need to be made to preserve the market conditions that foster lower lead usage in our economy. What is needed today, however, is the public will to apply pragmatic measures to a serious public health problem. Such steps will hasten the day when lead poisoning has become as rare a threat to this nation's children and adults as polio is today.*

After completion of the initial de-leading program aimed at abatement of the two million units with deteriorating paint, the abatement program could be either retired or extended to begin addressing the 38 million housing units with intact leaded paint -- a significant, though less immediate, source of concern. If the abatement program were terminated, the excise could either be repealed or the proceeds diverted to the General Fund of the Treasury. If the tax is discontinued, other constraints on increased use of lead may well be necessary to preserve the market conditions that foster lower lead usage in our economy.

These, however, are issues for another day. Our primary concern at present is the millions and tens of millions of children who can be protected from actual lead poisoning by removing peeling paint immediately.

Unlike the scientific breakthroughs that had to occur before conquering diseases such as polio, effective solutions to this problem are at hand. Now, the major impetus for change must come from the realization that the current situation is both intolerable and unnecessary. As a society, we must translate that realization into the only truly effective response to lead poisoning: its prevention. We can choose to start on the road toward that goal or, through continued inaction, we can consign ourselves to the human misery and economic costs of ubiquitous lead contamination. The opportunity is ours. But the benefits will extend to our children and our children's children for generations to come.

(continues) **45**

## The Local Prevalence of Lead Poisoning in the US

This appendix presents estimates of the extent of lead exposure in American children on a local basis; such figures have not been available previously. These estimates clearly reveal the seriousness of lead poisoning for communities in many urban areas of the country: **in many of these areas, a majority of children have unacceptable blood lead levels (i.e., above 10-15 ug/dl)**. And the prevalence of affected children is especially alarming among the urban poor in certain regions of the U.S. Urban areas (standard metropolitan statistical areas, or SMSAs) with older central cities and relatively small amounts of new housing -- the "Rust Belt", the Northeast, and much of the Midwest -- typically have higher prevalence rates. By contrast, urban areas in the "Sun Belt" and the West Coast generally have lower prevalence rates, because of their higher proportion of newer housing.

Tables A-1 and A-2 contain summaries of the data on a national basis. The importance of key parameters -- family income, child's age, race, and residence location -- are clear. Overall, in the central cities of large urban areas, almost 70% of poor Black children and about 35% of poor white children are at risk of lead toxicity. Tables A-3, A-4, and A-5 present a breakdown of these data on a local basis, by SMSA.[1]

## How the Estimates were Derived

These estimates are based upon information collected and analyzed for the Agency for Toxic Substances and Disease Registry (ATSDR), a component of the US Public Health Service. In 1988 the ATSDR published a comprehensive Report to Congress on childhood lead poisoning in the United States.[2] That report . . .

---

[1] Table A-3 contains estimates for SMSAs of over 1 million people. Tables A-4 and A-5 both contain estimates for SMSAs of under 1 million people, but those in Table A-5 tend to be smaller. As an artifact of the census data from which the estimates were calculated, the residence location could be distinguished as "central city" versus "not central city" for SMSAs included in Table 4 but not in Table 5. As a result, the calculations in Table 5 are likely to be proportionally underestimated.

[2] Agency for Toxic Substances and Disease Registry (1988), The Nature and Extent of Lead Poisoning in Children in the United States: A Report to Congress (Atlanta: U.S. Department of Health and Human Services/Public Health Service), Doc. No. 99-2966.

A  1

Table A-1. Estimated percentages of children 6 months to 5 years who are projected to exceed selected blood lead levels by strata and residence in SMSAs of 1 million or more.

| Strata | In Central City >10 (ug/dl) | >15 | Not In Central City >10 (ug/dl) | >15 |
|---|---|---|---|---|
| **> $6,000** | | | | |
| White | | | | |
| .5 - 2 years | 79.9 | 36.6 | 69.7 | 28.4 |
| 3 - 5 years | 80.8 | 35.5 | 70.2 | 27.1 |
| | | | | |
| Black | | | | |
| .5 - 2 years | 95.9 | 67.0 | 91.6 | 57.4 |
| 3 -5 years | 97.1 | 68.5 | 93.3 | 58.1 |
| | | | | |
| **$6,000 - 14,99** | | | | |
| White | | | | |
| .5 - 2 years | 65.1 | 23.5 | 53.1 | 17.4 |
| 3 - 5 years | 65.1 | 22.2 | 52.6 | 16.1 |
| | | | | |
| Black | | | | |
| .5 - 2 years | 91.4 | 53.6 | 84.6 | 43.9 |
| 3 - 5 years | 92.6 | 53.6 | 85.9 | 43.3 |
| | | | | |
| **> $15,000** | | | | |
| White | | | | |
| .5 - 2 years | 47.7 | 12.5 | 35.4 | 8.6 |
| 3 - 5 years | 46.7 | 11.4 | 34.2 | 7.7 |
| | | | | |
| Black | | | | |
| .5 - 2 years | 85.2 | 38.8 | 75.2 | 29.6 |
| 3 - 5 years | 85.7 | 37.7 | 75.4 | 28.3 |

**Young Black children living below the poverty level in urban areas may have a greater than 95% chance of having a blood lead level in excess of 10 ug/dl.**

A 3

(continues)

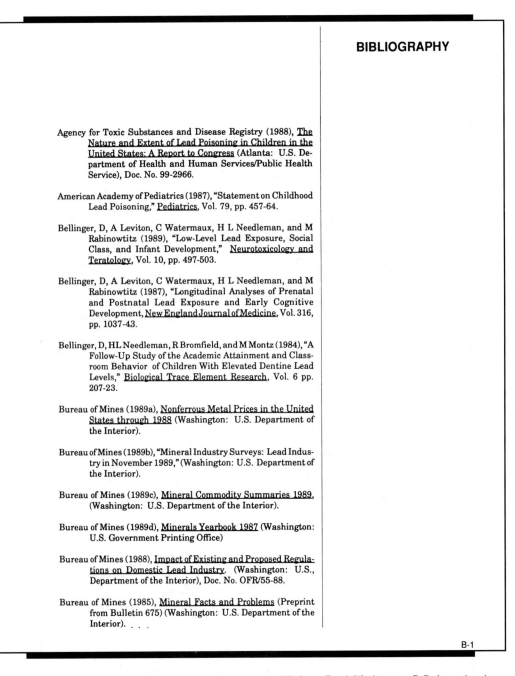

**BIBLIOGRAPHY**

Agency for Toxic Substances and Disease Registry (1988), <u>The Nature and Extent of Lead Poisoning in Children in the United States: A Report to Congress</u> (Atlanta: U.S. Department of Health and Human Services/Public Health Service), Doc. No. 99-2966.

American Academy of Pediatrics (1987), "Statement on Childhood Lead Poisoning," <u>Pediatrics</u>, Vol. 79, pp. 457-64.

Bellinger, D, A Leviton, C Watermaux, H L Needleman, and M Rabinowtitz (1989), "Low-Level Lead Exposure, Social Class, and Infant Development," <u>Neurotoxicology and Teratology</u>, Vol. 10, pp. 497-503.

Bellinger, D, A Leviton, C Watermaux, H L Needleman, and M Rabinowtitz (1987), "Longitudinal Analyses of Prenatal and Postnatal Lead Exposure and Early Cognitive Development, <u>New England Journal of Medicine</u>, Vol. 316, pp. 1037-43.

Bellinger, D, HL Needleman, R Bromfield, and M Montz (1984), "A Follow-Up Study of the Academic Attainment and Classroom Behavior of Children With Elevated Dentine Lead Levels," <u>Biological Trace Element Research</u>, Vol. 6 pp. 207-23.

Bureau of Mines (1989a), <u>Nonferrous Metal Prices in the United States through 1988</u> (Washington: U.S. Department of the Interior).

Bureau of Mines (1989b), "Mineral Industry Surveys: Lead Industry in November 1989," (Washington: U.S. Department of the Interior).

Bureau of Mines (1989c), <u>Mineral Commodity Summaries 1989</u>, (Washington: U.S. Department of the Interior).

Bureau of Mines (1989d), <u>Minerals Yearbook 1987</u> (Washington: U.S. Government Printing Office)

Bureau of Mines (1988), <u>Impact of Existing and Proposed Regulations on Domestic Lead Industry</u>. (Washington: U.S., Department of the Interior), Doc. No. OFR/55-88.

Bureau of Mines (1985), <u>Mineral Facts and Problems</u> (Preprint from Bulletin 675) (Washington: U.S. Department of the Interior). . . .

B-1

Reprinted with permission of Environmental Defense Fund, Washington, D.C. An updated edition of *Legacy of Lead* is under development. For further information, contact Environmental Defense Fund, 1875 Connecticut Ave. NW, Washington, D.C. 20009.

## Suggestions for Discussion and Writing

1. What problem does the proposal address? Try to state the problem in a single sentence. Compare your statement of the problem with that of the writers on page 187 in the Executive Summary. Why do you think the writers wait until the end of the problem section to state the problem? What is the authors' solution to the problem?

2. The proposal adheres to the conventional problem/solution format. What claims do the writers make in the problem section to persuade readers that a problem exists? For a claim to be valid, it must relate to readers' prior knowledge and experience and to their values. According to these criteria, do you, as an interested citizen, think the claims in this proposal are valid? Why or why not?

3. The writers make the following claim: Removing lead from the environment is more permanent and cheaper than treating lead poisoning (p. 27 of the proposal). What kinds of evidence do the writers use to support this claim? Would you, as an interested citizen, accept this evidence as valid? Why or why not?

4. Examine the layout for Chapter 1 of the proposal. What graphics and visual text do the writers use? How do these facilitate readers' comprehension and fluency? How do they help readers locate information?

5. Writers prepare an executive summary after writing a report. Determine where the writers got the information for the executive summary for this proposal.

6. Compare the information in the summary to the information in the report. In writing the summary, did the writers omit any information that readers needed to know to make a decision about supporting the proposed legislation? If so, what information was omitted? Why do you think readers needed it?

7. Examine Chapter 1 of the proposal to determine if the language and syntax are appropriate for the readers. What strategies do the writers use to make certain that novices understand the discussion? Give examples. How do the writers help readers understand discussions involving statistical results?

8. Strategies for defining terms include (1) classifying, (2) describing, (3) comparing, (4) exemplifying, and (5) visualizing. Find examples of these strategies in Chapter 1 of the proposal and bracket them. Identify the type of strategy used in the margin next to the bracket.

   Definitions can be brief or extended. Brief definitions can be appositions, parenthetical text, or addenda to sentences. Highlight examples of these forms in Chapter 1 of the proposal.

# SERVICE PROPOSAL: PROPOSAL TO PROVIDE SERVICES TO CLEAN UP LEAKING STORAGE TANK

Because much business by consulting firms is obtained through proposals, the continued employment of their employees depends on the success of their proposals. Environmental Science and Engineering, Inc., or ESE, is a consulting firm. The following proposal (Figure 5.2) in the form of a letter was written by an ESE engineer in response to a request for assistance in solving an environmental problem caused by a leaking storage tank. When an underground petroleum tank was removed from its premises, a midwest company discovered that fuel had leaked into the ground. Because the company was not equipped to carry out the cleanup required by the Illinois Environmental Protection Agency (IEPA), it requested ESE's assistance. ESE responded with the following proposal that outlines the services it would provide in cleaning up the spill and fulfilling IEPA requirements.

Because the administrator in charge of the cleanup for Midwest Petroleum has only a general knowledge of the field, the proposal writer does not provide a great deal of detail on the mechanical aspects of the solution. However, he does discuss the steps that need to be taken to fulfill IEPA requirements—the reader's major concern. The writer hopes to persuade the reader that ESE is qualified to carry out the appropriate procedures and that he should therefore accept the proposal. The writer is aware that other firms are probably also bidding on the project.

Throughout the proposal the writer uses technical terms and "buzz words" (words emanating from a professional community indicate he is up-to-date on new research results, methods, and trends) and that, therefore, the company is knowledgeable in its field. Because the Illinois Environmental Protection Agency uses such terms as *impacted, remediation, excavation,* and *off-site disposal* in its regulations, the writer also uses these terms: He discusses remediating (cleaning up) the impacted soil (soil into which the petroleum leaked) by excavating (removing) the contaminated dirt to off-site disposal (a special landfill). However, the writer also uses nongovernmental technical terms such as *perimeterize,* which is limited to use in his own field and therefore appears as jargon to a reader who is outside that field.

As you read the proposal in Figure 5.2, consider whether the writer adheres to the conventions of a proposal.

---

### Suggestions for Discussion and Writing

1. This is the only proposal in this chapter that is written to provide services to a reader. In what ways does this proposal differ from the others? In what ways is it similar?

2. To which conventions for proposals does this document adhere? From which does it deviate?

3. At the conclusion of the first paragraph, the writer indicates that he will discuss the steps involved in the assessment and clean up of the leak. However, the writer does not follow this statement with steps. The reader's fluency may be interrupted as he attempts to determine the various steps. Revise the text of the proposal and add visual clues related to the steps so the reader will read what he predicts.

4. The author uses a number of polysyllabic and Latin-based words that could be replaced with simpler ones (e.g., *remediation* could be replaced by *cleanup*). The writer also uses passive voice when active voice would be more direct (e.g., *Remediation of soils . . . is required by the IEPA* could be written as *IEPA requires that soils be cleaned up*). Study the first section of the proposal (down to the subhead Typical Response Schedule and Cost) to determine if other terms could be simplified and other sentences made more direct. Make the revisions.

November 15, 1991

**ESE** Environmental Science & Engineering, Inc.

8901 N. Industrial Road          (309) 692-4422
Peoria, Illinois 61615-1589    Fax (309) 692-9364

Mr. Robert Apple
Midwest Petroleum
Peoria, IL 61603

Re: **Leaking Underground Storage Tank Compliance Incident #42**

Dear Mr. Apple:

This letter is prepared in response to our telephone conversation of November 8, 1991 and our November 9, 1991 meeting regarding the apparent petroleum release at the Midwest facility. Based on the information to date, it appears that a release from at least one of the two diesel underground storage tanks (UST) removed from your facility has impacted the adjacent soils and potentially the local shallow groundwater. The Illinois Emergency Services and Disaster Agency was subsequently notified of the site status. The following paragraphs describe the typical steps involved in assessment and remediation of fuel releases in Illinois. As we discussed, we are prepared to assist you with all of these steps.

Remediation of soils which are in excess of the IEPA's objectives for benzene, toluene, ethylbenzene and xylene (BTEX) and polynuclear aromatic compounds (PNAs) is required by the IEPA. In Illinois the IEPA generally promotes excavation and off-site disposal of contaminated soil which does not meet clean-up objectives. Although this method does not appear justified in many cases and does not permanently solve the environmental hazards, it is difficult to vary from this process. Corrective action activities associated with the remediation of petroleum impacted soils (excavation

**Side annotations (left column):**

Latin designation for "Subject."

Formal style. Addresses reader by surname.

Purpose.

Places initials in parenthesis so reader will know to what they refer when they are used later.
Forecast.

Statement of problem. Paragraph focuses on IEPA requirements for corrective actions. Disagrees with IEPA but stresses need to comply anyway. Indicates alternative steps depend on conditions. Indicates to reader ESE is knowledgeable in IEPA regulations.

**Side annotations (right column):**

Discusses necessary steps to be taken and sequences steps logically.

**Figure 5.2  A proposal presented in letter form.**  Source: Environmental Science and Engineering, Inc.

Header: helps reader determine where they are in a document.

November 15, 1991
Page Two

and off-site disposal) do not require Agency approval of those activities in which the cost is less than $150,000. Corrective action may begin and continue up to the $150,000 without the need to submit and have approved a Corrective Action Plan.

Step 1: If total costs are less than $150,000 IEPA doesn't have to approve clean-up plan.

Paragraph focuses on need for subsurface investigation. Indicates someone will need to carry out an investigation.

Does not provide readers with explanation of the acronym LUST (Leaking Underground Storage Tank). This should be provided the first time it's used, even if reader knows it.

During the tank removal operation, shallow groundwater was apparently encountered in both tank cavities. A slight product sheen was noted on the water surface and strong petroleum odors were reported in the soils displaced from the northern tank excavation. In these cases, the IEPA requires a subsurface investigation to better define the condition of the shallow groundwater aquifer surrounding the LUST. It would appear the groundwater quality near the northern tank cavity will require investigation. Based on laboratory analysis of the soil closure samples collected from the southern tank excavation, this area may or may not require a groundwater investigation.

Step 2: Because of findings, further investigation (subsurface) is required.

Paragraph focuses on preparing a CAP if findings require additional clean-up activities.

Indicates additional steps and forms necessary. Implies someone will need to know how to carry out the additional clean-up steps and develop a CAP acceptable to the IEPA.

If groundwater surrounding one or both of the tank cavities is found to be in excess of the IEPA's generic clean-up objective for groundwater, the Agency will require remediation of the shallow aquifer. Generally, when impacted groundwater is identified, the IEPA requires an investigation which includes the installation of three to four (3-4) stainless steel monitoring wells and several soil borings. Unlike the soil remediation via excavation and off-site disposal, the IEPA must pre-approve each groundwater investigation step prior to its completion if State reimbursement for costs are to be sought. The pre-approval process involves the drafting and submittal to the IEPA of a Corrective Action Plan (CAP).

Step 3: If findings are high, more complete clean-up is required and further investigation (hydrogeologic) needed.

Step 4: Prepare a CAP.

(continues)

**Figure 5.2**    *(continued)*

November 15, 1991
Page Three

Paragraph focuses on a remediation plan. Implies someone needs to know how to develop this plan and carry it out.

The CAP will detail the proposed subsurface investigation including the specific locations of proposed boring and monitoring wells, sample collection and analysis, aquifer testing, etc. The purpose of this phase of the investigation is to perimeterize the plume of contamination present in the local shallow groundwater, and to determine the characteristics of the aquifer. With the information obtained from the above-mentioned hydro-geological investigation, a plan for remediation of the petroleum-impacted groundwater, which generally involves the installation of a groundwater pumping and treatment system, will be formulated for this site.

Step 5: Prepare a remediation plan.

Proposal. Main organizing idea for solution section. Having led reader to recognize that company needs someone to develop plans, carry out activities, and complete forms to fulfill IEPA requirements, writer now indicates ESE can do these things.

ESE can assist your company in the required correspondence for submittal to IEPA, including soil disposal permits, reimbursement application, 20-Day Certification Form, 45-Day Report Forms, Corrective Action Plans (CAP·) and Correction Action Report (CAR). We propose to perform these tasks for your review prior to submittal to IEPA. ESE will provide coordination of field investigations, including sampling and laboratory analysis in accordance with IEPA requirements. The following paragraphs summarize the typical steps in the assessment and remediation process.

**Typical Response Schedule and Costs**

Procedure Section. Uses subhead to indicate new section. Boldface type makes it easy for reader to see. Lists schedule of activities so reader can easily comprehend what must be done. Uses numeration so reader knows how many steps. Uses hanging indent so reader can easily see each step.

1) IEPA issues an operating and report procedures packet which includes reporting forms for Agency submittal.

2) ESE will provide guidance in reviewing and compliance with the required certification items on the Leaking Underground Storage Tank program 20-Day Certification Form.

3) ESE will assist with the completion of

**Figure 5.2** *(continued)*

November 15, 1991
Page 4

the Application for reimbursement of
Corrective Action Cost from the Underground Storage Tank Fund. ESE will
also provide detailed invoices required by the IEPA for reimbursement.

4) ESE will request an IEPA Generator
Number to allow removal and off-site
disposal of petroleum impacted soil
and/or water if required.

5) ESE will collect and analyze a composite soil sample to allow disposal
to a special waste landfill if selected.

6) Performance of a limited subsurface
investigation which includes the
completion of 5-10 soil borings. Each
boring will be continuously sampled
and field screened for evidence of
petroleum impact.

7) ESE will complete the 45-Day Report
which will present the findings of
the limited subsurface investigation.
ESE will also prepare a Corrective
Action Plan form for submittal to
IEPA which outlines a hydrogeological
investigation, including the installation of monitoring wells to better
define the shallow aquifer characteristics.

8) IEPA review and approval of CAP.

9) ESE will execute the CAP investigation.

10) ESE will prepare a revised CAP if
further investigation is warranted or
a Corrective Action Report (CAR) if

(continues)

**Figure 5.2**   *(continued)*

November 15, 1991
Page 5

> it is not warranted. A CAR summarizes
> the findings of the site investigation
> and proposes a remedial system for the
> site if required.

11) IEPA review and approval of the CAR.

12) ESE will design required remedial systems. ESE can maintain and operate the
remedial system until IEPA clean-up objectives for the site are met, at which
time IEPA closure of the site is requested.

13) IEPA grants closure of the site.

**Estimated cost associated with typical sites as out-
lined above:**

| ITEM | DESCRIPTION | COST |
|------|-------------|------|
| 1-5 | Consulting services and associ-<br>ated laboratoory fees | $500-$1,000 |
| 6 | Performance of a limited subsur-<br>face investigation which in-<br>cludes completion of 5-10 soil<br>borings. (Five borings are pro-<br>posed for each tank area if re-<br>quired) | $2,500-$5,000 |
| | ESE personnel to observe and<br>monitor soil boring operations,<br>collection and screening of<br>samples. Field time estimated at<br>1-2 days | $450-$900 |
| | Environmental laboratory analy-<br>sis of 5-10 soil and groundwater<br>samples for BTEX and PNA<br>    BTEX (Method 8020) $100/test<br>    PNÅ (Method 8310) $250/test | $1,750-$3,500 |

Budget Section. Uses
table format so reader
can easily see cost of
each item. Numbers
items so reader can
easily refer to each.

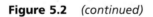

**Figure 5.2** *(continued)*

November 15, 1991
Page 6

| **ITEM** | **DESCRIPTION** | **COST** |
|---|---|---|
| 7-8 | Completion of IEPA 45-Day Report and development of a site specific Corrective Action Plan | $2,500-$3,000 |
| 9-10 | Implementation of CAP and development of a revised CAP or CAR for IEPA submittal . | $10,000-$15,000 |
| 11-13 | If groundwater is impacted, estimated costs associated with the design, installation and maintenance of a groundwater remediation system. | $75,000-$100,000 |

**Boilerplate request for a contract. (Always written in the same way and used on each letter responding to RFP.)**

Environmental Science & Engineering can provide engineering services as outlined in the enclosed Exhibit A - Work Order No. 1 and Short Form Contract. If your review of this agreement proves satisfactory, please sign and return both copies. We will subsequently return a fully signed copy for your files.

**Gracious conventional closing.**

We would welcome the opportunity to assist you with this proejct. If there are any questions, please call the undersigned or Sandra Rudolph at your convenience.

**Closing.**

Sincerely,

**Environmental Science & Engineering, Inc.**

**Signature.**

Michael J. Hoffman, P.E.
Sr. Environmental Engineer

**Word Processing ID.**
**Enclosures.**

SKR. BW - Ltr---
Enclosures

**Figure 5.2**   *(continued)*

# SOLICITED RESEARCH PROPOSAL: *EXXON VALDEZ OIL SPILL PROJECT DESCRIPTION*

Numerous projects—from activities to save the murres (birds) in Alaska to apprenticeships for high school students in high-technology jobs to research on the effect of noise on the hearing of residents within a ten-mile radius of an airport—depend on monies from governmental and private agencies. These monies are usually obtained through competitive proposals. Unless a proposal is accepted, a project will probably not be undertaken. The need to write a successful proposal can be extremely important if you believe in a project.

Successful proposals mean not only the initiation or continuation of a project but also the creation or continuation of jobs for people involved with a project. Thus, proposals have social and economic effects. You will discover how proposals meet societal needs and affect a local economy as you read the following proposal.

In October 1991 a federal judge imposed $1.025 billion in fines and damages on Exxon for the havoc wreaked on the state of Alaska by the Valdez oil spill. A six-member Oil Spill Trustee Council, composed of the Alaska attorney general and representatives from two state and three federal agencies, was appointed to determine how the money should be spent. The council decided to solicit proposals related to any of three broad areas: land purchases to protect vital habitats, scientific studies, and endowments from which interest could be used to finance restoration. An RFP that provided guidelines, including a size limitation of three pages, was issued to interested organizations.

More than 450 proposals were submitted to the council. Because the total amount of funds requested far exceeded the amount available, the council will only be able to approve some of the proposals (*Time* Lemonick, 1992).

Both economic and political factors will affect the council's decision regarding which proposals to fund. Although the public supports funding for projects involving habitat acquisition to protect what is left of the region's ecosystem from harm caused by logging and development projects, critics worry that the council may lean toward projects that favor the fishing industry or which involve scientists hired by the agencies represented on the council.

The following solicited proposal (Figure 5.3), which falls under the category of scientific studies, was submitted in 1992. It proposes to use funds from the council to conduct a feasibility study related to restoring the region's murre population. (A murre is a black and white seabird of the auk family that resembles a thin penguin with wings.) Local residents share a special fondness for this strange-looking bird.

**Project identification information. Follows guidelines of RFP.**

### EXXON VALDEZ OIL SPILL PROJECT DESCRIPTION

**Project Number:** 93022

**Project Title:**     Evaluating the Feasibility of Enhancing Productivity of Murres by using Decoys, Dummy Eggs, and Recordings of Murre Calls to Simulate Normal Densities at Breeding Colonies Affected by the *Exxon Valdez* Oil Spill, and Monitoring the Recovery of Murres in the Barren Islands

**Project Category:** Manipulation and Enhancement; Restoration Monitoring

**Project Type:** Birds

**Lead Agency:** Department of the Interior - Fish and Wildlife Service

**Cooperating Agencies:** None

**Project Term:**     January 1, 1993 to December 31, 1993

**INTRODUCTION:**

**PROBLEM SECTION**

A. Background on the Resource/Service

**Begins by appealing to readers's attitudes toward the birds to gain sympathy toward the project. Uses comparison/contrast organizational pattern. Begins with disussion of normal breeding conditions before spill.**

Murres were the species of higher vertebrates most heavily affected by the oil from the *Exxon Valdez* spill. These diving seabirds have continued to demonstrate abnormal breeding behavior and low reproductive output at several sites since the spill. Factors that normally result in increased breeding success of common murres are breeding in high-density concentrations and laying eggs in synchrony with  neighbors. Being one of a crowd apparently reduces vulnerability to avian predators. Within a colony, birds in groups that breed early tend to be more productive than birds breeding later, and older birds tend to breed earlier and be more successful than young birds. Prior to laying, murres tend to be flighty. In cases where a small percentage of murres in a cluster have begun to incubate before others have laid, incubators tend to leave their eggs exposed to predators, joining the flock when panic flights occur. Nevertheless, as more birds lay there is a tendency for incubators, now apparently feeling safer with company, to remain with eggs when non breeders flush.

**Contrasts with present breeding conditions after the spill.**

**(This sentence is actually part of solution section.)**

For reasons not yet fully understood, murres at colonies affected by the oil have not yet resumed normal breeding schedules. Apparently a relatively small proportion of birds have laid their eggs earlier than others, and egg predation by gulls has been high. Perhaps a substantial proportion of experienced breeders were killed in the spill so that the population now is composed of mostly young, inexperienced breeders. It is not well understood how crucial the presence of older birds is to the social facilitation of normal breeding, and it is possible that a shortage of experienced breeders is causing the abnormal timing and poor reproductive success. Another contributing factor could be reduced breeding densities, since populations were reduced by mortality of adults. The use of tape-recorded murre calls, placement of decoys, and dummy eggs could stimulate more normal breeding behavior.

(continues)

**Figure 5.3    A solicited proposal.**    Source: Alaska Department of the Interior— Fish and Wildlife Service.

Readers for this proposal are not experts on the topic. Therefore, the writer fully explains the problem, providing background information prior to presenting the problem in the Summary of Injury section. This is the conventional sequence for the problem section of a proposal.

Project Number: 93022

**B. Summary of Injury**

Over 100,000 murres were killed by the oil, and counts of birds at colonies within the trajectory of the oil indicated reduced populations after the spill. In the 3 years following the spill, remaining murres at colonies affected by the oil have initiated laying up to 1 month late, if they laid at all, and reproductive output has remained much lower than would be expected. Three consecutive years of poor reproductive success is very unusual based upon other studies.

**C. Location**

Experiments would be conducted at murre colonies in the Barren Islands, located between the Kenai Peninsula and the Kodiak Archipelago.

**WHAT**

**A. Goal**

The purpose of this project is to evaluate the feasibility of using artificial means to stimulate normal breeding behavior, as measured by nesting chronology and success, in murres at colonies affected by the oil spill.

**B. Objectives**

1. Determine the feasibility of enhancing the breeding success of murres by using decoys, dummy eggs, and recorded murre calls.

2. Monitor the recovery of murres in the Barren Islands.

**WHY**

**A. Benefit to Injured Resources/Services**

If murres can be induced to resume nesting at normal dates and if predation were reduced, reproductive success should increase. Increased recruitment from birds produced at injured colonies is likely to provide the best opportunity for populations to recover from reductions caused by the *Exxon Valdez* oil spill. Pioneering from other colonies outside the spill area is not likely to contribute in a major way in the near future since murres exhibit a high tendency to return to their natal colonies to breed, especially if there are available nest sites. There would be available nest sites at colonies with reduced populations. The monitoring phase is essential to understand the results of the feasibility study and to assess the recovery of the colony as a whole following the oil spill. The underlying causes of the abnormal nesting behavior (e.g., delayed laying) are not yet understood, and monitoring data will provide the basis for testing various hypotheses. Understanding the impact of the oil spill may make it possible to minimize damage in future spills by directing clean up efforts appropriately. Moreover, documentation of the response of murres in the aftermath of the oil spill will provide a basis for predicting the extent of the injury from future spills.

107

Heading.

Implied main organizing idea for problem section: Murre population was almost killed off by spill and is very slow in coming back because of low reproduction rates. Location section is usually placed in procedures section of a proposal, not here.

SOLUTION SECTION

Main organizing idea for solution section.

**Figure 5.3** *(continued)*

Even though the proposal is short, the writer uses visual text to help readers locate information and to perceive the different chunks of information quickly and easily. Notice, too, that the major headings in this proposal—Introduction, What, Why, How, and When—actually parallel the conventional proposal categories: background and rationale for problem

Relates to readers'
goals for allotting
funds.

## Project Descriptions

**B. Relationship to Restoration Goals**

This project meets the Trustee Council goal of restoring the spill area to its pre-spill condition by providing information that could be used to develop a management action. If one or more of the experimental treatments prove to be feasible, it should be possible to implement the technique extensively enough to generate improved success for a portion of one or more colonies. At least for these portions, more young should be produced and ultimately begin the process of recovery to former population levels.

**HOW**

**A. Methodology**

Treatment and control plots would be selected at East Amatuli Light Rock and on Nord Island in the Barrens. Decoys, and solar powered sound players would be placed in selected locations prior to the arrival of murres on cliffs. It would be necessary to use technical climbing gear to accomplish the objective on Nord Island. Time-lapse cameras would be used to monitor plots on E. Amatuli Rock because access after murres have laid would disturb the birds.

**B. Coordination with Other Efforts**

The two subprojects included here are complimentary. Data from the monitoring program will be used to assess the effectiveness of this project, and a single project leader would guide both projects.

**ENVIRONMENTAL COMPLIANCE**

This is a non-intrusive project which appears to qualify for categorical exemption under NEPA.

Time Line.

**WHEN**

| Jan. - April 1993 | Plan and arrange logistics (e.g., boat charters), recruit seasonal employees, develop detailed study protocols, assemble field gear, purchase equipment |
| May 1993 | Place decoys, players, dummy eggs, and time-lapse cameras in field |
| Jun. - August 1993 | Conduct field studies |
| Sept. - Oct. 1993 | Analyze data |
| Nov. - Dec. 1993 | Write progress report |
| Dec. 15, 1993 | Submit progress report |

Budget.

BUDGET ($K)

| | USFWS |
|---|---|
| Personnel | $ 84.5 |
| Travel | 9.0 |
| Contractual | 126.0 |
| Commodities | 15.0 |
| Equipment | 25.0 |
| Capital Outlay | 0.0 |
| Sub-total | $ 259.5 |
| General Administration | 21.5 |
| Project Total | $ 281.0 |

**Figure 5.3**    *(continued)*

(introduction), objectives (what), explanation of solution (why), methodology and procedures (how), and schedule (when).

As you study the proposal in Figure 5.3, consider the writer's claims to support the argument that artificial means to increase murre fertility should be tested in order to bring the murre population back to what it was prior to the oil spill.

---

### Suggestions for Discussion and Writing

1. In the problem section, the writer discusses the problem with the murres after the Valdez oil spill but does not discuss the spill itself. Why do you think the writer decides not to discuss the spill?

2. What claims does the writer make to persuade readers (the Oil Spill Trustees Council) to fund the proposal? If you were a member of the council, what claims would persuade you to approve this project?

---

# IN-HOUSE PROPOSAL FOR A CHANGE: PROPOSAL TO ADD A VALVE ON CIRCUIT BREAKERS TO SIMPLIFY REPAIRS

Traditionally employees have had little opportunity to provide suggestions to their companies. Today that is changing, and many companies encourage workers to submit proposals for improving the quality of the workplace and its output. As the result of a successful proposal, employees have the satisfaction of seeing their ideas transformed into practice and often improve the workplace for themselves and their coworkers while simultaneously improving profits for the company. The following proposal (Figure 5.4) does just that.

A supervisor and a mechanic with Appalachian Power Company collaborated to write this proposal to persuade management to add a valve on circuit breakers to simplify repairs. Appalachian Power encourages employees to propose operational changes that will save money and improve the quality of its products and services. At the end of the year, the parent company, American Electric Power, publishes the proposals that have been the most helpful in a magazine and awards a monetary bonus to the individuals who made the suggestions. The following document is one of the award-winning proposals.

The proposal is written on a form that basically follows the conventional format of a proposal. Although the writers only need to fill in the blanks, they still "invent" their text for the problem, solution, and

PROBLEM SECTION.
Uses comparison/
contrast organizational
pattern. Begins by
describing difficulties
in repairing a check
valve problem at
present. Describes
process in sequential
order.

SOLUTION SECTION.
Organizing idea for
section. Contrasts steps
involved in procedures
using new type of
valves with procedures
described in problem
section. Also contrasts
results of new valve
procedure with old
procedures so readers
recognize differences.
Relates to values of
company.

Uses chart to show
advantages. Provides a
quick summary for
readers.

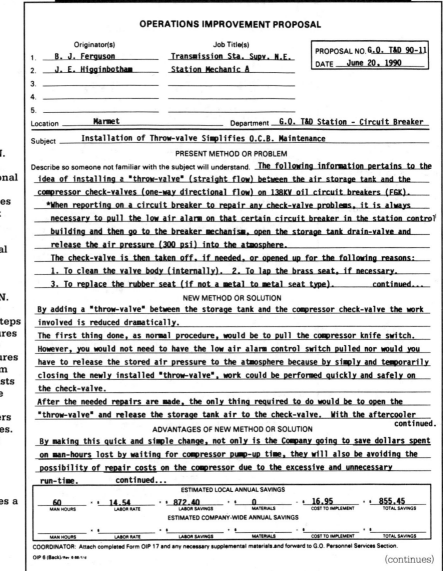

**Figure 5.4    A proposal to change a process.**   Source: Reprinted with
permission of Appalachian Power Company, B. J. Ferguson, J. E. Higginbotham.

Because of the preformatted form, sections rather than pages are continued, making the reader jump back and forth between pages to complete a section.

Indicates specific problems related to procedure.

Main organizing idea of Problem Section.

---

DM 20

**OPERATIONS IMPROVEMENT PROPOSAL**

FILE NO. Ƶ-3010

Originator(s) B. J. Ferguson / J. E. Higginbotham     Job Title(s) Transmission Sta. Supv. NE / Station Mechanic A

Location Marmet     Department G.O. Station - Circuit Brk.

Proposal No. G.O. T&D 90-11     Date June 20, 1990

Subject: Installation of Throw-valve Simplifies O.C.B. Maintenance

(Start with paragraph briefly summarizing your idea and what it accomplishes. Describe problem or old method. Follow with details of solution or new method. Answer questions that may be asked by someone using your idea: Include how—to—do—it procedures, materials and equipment involved, actual operating experience, specific benefits, costs, man-hours saved. Tell who did what and when. Please double-space typing.)

**PRESENT METHOD OR PROBLEM** (continued)

The compressor is then reconnected to the air storage tank by the aftercooler line, the drain-valve is closed on the storage tank, and the compressor knife switch placed back in the closed position (which had been opened prior to any work being performed) in order to get the proper amount of operational air pressure in the storage tank (300 psi).

The pump-up time for the compressor (depending, of course, what condition it is in to begin with) can range around or within one hour.

After the compressor has fully charged the storage tank and shut down, the next step is to loosen the fitting on the aftercooler line to see if, in fact, the valve seat is sealing properly. If the slightest leak is detected, it will then be necessary to repeat the entire procedure as outlined above, starting with deactivating the compressor knife switch and, gain, releasing the stored air to the atmosphere via the drain-valve, because in most cases, the leak will worsen.

The air will continue to escape up through the aftercooler line and into the compressor, usually venting through the head gaskets (low or high pressure side) or the compressor unloader body. It should also be noted that during this leakage, the motor, when called upon by the governor switch, will be starting under load. This can lead to blown fuses or possibly cause damage to the motor itself.

After a second repair is made to the check-valve, the compressor has to, again, bring the storage tank air to normal pressure. This excessive running time on the compressor, if for just one repair on the check-valve, is completely unnecessary. It can and will at times, lead to other problems within the compressor due to the heating it will be subjected to.

**NEW METHOD OR SOLUTION** (continued)

line still disconnected, you can now detect whether or not the seal is proper and if it's not, there is just the matter of reclosing the "throw-valve" and making the proper adjustments.

**ADVANTAGES OF NEW METHOD OR SOLUTION** (continued)

It should be noted too, that using this technique is a much better way of getting the check-valve to initially seal. Because, with the storage tank having the proper air pressure in it and when the "throw-valve" is opened up into the check-valve, this pressure forces the two seating surfaces together helping to form a tight seal.

(ATTACH SUPPLEMENTARY MATERIAL IF NECESSARY FOR CLARITY)

---

**Figure 5.4** *(continued)*

**Appendix.**

**Includes details of calculation so readers can check if they wish. This information is often placed in an appendix rather than the main text.**

**Uses a diagarm so readers can see what writer is discussing Facilitates readers' comprehension by emphasizing visually the two parts that must be added.**

**Diagram is simple so readers can follow it easily.**

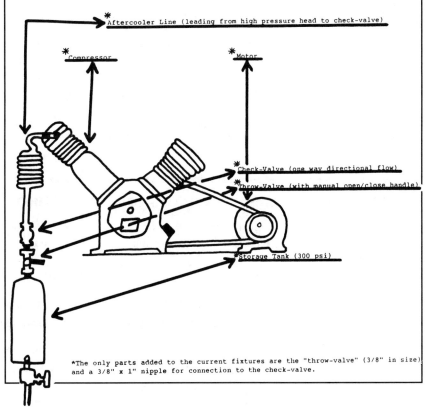

SAVINGS CALCULATION

It is estimated that a throw valve would save approximately two hours per each check valve maintenance procedure.  There are approximately 15 such procedures done per year.  Assuming a crew of two men, the savings would be as follows:

    2 X 2 X 15 X $14.54 = $872.40/year


Cost to implement:

    Cost of throw valve = $8.00/each
    Labor to install (1 MH) = $14.54

    Amortize this over a 20 year breaker life, which is conservative:

    (8 + $14.54)/20 = $1.13 per breaker

        15 X $1.13 = $16.95

Aftercooler Line (leading from high pressure head to check-valve)

Compressor

Motor

Check-Valve (one way directional flow)

Throw-Valve (with manual open/close handle)

Storage Tank (300 psi)

*The only parts added to the current fixtures are the "throw-valve" (3/8" in size) and a 3/8" x 1" nipple for connection to the check-valve.

**Figure 5.4**   *(continued)*

advantages sections, and they have to determine whether to include graphics. Notice that they include a diagram to help the reader see exactly where the new valve should be installed. If the reader approves the proposal, the diagram can be given to the secondary reader responsible for installing the new valve. The writers also include an appendix page, which presents the calculations for savings to support their claims that the change will save the company money.

Although the employees direct their proposal to their supervisor, who has to approve the alteration to the circuit breaker, they are aware that if the supervisor approves the change, their proposal will also be read by the committee that selects the prize-winning suggestions for publication.

As you read the proposal in Figure 5.4, consider how the employees appeal to the economic factors readers will consider in determining whether to approve the proposal.

---

### Suggestions for Discussion and Writing

1. What claims do the employees make to persuade their audience to accept their proposal? Based on the situation described in the introduction to this proposal, why do you think the audience accepted the claims?

2. To which conventions of a proposal do the writers adhere? From which conventions do they deviate?

---

## CHAPTER SUGGESTIONS FOR DISCUSSION AND WRITING

1. The proposal related to lead poisoning contains an appendix. Why do you think the writer includes the information contained in it?

2. The proposal related to the leaking storage tanks hardly discusses the problem, whereas the murre proposal provides a great deal of detail about the problem. Why do you think these proposals differ in their treatment of the problem sections? What changes, if any, would you make?

3. Consider the language and syntax used in the description of the oil spill project and in the explanation of the leaking storage tank solution. Compute a Fog Index for each, using the following formula.

a. Take a sample of at least 100 words. Count the number of sentences included in the sample. Determine the average number of words per sentence by dividing the number of words by the number of sentences. Treat independent clauses as separate sentences. Example: "We read. We learned. We improved." This statement should be counted as three sentences even if semicolons or commas were used instead of periods.

b. Count the number of polysyllables (words of three syllables or more) per 100 words. Omit from this count capitalized words, combinations of short, easy words like manpower and insofar and verbs made into three syllables by adding -es or -ed.

c. Add the average number of words per sentence to the polysyllable percentage, and multiply the sum by 0.4. Ignore digits after the decimal point. Because few readers have more than 17 years of schooling, assign any passage that tests higher than 17 a fog index of 17+.

Is there very much difference between the Fog Indexes of the two proposals? The Fog Index is supposed to indicate the grade level at which a text is written. Do you think the index you've calculated for each proposal accurately reflects the grade level a reader will need to read it? Why might someone with a lower grade level be able to read these proposals? What changes would you make to the language or syntax of each proposal to make it more readable? Make the revisions.

## Note

Lemonick, Michael D. *Time*. 1992. 28 September, 60–61.

**CHAPTER**

**6** Reports

## INTRODUCTION

**M**ost reports simply document routine business operations. Occasion-ally, however, a report galvanizes society into action with its findings or recommendations. The report on the Challenger disaster resulted in a complete overhaul of the shuttle program. The report on the Three Mile Island nuclear accident resulted in the creation of a nuclear utility watch-dog agency (the Institute for Nuclear Power Operations), an improved sys-tem of communications among nuclear utilities, and an effective procedure for reporting problems to the Nuclear Regulatory Commission that ensures prompt and appropriate corrective actions. An environmental impact state-ment related to the New York borough of Manhattan resulted in the saving of an ancient slave burial ground as a historic landmark. You'll have an op-portunity to read excerpts from these reports in Part III. In this chapter you'll read part of a report on waste reduction at McDonald's that resulted in significant environmental savings by prompting changes in procedures related to fast-food production and service.

### Routine Reports

Reports are constantly written to keep supervisors and managers up to date on developments in their organizations, on the progress of their pro-jects, and on the results of studies or programs. The larger a business, the more reports are generated. Some reports, such as progress or periodic re-ports, are written on an interim basis—daily, weekly, monthly, or annually. Some, like trip reports and minutes to meetings, are written sporadically. Final reports assume a variety of forms, the most common of which are in-formation, evaluation, feasibility, and research reports. Some reports re-quire complete narrations, but others are simply forms to be filled in. A report may be integrated into a letter or memorandum, or it may stand by itself.

## Historical Reports

People have been reporting events and new developments since very early times, when men reported their hunting activities through the medium of cave paintings. It is through historic reports that we know today about plagues, such as the plague of Athens (5 B.C.) and the great plague of London (A.D. 1665).

In the early technical reports, writers such as Frontinus, who wrote *De Aquus (Aqueducts of Rome)* in A.D. 97, set their reports within their historical and social context. The writers were as interested in providing the audience with an understanding of how the topics on which they were reporting both influenced and were influenced by the situation and concepts of the period as they were in describing the topics. However, by the early twentieth century, writers increasingly divorced their topics from the social and historical contexts in which they occurred, and their documents began to resemble the impersonal reports we read today.

# CONTEXT

## Situation

Most reports are reader-initiated. Readers request that writers provide them with information about a specific topic, program or project, or situation. Sometimes writers already have the information; other times they need to gather it. They may write an entire report alone, or they may do it collaboratively. Often people other than the writers have some of the information that needs to be included in a report. Writers may interview these people orally or ask them to submit the information in written form to be incorporated into a main report.

Reports often involve economic, political, and social factors. Evaluations of a situation may indicate that people have failed to perform appropriately or that a program is not fulfilling its objectives. Evaluation reports can result in people losing their jobs, projects being canceled, and the profits of organizations being reduced. Both the Kemeny report on the Three Mile Island Nuclear Accident and the Rogers report on the Challenger disaster (Part III) precipitated these results. Knowing that a report may have such an effect, writers may wish to circumvent or completely avoid actually blaming anyone, especially if they need to continue working with the people. Instead of using active voice and stating, "The operator forgot to turn the valve off," a writer may use the passive voice and state, "The valve was not turned off." Writers may also be careful that their language does not place an undue burden on a person. Rather than state "X continues to fail to hold a heading within five degrees while flying straight and level," a writer may state, "X continues to face the challenge of maintaining a heading while engaged in straight and level flight."

Making recommendations can also become problematic if writers know that the recommendations are contrary to readers' expectations or hopes. This situation is further complicated when readers sponsor and finance a report. If the results do not support the views of the sponsoring agency, the agency may not hire the writer's organization for future work. When McDonald's asked the Environmental Defense Fund to study the feasibility of engaging in a waste reduction program, the two companies agreed on certain conditions before signing the contract to avoid such a situation. Among those conditions, EDF agreed not to recommend changes that would affect the company's administrative operations because that would be too costly.

Establishing such preconditions can often prevent problems resulting from conflicting loyalties. Another method writers use to avoid potential political and economic consequences is to maintain a neutral stance in relation to their topics. However, readers may select information in a report to support their own views. Both proponents and opponents of McDonald's waste reduction program used the information selectively to support their respective arguments.

Sometimes a topic itself is controversial. Under these conditions, writers may adapt the conventions of a scientific research report to maintain objectivity. The research design for the Foley Square federal courthouse that is included in Part III follows these conventions. The report revolves around the question of what to do with a slave burial site. Some members of the African-American community wanted the bodies to remain interred and the site to remain as it was; others were willing to allow the bodies to be excavated but demanded that the site be declared a historical landmark. Anthropologists wanted to study the remains, and the federal government wanted to build a courthouse on the site.

## Purpose

Although a writer's primary purpose in writing a report is to provide information, writers also have secondary purposes. They want to persuade readers to accept their data. They may also want to persuade readers to accept their point of view and recommendations. To do so they must appeal to their readers' needs and values. In addition they need to use sources, methods, and statistical analyses that meet the criteria of their readers' communities and to present these so that readers can follow their line of reasoning.

## Audience

Readers, regardless of their organizational position, assume the role of decision makers when they read a report. They may need to determine if they want to accept a researcher's findings, if they agree with the results of an evaluation report, or if the information changes their point of view concerning a project.

Reports are read by a wide variety of readers who read different parts of a report and who engage in different reading behaviors (Figure 6.1). Reports may be read by experts, but they are also often read by generalists and novices who have requested that experts provide them with information for making a decision. Generalists and novices seldom read an entire report. They usually read only the abstract or the executive summary. They

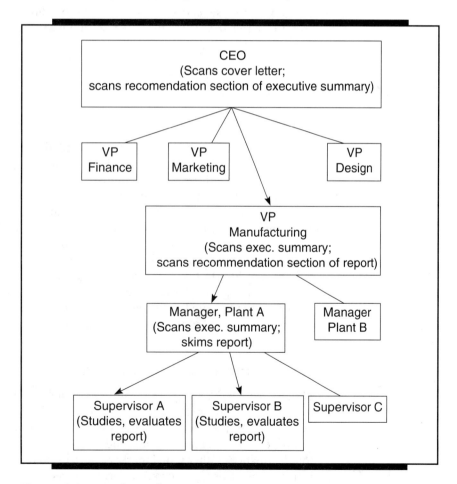

**Figure 6.1   Readers of a report.**

may also study the conclusions and recommendations. Sometimes they scan the introduction to obtain an overview of the project or situation.

A chief executive officer (CEO) may scan a cover letter and possibly the recommendation section of an executive summary. This executive may then pass the report to the appropriate subordinates, who will probably scan the executive summary and the recommendations related to their specific areas before passing it to those in their divisions responsible for the areas discussed. These readers will probably not only scan the executive summary, conclusions, and recommendations section but also skim the entire report before passing it to their subordinates, who will be expected to study and evaluate the information. Of course, if the information is sufficiently important, the CEO may study the entire report, evaluating the information in relation to the organization's goals.

Sometimes supervisors request their subordinates to read an entire report and summarize it for them. These summaries are often transmitted in memo form.

# DISCOURSE STRATEGIES

## Persuasion

All reports are persuasive. Writers try to persuade readers to accept their results, conclusions, and recommendations. Writers support their results with claims that their information, methods, procedures, and analyses are valid. The sources of information they cite, and the methods, procedures, and statistical analyses they describe, serve as evidence to support their claims. If the claims relate to readers' values, needs, and attitudes, and if they are based on evidence that is considered valid by readers' organizational and professional communities, then readers may accept the claims. Readers must also be able to follow a writer's line of reasoning if they are to be persuaded to accept the conclusions and recommendations. The conclusions must be directly related to the results, and the recommendations must be directly related to the conclusions (see Figure 6.2).

## Navigation and Location Aids

Because reports, like proposals, can be extremely long, readers may skim or read only parts of them. Tables of contents and lists of tables and figures and indices help readers locate relevant information. Chapter titles, headers, page numbers, and headings and subheadings also help readers search for information. Other visual markers, such as boldface type and boxes, can help readers identify important information as they skim a document.

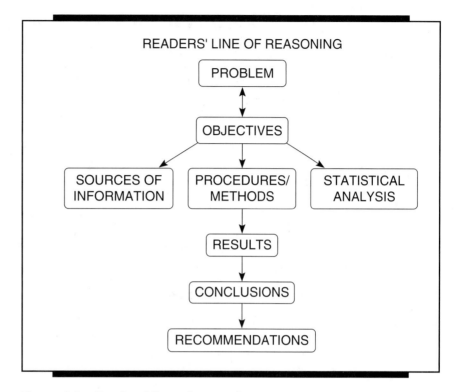

**Figure 6.2    Readers' line of reasoning.**

## Ancillary Documents

As we have already indicated, preview documents are extremely important for reports. Cover letters, title pages, and executive summaries and abstracts may be the only parts of a document read; readers may never look at the main text.

Supplemental documents often provide detailed information relating to writers' sources of information, methods, procedures, and statistical analyses. Some readers want to see these details before they accept a writer's claims as valid. Supplemental documents are also often necessary for secondary readers, who are usually the ones who must implement the recommendations. The detailed information that these readers need to carry out the recommendations can be placed in appendices. Generalists and novices may also need a glossary to help them understand a report. In addition, experts may want a list of references to read additional information on a topic, to study some of the primary documents, and to investigate questionable information.

## Organizational Pattern and Sequence

The organizational pattern of a final report is analytical, consisting of the major areas involved in a study: (1) problem, background, or rationale; (2) objectives; (3) methods/procedures; (4) results; (5) conclusions; and (6) recommendations. Although reports generally discuss each of these topics in a different section, a report may combine several in a single section. In reports that are concerned with several parts of a single project, each part may be subdivided into the various topics as the reports in this chapter on waste reduction and lean beef are organized.

## Graphics

Graphics can help generalists and novices comprehend the concepts presented in a report. Diagrams and tables can provide experts with detailed aspects of a mechanism or process or with lists of information that they need.

## Conventional Format

Reports are usually written in an impersonal and formal style. Writers consult organizational or professional style guides to determine the specific styles for citing and listing references, punctuating sentences, and formatting a report.

# THE DOCUMENTS

This chapter includes a feasibility study and a research report. You will have an opportunity to consider the similarities and differences in these two types and to examine how each adheres to the general conventions for reports.

In Part III you will read other types of reports. The excerpts from the Kemeny Report on the Three Mile Island nuclear accident and the Rogers report on the Challenger disaster exemplify evaluation reports. The chapter on the Challenger also includes an example of an activity report. Finally, the Foley Square research design will give you an opportunity to study the information section of an evaluation report.

In this chapter you will also study several historical reports that serve as the predecessors of today's documents. As you read these, you will see how reports of the twentieth century have evolved from those that date as far back as the fifth century B.C.

As in the previous chapters, we will examine the first report in each category closely. You can then apply what you've learned to the other reports.

## Feasibility Study: *Waste Reduction Task Force Final Report,* McDonald's Corporation/Environmental Defense Fund

At the beginning of the 1990s, environmentally concerned consumers began pressuring fast-food restaurants to eliminate the use of plastic containers and to reduce the amount of overall waste generated. Some companies acceded to consumer demand by replacing plastic containers with paper wrap. Other companies continued to use plastic to package their foods, arguing that plastic comprised an extremely small portion of municipal solid waste.

To determine what to do, McDonald's requested the Environmental Defense Fund, a not-for-profit environmental action organization, to collaborate on a study examining feasible ways to reduce solid waste. The purpose of the study was to determine the extent to which McDonald's could become "environmentally correct."

Writers were aware of the secondary purposes related to the economic and political factors associated with the report. McDonald's executives hoped to use the report's recommendations to increase sales and thereby increase profits. They also expected to use the information in the report as supporting evidence to counter opponents' arguments.

EDF also had secondary purposes for writing the report. EDF hoped that if McDonald's adopted its recommendations, other quick-service restaurants as well as other industries would adopt the approach. In addition, EDF hoped that publicity related to McDonald's acceptance of its recommendations could increase public awareness of the organizations's projects.

Although the primary audience for the report was McDonald's executives, peripheral readers would also see the report. Results of the study were not only sent to executives of the two organizations involved, but were also made available to the news media and interested citizens.

The writers were aware of the economic, political, and social factors involved. They knew that McDonald's was not likely to accept the recommendations if profits would be reduced or the company's business practices would have to be changed. Therefore, the writers supported many of their claims with evidence that business operations would not be disrupted and that costs would not increase (and in many instances, would actually decrease). Because the methods and procedures included an in-depth study of McDonald's business operations, the readers' organizational community accepted the writers' conclusions as valid.

A portion of the report follows. The cover page, table of contents, and executive summary provide you with an overview of the complete report. The excerpts reprinted here from the first two chapters are fairly representative of the succeeding chapters. The excerpts give you an opportunity to study how the information is organized between chapters and within chapters and to consider the point of view and focus, style, graphics, visual text, and format of the report. You also have an opportunity to study how the writers develop their arguments.

**Persuades Readers to Accept Conclusions and Recommendations**     The writers follow several strategies to persuade readers to accept the conclusions of the report. Throughout the document, they appeal to readers' attitudes and values. At the very outset in Chapter 1, the writers indicate in their list of objectives that their recommendations are to be consistent with McDonald's business practices and future growth. Among their criteria for judging the merits of a recommendation are the practicality and economic costs and benefits of the changes. Furthermore, the writers base their methods and procedures on criteria accepted by their readers' organizational community as well as their own professional communities. These procedures include interviews with experts in the various areas of waste and observations at exemplary waste reduction centers around the country. Finally, the writers base their conclusions on the results of their study. Readers can follow writers' line of reasoning from their results to their conclusions.

**Provides for the Needs of Multiple Readers**     The report provides easy access to readers with differing reading behaviors and in different roles, fields, and communities. An executive summary, on pages 249 through 255, includes a brief overview of the major sections in the report and the main conclusions and recommendations. A CEO can quickly read these pages and make decisions concerning the company's environmental stance. Peripheral readers such as reporters, interested citizens, and McDonald's franchise owners can also use this summary to learn the policies and actions recommended for the company. Writers with McDonald's and EDF can also use the summary information in other documents that they write to support the company's decisions and to counter their opponents' arguments.

For readers who wish to obtain additional information, especially in relation to background data and the study itself, a table of contents includes chapter titles and two levels of headings to provide readers with a map for navigating the main text. Pages are numbered, and chapters include headings and subheadings to help readers who are searching for specific topics.

Within the main report, headings and subheadings presented in boldface and underlined type help readers skim or scan a chapter. Additional spacing around headings and subheadings, the use of lists, bullets, and hanging indentation to set them off, and the placement of new sections on new pages rather than crowded at the end of a page, make the document very legible. Numerous tables and graphs provide readers with a summary of data, easy access to specific data, and help in comprehending information.

Introductions to each of the chapters include a forecast so readers can accurately predict what they will read. Because the forecast includes a listing of the topics that will be covered, it also offers readers a choice in determining the sections they want to read.

**Uses a Logical Organizational Pattern**    Both between chapters and within chapters, the document is organized from general to specific information. Readers begin in Chapter 1 with an overview of the entire operation and then move to a detailed examination of specific topics in Chapters 2 through 5. These four chapters are divided according to the major topics related to waste reduction: source reduction, reuse, recycling, and composting. Each of these four chapters is organized chronologically, presenting background information and present conditions and then discussing future actions.

**Follows a Conventional Format**    A feasibility report follows the conventions of a general report. However, additional conventions govern the discussion of results. In a feasibility study the discussion is subdivided into two major categories: (1) present conditions and (2) alternatives for changing those conditions or future actions.

The waste reduction report adheres to the conventions of a feasibility study. Sections I and II of Chapter 1 provide the background, rationale, and objectives of the study. Section III discusses the methods and procedures used to gather the data. Results and conclusions for the overall general operation of McDonald's are presented in Sections IV through VI of Chapter 1, and the results and conclusions of the four specific areas of environmental concern are presented in Chapters 2 through 5. Each results section is presented in terms of present and future status.

Although many feasibility reports maintain a neutral tone, this one does not. A bias toward making changes to improve the environment is obvious from the first sentence of Chapter 1 in which the writers, in marked hyperbole, comment, "This report marks the culmination of an historic project" to the conclusion of the introductory section when the writers gush, "The results of the task force far exceeded either EDF's or McDonald's expectations."

However, the writers adhere to the conventions for maintaining a formal and impersonal tone; they use the names of their organizations and do not refer to themselves in first person plural.

As you read the report, consider the writers claims and conclusions as they develop their arguments.

**Title Page**

**Sponsoring Organizations**

McDONALD'S CORPORATION

ENVIRONMENTAL DEFENSE FUND

Waste Reduction Task Force

**Title of Report**

# final report

**Date**

APRIL 1991

**Writers**

This report was collectively written by the members of the EDF-McDonald's Waste Reduction Task Force, for McDonald's, Terri Capatosto, Robert Langert, Keith Magnuson and Dan Sprehe, and for EDF, Richard Denison, Jackie Prince and John Ruston.

© 1991 ENVIRONMENTAL DEFENSE FUND and McDONALD'S CORPORATION

ALL RIGHTS RESERVED. THIS DOCUMENT MAY NOT BE REPRODUCED IN WHOLE OR IN PART IN ANY FORM WITHOUT THE WRITTEN PERMISSION OF AN OFFICER OF THE ENVIRONMENTAL DEFENSE FUND OR THE McDONALD'S CORPORATION.

PRINTED ON RECYCLED PAPER

**Begins with Executive Summary**

# WASTE REDUCTION TASK FORCE
# FINAL REPORT

# EXECUTIVE SUMMARY

**Parallels Introduction for Chapter 1.**

## INTRODUCTION

On August 1, 1990, the McDonald's Corporation and the Environmental Defense Fund (EDF) joined forces in a unique collaborative project. In this effort, the nation's largest quick-service restaurant business and one of America's leading environmental research and advocacy organizations worked together to find ways to reduce McDonald's solid waste through source reduction, reuse, recycling and composting.

The results of the task force far exceeded our expectations and original goals. What started as a cautious venture for both parties ended in a successful partnership. While the task force originally intended only to recommend options for McDonald's to consider, the six-month process instead produced a corporate Waste Reduction[1] Policy (Appendix 1) and a comprehensive Waste Reduction Action Plan (Appendix 2), both of which are already in effect.

Building on McDonald's past achievements, the Waste Reduction Policy will guide long-term activities throughout the McDonald's family of restaurants, distribution centers and suppliers. The Waste Reduction Action Plan contains 42 discrete initiatives, pilot projects and tests in the areas of source reduction, reuse, recycling and composting that McDonald's will undertake within the next two years.

**Specifies terms so all readers interpret terms the same regardless of their community.**

---

[1]In this report "waste reduction" refers to actions that reduce the quantity or toxicity of solid waste requiring disposal in landfills or incinerators. "Source reduction" is construed more narrowly, referring to measures that reduce the weight, volume or toxicity of products and packaging prior to their use.

(continues)

Combines information
from the Introduction
and Origin sections of
Chapter 1.

The work of the task force took place against a backdrop of high public concern over solid waste disposal. Disposable packaging has taken center stage in discussions on solid waste management, in part because of its high visibility and short useful life. Disposable food-service packaging has emerged in response to changing American lifestyles, allowing people to eat meals at their convenience, whether at work, on the move, or at home. Quick-service restaurant packaging accounts for only a small fraction of municipal solid waste, so any complete solution to our solid waste problems will require actions across industries and by all consumers. Given this reality, EDF in particular desired to create a model approach that could be used by other companies. And as McDonald's new policy recognizes, being a business leader carries the responsibility of being an environmental leader as well.

Writers' bias.

Relates to readers'
attitudes and values.

The most promising solutions for diverting wastes from disposal lie in reduction, reuse, recycling and composting.[2] The task force sought to produce the maximum possible reduction in McDonald's solid waste in a way that was consistent with McDonald's business practices. In considering McDonald's packaging and other solid wastes, we looked at every aspect of McDonald's operations, including materials discarded behind-the-counter and by customers in the restaurant lobby, disposable packaging used in McDonald's take-out business, and the distribution and supply system.

Summarizes
Conclusions section of
Chapter 1.

Claim
↓
Evidence

Waste sorting studies conducted in part for the task force show that almost 80% of McDonald's on-premise waste, by weight, is generated "behind the counter," in the preparation area and restaurant supply system. Many of these waste materials, such as corrugated boxes, are candidates for source reduction and can also be identified and separated for recycling by the restaurant crew. The behind-the-counter component of McDonald's solid waste can also be reduced by substituting reusable bulk storage systems for disposable ones. For example, McDonald's already utilizes reusable plastic trays for holding sandwich buns in place of corrugated containers and large steel canisters for storing Coke in place of cartons.

Packaging materials used by take-out and drive-thru customers (50–70% of the business, depending on a restaurant's location) cannot easily be corrected by in-store recycling programs initiated by McDonald's. For these items, source reduc-

---

[2]One of the reports on solid waste management reviewed by the task force was Recycling and Incineration: Evaluating the Choices, edited by EDF task force members Richard Denison and John Ruston (Washington DC: Island Press, 1990, 320 pages). To obtain a copy or receive more information, write for the EDF Solid Waste Publications List; EDF, 257 Park Ave. South, New York, NY 10010.

tion steps and design changes that allow packaging to fit into evolving community recycling programs will deliver the greatest environmental benefits.

The task force concluded that there is no single method for minimizing solid waste at McDonald's. Rather, there are a number of specific solutions that, collectively employed, will achieve significant waste reductions.

## ORIGIN OF THE TASK FORCE

**Parallels same section in Chapter 1.**

The origin of the task force dates to October 10, 1989, when Ed Rensi, president of McDonald's U.S.A., met with Fred Krupp, EDF's executive director, at EDF's request. McDonald's and EDF staff held several follow-up meetings and tours of McDonald's facilities. This built a working relationship and led EDF to propose a joint task force, which was formally announced on August 1, 1990.

At the outset, EDF recognized McDonald's substantial existing initiatives in recycling, and McDonald's interests in going further. McDonald's recognized EDF's expertise in solid waste management and the importance of seeking expert opinions. In undertaking this project, McDonald's committed itself to an unprecedented level of scrutiny by an outside organization. In order to maintain the independence of the parties, the agreement that established the task force included several protections, including a provision that each side pay its own expenses.

The task force was comprised of four members from McDonald's and three from EDF. The McDonald's team included specialists from McDonald's own Operations and Environmental Affairs departments, and from the Perseco Company, McDonald's packaging purchaser. The EDF task force members were trained in biochemistry, chemical engineering, economics and environmental science. Among them, the EDF staff held 17 years of experience in analyzing solid and hazardous waste management issues. Short biographies of the task force members follow the Executive Summary.

## THE TASK FORCE WORK PROCESS

**Parallels same section in Chapter 1.**

The task force examined in detail only McDonald's materials use and solid waste issues in its U.S. operations, including its restaurants, distribution centers and suppliers. We took broader environmental impacts into consideration, in part to ensure that changes resulting in solid waste reductions would not create or exacerbate other negative environmental impacts.

(continues)

The task force devoted considerable effort to understanding McDonald's business. Numerous hours were spent in various McDonald's restaurants, and the EDF task force members each worked in a restaurant for a day. The task force benefitted from the willingness of McDonald's and its suppliers to open their doors for a review of their operations. We toured the facilities of two McDonald's food suppliers, five packaging suppliers and one of McDonald's largest distribution centers. The task force also visited Plastics Again, a polystyrene recycling facility in Massachusetts, and Resource Conservation Services, a composting facility in Maine. Most of these visits included tours, formal presentations and extensive question and answer sessions with top management and technical experts.

During the course of this project, McDonald's brought in experts from various departments to discuss issues in depth. Likewise, additional EDF staff provided expertise on environmental issues beyond solid waste, as did experts from other environmental organizations.

———

**Summarizes information on the two organizations collaborating on the report.**

There are more than 8,500 McDonald's restaurants in the United States. More than seventy-five percent of the U.S. restaurants are owned and operated by local, independent franchisees. The McDonald's Corporation establishes and maintains strict operating standards for the entire system. At the same time, decision-making is decentralized, allowing the flexibility to adapt to local conditions.

McDonald's is served by a network of almost 600 independent food, packaging and equipment suppliers. McDonald's does not own any of its suppliers, and is not a manufacturer. McDonald's relationships with its franchisees and suppliers is a partnership that entails mutual risks and rewards. The success of the McDonald's system is dependent upon the success of the local licensee. As a low margin, "penny-profit" business, McDonald's success depends on high sales volumes.

On the other side of the task force, the Environmental Defense Fund is a national non-profit organization that links science, economics and law to create innovative, economically viable solutions to environmental problems. EDF was founded in 1967 by scientists on Long Island to fight the spraying of the pesticide DDT. With its headquarters now in New York City, EDF has grown to seven offices nationwide. EDF's professional staff of over 110 scientists, economists, attorneys, engineers and administrators are supported by more than 200,000 members and over 100 private foundations.

## HIGHLIGHTS OF McDONALD'S ACTION ITEMS

**Relates to Chapter 2. Summarizes information from paragraph 2.**

SOURCE REDUCTION. In a major source reduction step endorsed by the task force, McDonald's is completing a switch from polystyrene foam "clamshells"

iv

to paper-based wraps for packaging its sandwich items. The wraps provide a 70–90% reduction in packaging volume, resulting in significantly less space consumed in landfills. Compared to the polystyrene foam boxes they replaced, the new sandwich wraps also offer a substantial savings in energy used and substantial reductions in pollutant releases measured over full lifecycle of the package, according to Franklin Associates, an independent consulting firm.[3]

**Relates to topics listed in forecast cost, paragraph 5.**

Source reduction can similarly reduce environmental impacts in the production of paper packaging items. For example, McDonald's is systematically reviewing its use of chlorine-bleached paper products with the intention of reducing the use of this type of paper wherever feasible. Because the practice of chlorine bleaching in the manufacture of white paper is a significant source of water pollution from paper mills, considerable environmental benefits can result from a shift to brown, unbleached paper, or paper bleached with more benign chemical processes. McDonald's is currently phasing in recycled brown paper carry-out bags and oxygen-bleached coffee filters in all of its restaurants. The new wrap for the Big Mac is also made with unbleached paper.

As a matter of policy, McDonald's will continue to evaluate and improve the environmental aspects of its packaging, guided by a detailed set of Waste Reduction Packaging Specifications developed by the task force (the specifications are listed in Chapter 1 of the Final Report). These specifications are being communicated to all of McDonald's nearly 600 suppliers, and will be given weight equivalent to McDonald's existing packaging specifications for functionality, cost and availability.

**Parallels Chapter 3.**

REUSE. Behind-the-counter restaurant supply operations provide the greatest near-term opportunities to cut waste by replacing disposable containers or products with reusable bulk storage systems. For example, in 1991 McDonald's will test reusable shipping containers for condiment packets, bulk storage containers for cleaning supplies, durable shipping pallets in distribution centers, and reusable coffee filters. For in-store customer service, McDonald's will test pump-style condiment dispensers in place of individual packets, a system for fill-

---

[3]The new wraps offer significant reductions in energy use and pollutant releases relative to polystyrene foam packaging even under the highly hypothetical assumption that 50% of the polystyrene is recycled. A summary of the Franklin Associates report documenting these benefits is available from McDonald's Environmental Affairs Department, McDonald's Plaza, Oak Brook, IL 60521.

v                                                                    (continues)

**Parallels Chapter 4.**

ing customers' reusable coffee cups, and reusable lids for salads and breakfast entrees.

RECYCLING. According to waste composition studies, 34% by weight of the solid waste generated at a typical McDonald's restaurant consists of corrugated shipping containers. By the end of 1991, McDonald's restaurants will be recycling corrugated boxes nationwide.

Coated and uncoated paper food-contact items such as sandwich wraps, french fry cartons and cold drink cups make up 11% of on-premise waste. In 1991, McDonald's will assess the recyclability of this type of paper packaging in a 30-restaurant test. McDonald's will also continue to evaluate a processing system developed in southern California to recover, for later recycling or composting, the majority of materials discarded by customers in the restaurant lobby and in behind-the-counter operations.

McDonald's will also initiate a pilot program in approximately 200 restaurants to collect and recycle polyethylene used for inner wraps and shipping packages (e.g., the plastic bags that contain sandwich buns) and for jugs and other containers. These polyethylene materials account for another 3% of the weight of McDonald's on-premise waste. To facilitate polyethylene recycling, McDonald's will "homogenize" its use of this material, using only one type of plastic film for wraps and bags.

In April of 1990, McDonald's announced that it would continue to support recycling by helping to "close the loop" through purchasing products made from recycled materials. Through the McRecycle U.S.A. program, McDonald's has committed to purchasing over $100 million worth of recycled materials for restaurant construction and renovation annually, a goal that will be met in 1991.

McDonald's is currently the largest user of recycled paper in its field. Building on this experience, McDonald's has directed its nearly 600 suppliers to use corrugated boxes that contain at least 35% recycled materials, a target considerably higher than the industry average. McDonald's has directed its suppliers to maximize the percentage of recycled "post-consumer" materials (materials that have served their end use and would otherwise be discarded into the waste stream, and are in need of expanded markets, as opposed to manufacturing scrap, which is already commonly recycled). McDonald's brown, 100% recycled carry-out bags contain 50% post-consumer content, and the 100% recycled napkins that will be used in all restaurants by August of 1991 will contain at least 30% post-consumer recycled content. Although McDonald's ability to use post-consumer materials in food-contact packaging is limited by federal regulations and health concerns, McDonald's will work with its suppliers to incorporate post-consumer content in such packaging wherever possible and allowable.

vi

**Parallels Chapter 5.**

COMPOSTING. About 34% of McDonald's on-premise waste consists of organic materials such as eggshells, coffee grounds and other food scraps. Used paper items such as discarded napkins represent another essentially organic component of McDonald's waste. Two tests are now underway to evaluate whether this waste fraction can be composted into a high quality soil or humus product. These tests include a series of controlled experiments to evaluate the compostability of various paper packaging items, and a pilot program involving ten stores in the Northeast that are sending their partially separated organic materials to a composting facility in Maine. The switch to paper-based sandwich wraps produced and enhanced composting possibilities.

*By the end of 1991, McDonald's will have in place programs to recycle or to test the recycling or composing potential of more than 80% by weight of all on-premise waste. McDonald's will continue to investigate ways to reduce or recycle remaining waste materials.*

**Parallels Chapter 6.**

INSTILLING THE WASTE REDUCTION COMMITMENT. As described in Chapter 6, the last but most important chapter of our Final Report, McDonald's Waste Reduction Policy will be adopted throughout the McDonald's system, including the home and regional offices, restaurants, distribution centers and suppliers. The Waste Reduction Action Plan clearly defines the company's waste reduction activities and initiatives, identifies their status, the departments responsible for implementation and the management mechanisms that ensure integration into McDonald's standard operating procedures. McDonald's senior environmental affairs officer will regularly report to the board of directors on progress toward the company's waste reduction goals, which will also be communicated to McDonald's customers, shareholders, suppliers and employees. . . .

vii

(continues)

# TABLE OF CONTENTS

**Indentation indicates hierarchy of information. Dotted lines from text to page numbers facilitate readers' locating the appropriate page number.**

# CHAPTER 1:
# SETTING THE STAGE
# FOR WASTE REDUCTION

**Introduction headings**

## I. INTRODUCTION

This report marks the culmination of an historic project between McDonald's Corporation, the largest quick-service restaurant in the United States, and the Environmental Defense Fund (EDF), a national non-profit organization. The EDF-McDonald's Waste Reduction Task Force held its first official meeting in August of 1990, a year after discussions between us had begun. For over six months, we identified and investigated opportunities to reduce, reuse, recycle and compost the materials used and the wastes generated by McDonald's USA restaurants, distribution centers and suppliers.

Members of the McDonald's-EDF task force brought a variety of expertise to the table. McDonald's team included specialists from McDonald's Operations and Environmental Affairs Departments as well as an expert in packaging development and source reduction from Perseco, McDonald's packaging purchaser. These individuals, who represent core areas of policy and operations within the McDonald's system, were empowered to make the necessary decisions affecting the basic operations of the company to implement the findings and recommendations of the task force.

**Purpose for report**

EDF's team of solid waste experts consisted of a Ph.D. scientist in biochemistry, an economist with a background in policy analysis and a chemical engineer with business and environmental studies degrees. The task force effort involved intensive research and face-to-face discussion by EDF and McDonald's staff. This report reflects the consensus achieved by the task force.

**Foreacast**
  **Purpose of chapter.**
  **Sequence of data.**

This chapter covers several topics essential to understanding the *Corporate Environmental Policy* and the *Waste Reduction Action Plan* set out in the Executive Summary. First, the chapter describes the origins, activity and work products of the McDonald's-EDF Task Force. Second, it explains how the McDonald's system functions and the role of packaging in McDonald's operations. Finally, data gathered by the task force on the composition of McDonald's packaging and waste are presented and discussed.

1

(continues)

**Definition**

We distinguish between waste reduction and source reduction for purposes of this report. "Waste reduction" includes any action that reduces the amount or toxicity of municipal solid waste prior to disposal in a landfill or incinerator. For example, waste reduction can include making packaging more recyclable or compostable. "Source reduction" is construed more narrowly and refers to measures that reduce the weight, volume or toxicity of products and packaging prior to their use. Source reduction reduces both the environmental effects of raw materials extraction and manufacturing and the amount of material entering the waste stream. This is why source reduction is the most preferred option in the waste management hierarchy.

**Objectives**

The task force established three major goals. First, we set a goal of *maximum reduction in materials use and waste* for McDonald's. Second, we sought to develop a recommendations and options for waste reduction *consistent with McDonald's business practices and future growth.* These two goals must be realized at all levels of the McDonald's system, to ensure significant progress and to spur the development of waste reduction initiatives throughout the system—among restaurant operators, the corporate office, regional offices, suppliers and distribution centers. Members of the task force agreed that a sound environmental policy ultimately is good business.

A third goal of the task force, particularly its EDF members, was to create a *model approach to waste reduction* for others facing similar issues in the quick-service restaurant industry or other industries. We believe that this report, and supplemental information describing the task force process in more detail, serve this purpose.

The results of the task force far exceeded either EDF's or McDonald's expectations. At the onset, the goal was merely to develop recommendations and options for McDonald's to consider. Instead, we have developed a comprehensive policy and action plan, which are already in effect at McDonald's.

**Headings are numbered for easy identification.**

## II. <u>ORIGIN OF THE TASK FORCE</u>

The use of disposable packaging is standard in a variety of institutions such as schools, hospitals, grocery stores and quick-service restaurants. Disposable packaging has developed along with lifestyle changes and has transformed the dining habits of Americans by allowing us to eat restaurant-prepared meals at our convenience, whether at work, on the move or at home. However, as Americans have become aware of the nation's growing solid waste problem, quick-service restaurants have been targeted for their use of certain types of disposable packaging. Although packaging from quick-service restaurants is a small fraction of municipal solid waste, such packaging is highly visible in part because of its short useful life.

Environmental Protection Agency statistics show that the United States discards over 40 million tons of packaging/containers of all kinds each year, more than 30% of the municipal solid waste stream.[1] Thus, to effectively reduce materials consumption and solid waste generation, changes must be made in all sectors of our economy, by all manufacturers and users of all types of packaging, including the consumer.

The task force process originated in July of 1989, when EDF proposed to Michael Quinlan, president and chief executive officer of McDonald's Corporation, a meeting to discuss the company's use of materials and packaging and the solid waste it generates.

Ed Rensi, president of McDonald's USA, met with Fred Krupp, EDF's executive director, on October 10, 1989 to discuss EDF's inquiry. Afterward, a series of staff-level meetings were held. EDF saw enormous potential in McDonald's commitment to assuming a leadership position on environmental quality in general and solid waste in particular. Likewise, McDonald's recognized EDF's expertise in waste reduction and solid waste management. In December of 1989, EDF proposed that McDonald's and EDF form a joint task force to prepare a report on options for reducing, reusing, recycling and composting the materials used and the wastes generated by McDonald's restaurants and supporting operations.

McDonald's accepted the proposal. In doing so, the company committed itself to an unprecedented level of scrutiny by an outside organization. EDF brought to McDonald's a new perspective on finding solutions to the environmental issues it had been grappling with for years. McDonald's had already developed several major initiatives, in source reduction, recycling of corrugated boxes and polystyrene foam and in the purchase of recycled products. McDonald's saw the task force process as providing a framework, a systematic approach and a strong scientific basis for McDonald's solid waste decisions.

McDonald's also recognized an economic incentive to improving its materials and waste management practices. In 1990, McDonald's spent an estimated $53

---

[1] Franklin Associates, *Characterization of Municipal Solid Waste in the United States: 1990 Update* (Washington, DC: EPA Office of Solid Waste, June 1990).

(continues)

Headings use all capital letters, are in larger type than the main text, and are underlined. There is space between them and the next line.

million on waste hauling and disposal, and had projected an increase of as much as 20 percent in 1991. . . .

## III. <u>WORK OF THE TASK FORCE</u>

### A.   <u>Task Force Activities</u>

The task force held over 30 meetings (some via conference calls), with virtually all members attending every meeting. Typically, meetings lasted one or two days. We negotiated written documents and discussed guiding principles, progress on work assignments and specific proposals and packaging alternatives. The meeting locations alternated between McDonald's corporate office in Oak Brook, Illinois and EDF's offices in New York or Washington, D.C.

Procedures.

Subheadings indicate subcategories. Use capital and lower case letters. They are underlined. There is space between them and the next line. Subheadings are indicated by a capital letter. Subheadings are in a larger type than the main text.

The task force devoted considerable effort to understanding McDonald's business, including its restaurant operations, distribution system, materials use and waste composition. Numerous hours were spent in different McDonald's restaurants and each EDF member of the task force worked for a day in a McDonald's restaurant.

The task force also met with many of McDonald's suppliers and associated businesses. These field visits, which occurred over six months, included meetings with two food suppliers and five packaging suppliers, a tour of one of McDonald's largest distribution centers and visits to Plastics Again, a polystyrene recycling facility in Massachusetts and Resource Conservation Services, a composting facility in Maine. Most visits included tours, formal presentations and question-and-answer sessions with top management and experts from the companies. The willingness of McDonald's suppliers to open their doors and discuss their operations was critical to this report. . . .

### B.   <u>The Task Force Work Process</u>

The task force examined in detail only McDonald's materials use and solid waste issues in its USA operations. Where appropriate, the task force took broader environmental impacts into consideration, in part to ensure that changes resulting in less solid waste would not exacerbate other environmental impacts. Prior to selecting any options, McDonald's EDF agreed on five criteria that would be used to judge an option's merits. These criteria—consistency with the waste management hierarchy, magnitude of environmental impact, public health and safety, practicality, and economic costs and benefits—are described . . .

4

## C.  Guiding Principles in Selecting and Evaluating Options

In addition to the original options evaluation criteria presented in Figure 1.2, the task force identified several other important considerations and basic concepts. Since these cover issues that arise in all of the following chapters, we summarize them here.

**Level III heading. It is in smaller type than the other headings. It is in capital and lower case letters and it is underlined.**

### 1. Consistency with the "Preferred Packaging Guidelines"

One of the evaluation criteria refers to the Coalition of Northeast Governors' Source Reduction Council's "preferred packaging guidelines."[2] The Council's preferred packaging guidelines directly apply the waste management hierarchy to the design, composition and function of packaging materials and to the management of such materials once they become waste. We used these guidelines to provide a consistent methodology for evaluating options. In order of preference, the guidelines are:

1. Elimination of the package or packaging material where such items are not necessary to the protection, safe handling or function of the package contents.
2. Reduction of the package or packaging material if reduction does not result in a net increase in negative health or environmental or solid waste impacts.
3. Design and manufacture of packages and packaging material to be returnable, refillable and/or reusable.
4. Where source reduction itself would lead to net negative health, environmental or solid waste impacts, reduction of packaging waste by, first, designing and manufacturing packages and packaging materials to be recyclable and, second, increasing the recycled content of these materials. A package that is recyclable and is made of post-consumer recycled material is preferred.

---

[2]Coalition of Northeastern Governors' Source Reduction Council, *"Preferred Packaging Guidelines"* (Washington, DC) September, 1989. [For additional information, contact CONEG Research Policy Center, Inc., 400 North Capitol Street, NW, Suite 382, Washington, DC, 20001. (202) 624-8450]

5

(continues)

### 2. Accounting for "Lifecycle" Impacts

Strict adherence to the waste management hierarchy in the limited context of solid waste may not always provide an adequate basis for analysis. Recognizing that changes aimed mostly at reducing solid waste may have other positive and negative environmental impacts, a consideration of these broader "cradle-to-grave" impacts is necessary to ensure that a given option's benefits are not offset by negative effects (see Figure 1.4). Where sufficient data were available and the task force determined that consideration of lifecycle impacts was warranted, such factors were considered in prioritizing and evaluating options.

### 3. Accounting for Tradeoffs Among Solid Waste Reduction Options

Even within the waste management hierarchy, there is considerable interplay among its tiers. For example, a source reduction measure applied to a package may also have an impact—positive or negative—on the recyclability of the package. As explained in Chapter 2, McDonald's widely publicized decision to replace its polystyrene foam clamshells with thin paper-based wraps was based on the significant source reduction benefits—in reducing both the volume of solid waste and environmental impacts from production.

Even when packages made of the same basic material are compared, tradeoffs among the tiers can be encountered. It cannot simply be assumed, for example, that a source reduction measure is always to be preferred over a recycling measure. Chapter 2 discusses tradeoffs between source reduction and recycled content that McDonald's considered in its selection of new carry-out bags. It is critical that such interplay among the tiers of the hierarchy be carefully weighed; only by considering the specifics of individual packages can a clear conclusion be reached.

**Discusses overall operating procedures.**

## IV. HOW THE MCDONALD'S SYSTEM OPERATES[3]

The operating procedures and materials used in a McDonald's restaurant present a unique set of waste reduction opportunities and challenges. The uniform standards of the McDonald's system extend these opportunities to over 8,500 United

---

[3] This section was written largely by EDF based on its observations of the McDonald's system developed over the course of the task force's work. For more information about McDonald's and the quick-service restaurant industry, see:

Love, John F. *McDonald's: Behind the Arches* (New York, NY: Bantam Books, November 1986).

Anderson, Robert. *Grinding It Out* (Chicago: Contemporary Books, 1977).

Emerson, Robert L. *The New Economics of Fast Food* (New York, NY: Van Nostrand Reinhold, 1990).

6

States restaurants and to the McDonald's distribution and supply network. Any serious discussion of waste reduction and management at McDonald's must begin with a full understanding of how this extensive system works.

## A. <u>History and Institutions</u>

The first McDonald's restaurant was founded in San Bernadino, California, in 1948, when Dick and Mac McDonald renovated their drive-in restaurant and took their business in a new direction. The brothers cut their menu from 25 to nine items, revolving around a 15-cent hamburger, 10-cent fries and 15-cent shakes, low prices even then. Inside the small, octagonal glass and stainless steel restaurant, food was prepared in high-speed, assembly-line fashion. The McDonald brothers fashioned their windows to serve walk-up customers and replaced reusable dishware with paper bags and wraps. A limited menu and a high sales volume thus characterized the successful new restaurant, and disposable packaging became central to its operations and speed of service.

The McDonald's system was refined and transformed into a national institution through the efforts of Ray Kroc, who first visited the brothers in the summer of 1954. Within a year he had gained the national franchise rights to the McDonald's System, Inc., and had opened his own showcase restaurant in Des Plaines, Illinois.

In the 1950s and early 1960s, Kroc and his management team established the fundamental operating philosophy of the McDonald's system: Quality, Service, Cleanliness and Value ("Q.S.C. & V."). These goals were upheld by operating standards and specifications established by the corporation for the individual restaurants and enforced through a system of field evaluations. During these years, McDonald's also established the basic business relationships in its system, a system of mutual risk and reward that leads independent suppliers, restaurant operators and the McDonald's corporation to work together for the good of the entire system. From this foundation, McDonald's has built an $18 billion annual business encompassing over 11,000 individual restaurants worldwide, all based on high volumes and "penny profit" margins on individual menu items.

## B. <u>Management within the McDonald's System</u>

### 1. <u>The Corporation</u>

McDonald's CEO Michael Quinlan defines the company's management style as "tight-loose." The corporation sets stringent quality standards and procedural specifications that guide individual restaurants, suppliers and distribution centers. At the same time, management believes in giving field management and the international partners the flexibility to make decisions connected to local conditions.

(continues)

The company has 37 regional offices in the United States, as well as international regions that reach into 53 other countries. The corporate office gives these regional and international entities considerable autonomy. Planning, budgeting, local expansion and real estate decisions and almost all key functions of the company flow from the bottom to the top. Each regional office supports approximately 225 restaurants; a staff of 60 people assist franchisees and managers of company-owned restaurants in the operations and development of each restaurant.

McDonald's standards and procedures are instilled in managers and crew members through constant education. Basic and advanced operations courses have been taught to over 40,000 employees at "Hamburger University" near McDonald's home office in Oak Brook, Illinois.

Within the corporation, new menu items, equipment and operating procedures are tested extensively by McDonald's Operations Development, Product Development, Purchasing, Quality Assurance and Marketing departments. Bringing a new food item to market typically takes several years, beginning with tests by suppliers and McDonald's own labs and test kitchens, then small in-store tests and finally one or more larger regional tests before nationwide introduction.

### 2. Restaurant Owner-Operators

More than 75% of McDonald's restaurants in the United States are owned and operated by local franchisees. Twenty-year franchises are approved individually. The corporation does not earn any significant revenues from the sale of a franchise. Instead, franchisees pay a percentage of their monthly sales to the corporation as rent and service fees. Service fees cover about 75% of what the corporation actually spends on market research, R&D and franchise field services. The corporation, therefore, does not make money unless the franchises do. This relationship instills mutual risk and reward and a shared interest in the success of each local franchise.

Through its McOpCo (McDonald's Operating Company) subsidiary, the McDonald's Corporation also runs its own restaurants, accounting for about 25% of the McDonald's restaurants in the United States.

Reflecting their knowledge of local conditions and their vested interest in the success of the overall McDonald's system, individual owner-operators have always been a major source of innovations. For example, the Filet o' Fish Sandwich, the Big Mac and the Egg McMuffin are all notable inventions in a long list of menu items devised by local restaurant owner-operators. Through regional and national advertising cooperatives, to which each restaurant contributes a percentage of its

8

monthly sales, McDonald's owner-operators also determine the direction of much of McDonald's advertising.

### 3. Suppliers

Suppliers of food items, packaging and restaurant equipment to McDonald's restaurants are all independent companies. The McDonald's Corporation does not own any of its suppliers and is not a manufacturer. Although written documents are used by the corporation to define certain aspects of the relationship with a supplier, basic supply agreements have always been established on a handshake. Suppliers operate with open books and are paid essentially on a cost-plus basis.

This system allows the McDonald's Corporation to concentrate on the core aspects of the business—restaurant operations—and to select the best available suppliers. McDonald's works closely with its suppliers to develop new products and improve the quality of existing ones and expects that innovations developed by one supplier will be shared with other McDonald's suppliers who provide the same item. While McDonald's can in principle switch suppliers very quickly, in practice it has for the most part built mutually beneficial long-term relationships with its suppliers. In doing so, McDonald's has altered agricultural and food-processing technologies on a broad scale, one example being the growing and preparation of potatoes for french fries.

McDonald's food items are supplied through a combination of national suppliers with regional production plants (with purchasing done through McDonald's home office Purchasing department), and local suppliers for more perishable food products such as dairy items and buns (which are purchased through McDonald's regional purchasing offices).

### 4. Packaging Supply and System Distribution

The purchase of all paper and plastic food-service packaging[4] for McDonald's U.S. restaurants is coordinated through Perseco. Perseco is an independent and

---

[4] In this report, "food-service packaging" includes direct food packaging as well as food serviceware such as utensils, cups, napkins and carry-out bags.

(continues)

privately owned company that serves exclusively as McDonald's shopper for packaging. Perseco is a material-neutral buyer that selects and purchases packaging from more than 100 suppliers, including the most visible portion of McDonald's packaging: Customer-related packaging used to serve meals. Some packaging purchasing activities are also coordinated through the Home Office . . . for customers leaving the restaurant is packaged in paper bags, while in-store customers are served on reusable trays.

McDonald's ideal service standard is to take a customer's order within 2 minutes after he or she enters the restaurant and deliver the completed order within another minute. To serve customers freshly prepared food quickly and to efficiently assemble menu items in batches, the restaurant crew person calling production (the type and number of items to be prepared) attempts to anticipate the flow of customers just moments before their actual arrival. In most restaurants, this process is assisted by projections of sales volumes provided by computers that receive and compile data on past sales directly from the cash registers.

McDonald's standard for holding hot items in the transfer bin is 10 minutes, or less. Restaurants seek to minimize excess products because of the cost. The kitchen staff must therefore anticipate and respond to demand within tight time constraints while still avoiding preparing too little or too much food. The restaurants face a similar challenge in seeking to balance offering a widening range of menu items to meet customer demands and being able to handle and prepare food within McDonald's standards for speed and efficiency.

## D. Customers, Service and Expectations

More than 18 million people visit McDonald's restaurants daily in the United States, making it hard to generalize about the "average" McDonald's customer. Many customers make their decision to eat at McDonald's within hours or even minutes of arriving at the restaurant. They usually face a wide variety of dining choices in a given "trading area," which may be a few city blocks in an area like Manhattan, a highly transient area near a freeway interchange, or a suburban McDonald's location.

A typical McDonald's restaurant might serve 2,000 people per day and in a restaurant with a drive-through window, 60–70% of customers may take their food outside the restaurant. Even at peak volume periods such as lunch, McDonald's strives to provide the same speed of service as during the rest of the day. McDonald's customers *are* concerned with convenience, defined as both speed of service and the ability to select where they eat their meal: in the restaurant, in the car, at work or at home.

## V. THE ROLE OF PACKAGING IN MCDONALD'S BUSINESS

McDonald's purchases packaging to fit the food-service needs of its customers and to ensure that its menu items meet strict standards for Quality, Service, Cleanliness and Value. The company is neutral as to the types of packaging used as long as the package meets its "Q.S.C. & V." criteria, but recognizes that different packaging choices can have different environmental impacts and that McDonald's has a responsibility to minimize these impacts.

Historically, McDonald's has selected its packaging guided by a set of specific packaging criteria which take into account three major factors:

- Availability
- Functionality
- Cost

Some aspects of functionality related to sandwich packaging are as follows:

Insulation to keep food warm for a specified time in the holding bin (the bin has auxiliary heating, but different food items have different insulation needs, thus the package needs to meet individual product insulation requirements).

Breathability to keep food tasty and moist, while avoiding soggy food or food that becomes dry.

Handling ability to maintain the integrity of the sandwich, facilitate handling by employees and customers and ensure sanitation.

Appearance, including the ease of printing and appearance of graphics, for quick crew recognition and handling and merchandising to the customer.

In the area of functionality, extensive research and testing is required to determine the impact of the packaging on the specific presentation of food products and on the system in general. Testing of packaging begins at the supplier's research center, then at Perseco, then at McDonald's home office test kitchen. Finally, the most favorable options are tested in actual restaurants. As with other tests of McDonald's menu items and equipment, several McDonald's departments are involved in the overall evaluation of packaging. Several samples of food items packaged in each packaging option are analyzed and extensive data recorded, including data on the following components:

11                                                          (continues)

**Uses bullets to list items.**

- Internal food temperatures
- Temperatures taken at different time intervals
- Blind taste-testing
- Moisture analysis
- Grease resistance
- Product appearance and functionality in the package
- Locking mechanisms and folding characteristics of the package

At the same time, other factors within the overall areas of cost, functionality and availability are analyzed, including production and printing capabilities, availability of raw materials and impacts in the distribution system.

As a result of the work of the task force, a fourth primary consideration—waste reduction—will be added on a par with the three traditional criteria discussed above. This new criterion will be given substance and definition through the application of a set of waste reduction packaging specifications, shown in Figure 1.5. Use of these specifications is intended to bring about continuous packaging improvements in each of the following areas:

- reduction in materials use
- reduction in production impacts
- use of reusable materials
- use of recyclable materials
- use of recycled content
- use of compostable materials

## VI. <u>UNDERSTANDING MCDONALD'S PACKAGING AND WASTE</u>

The first step in defining opportunities for waste reduction in the McDonald's system is to clearly characterize McDonald's packaging and waste. In order to do this, the task force examined data from five major sources, including studies that were initiated at the task force's request. Taken as a whole, the data characterize the entire McDonald's system: supply, distribution, use and disposal. The data are presented on a per-restaurant basis both to recognize that waste disposal is handled through individual restaurants and to reflect the role of McDonald's restaurants in local communities. The data are presented as percentages, with total amounts shown on each chart, to provide an easily accessible picture of what McDonald's waste stream looks like. The five major sources of data are summarized below.

**FIGURE 1.5**

**WASTE REDUCTION PACKAGING SPECIFICATIONS**

**INTRODUCTION**

The following specifications are intended to provide guidance and direction to McDonald's suppliers in reviewing existing primary and secondary packaging and foodservice products and in designing and choosing appropriate materials for new packaging and products.

The order of the specifications is not random; rather, it reflects the following widely accepted hierarchy of preferred packaging options:

1. Elimination of the package or packaging material where such items are not necessary to the protection, safe handling or function of the package contents.

2. Reduction of the package or packaging material if reduction does not result in a net increase in negative health or environmental or solid waste impacts.

3. Design and manufacture of packages and packaging material to be returnable, refillable and/or reusable.

4. Where source reduction itself would lead to net negative health, environmental or solid-waste impacts, reduction of packaging waste by first designing and manufacturing packages and packaging materials to be recyclable and, second, by increasing the recycled content of these materials. A package that is recyclable and is made of post-consumer recycled material is preferred.

5. Where recycling is not a viable option, design and manufacture of packages or packaging materials to be compostable.

While this hierarchy serves as a primary directive for choosing among various options for improving the environmental characteristics of packaging and foodservice items, tradeoffs among its tiers can frequently arise. In such cases, the over-arching objectives of reducing virgin materials use and minimizing solid waste should be invoked. For example, consider the following example of deciding whether to increase the recycled content of a package or to decrease its basis weight (where both are not possible):

A bag made with 50% recycled content weighs 15 grams, half of which (7.5 grams) is recycled material. Thus, this bag uses 7.5 grams of virgin paper. Another bag made with virgin paper uses 17% less total material, so it weighs 12.5 grams — 2.5 grams less than the other, but contains 5 grams more virgin material.

In this case, the bag with recycled content is preferred because it uses only 60% (7.5 grams/12.5 grams) as much virgin paper as the lighter weight bag.

Finally, while the preferred packaging hierarchy should generally be adhered to in prioritizing efforts and resolving conflicts or tradeoffs, it is critical for suppliers to move forward on all fronts, seeking to optimize packaging and products with respect to all of the improvements discussed below.

13

(continues)

**Uses underlining to
catch readers' attention.**

## SPECIFICATIONS

**REDUCTION IN MATERIALS USE:**

1. <u>Strive to eliminate all unnecessary secondary packaging.</u>

Through changes in the mode of packaging, bulk packaging, or other means, eliminate individual wrapping of items, as well as inner bags, packs, and dividers wherever possible.

2. <u>Minimize use of materials in all packages and foodservice items.</u>

Examine all products and packaging, seeking ways to reduce the amount (size, weight, or both) or material used. This examination should include both primary and secondary packaging, and should consider reductions in the basis weight or gauge of materials, reduction in the dimensions of shipping boxes through more space-efficient packing of contents, etc.

**REDUCTION IN PRODUCTION IMPACTS:**
Choose packaging materials that are produced utilizing processes that use as little energy and other natural resources and release as little pollution as possible.

1. <u>Substitute unbleached or non-chlorine bleached paper for bleached paper wherever feasible.</u>

In designing or selecting paper products and packaging, utilize the following "bleaching hierarchy" (best to worst):

a.  Unbleached paper.

b.  Paper bleached and delignified with non-chlorine reagents (e.g., oxygen, hydrogen peroxide, ozone).

c.  Paper bleached with chlorine reagents other than elemental chlorine (e.g., chlorine dioxide, sodium hypochlorite).

d.  Paper bleached with elemental chlorine.

Review all existing paper products and primary and secondary paper packaging to determine the feasibility of replacing current bleached paper components with those higher up on the hierarchy. Utilize

the highest option possible in designing any new paper packaging.

2. <u>Select plastic resins made from relatively safer chemical feedstocks.</u>

Examine all current plastic items or components in products and packaging to determine whether less environmentally damaging resins could be substituted. In designing new products or packaging, select the most benign resins available.

3. <u>Use materials that are produced utilizing processes that minimize energy use and water consumption.</u>

4. <u>Ensure that all inks and pigments are of as low environmental concern as possible, and minimize use of such additives.</u>

Consider replacing petroleum solvent-based inks with soy-based inks, which do not contain smog-producing volatile organic chemicals (VOCs). Reduce the use of inks of all types through sparser printing.

**REUSABLE MATERIALS:**

1. <u>For each packaging item, including shipping packaging, consider the option of substituting a reusable container for a disposable one.</u>

Seek to substitute reusable shipping containers for corrugated boxes wherever feasible. As a supplement to existing disposable products and packaging, develop reusable counterparts that provide the equivalent function(s). This should include both behind-the-counter and over-the-counter shipping, storage, and foodservice products and packaging.

2. <u>Substitute bulk packaging for single-portion packaging wherever possible.</u>

14

**RECYCLABLE MATERIALS:**

1. For all existent materials and all new materials being introduced, seek to consolidate the range of materials used in order to maximize opportunities for recycling.

Utilize materials wherever possible for which viable recycling markets already exist; for example, substitute clear, low-density polyethylene for other films (e.g., bags and inner wraps); substitute clear, high density polyethylene for other container materials (e.g., bottles and jugs). Avoid multi-component packaging wherever possible.

2. Seek to design all disposables using materials which can be recycled into high-grade products.

Utilize materials that can be collected and processed in pure form to produce the same or similar products (i.e., use materials that allow for as close to closed-loop recycling as possible).

3. Minimize the use of inks, pigments and other additives.

Where similar packaging items need to be distinguished from one another, seek to do so through sparsely printed graphics, rather than coloring the entire item.

4. For disposables used for take-out/drive-thru packaging, increase use of materials that are included in municipal collection programs (e.g., curbside collection programs).

**RECYCLED CONTENT:**

1. Seek to increase recycled content in all primary and secondary packaging and foodservice items.

2. Seek to increase the use of post-consumer materials as part of the recycled content in all primary and secondary packaging and foodservice items.

Maximize post-consumer material in packaging that does not come into direct contact with food. Design all multi-layered packaging so as to facilitate incorporation of post-consumer content into the middle and/or outer layer(s). Aggressively seek reliable and consistent sources of post-consumer material that could be utilized in direct food-contact applications, and seek approval for their use.

**COMPOSTABLE MATERIALS:**

1. Where viable recycling options do not exist, design paper packaging to be compatible with organic waste composting.

Avoid multi-layer paper/plastic packaging wherever possible. Where multi-layer packaging is required, substitute compostable (cellulose-based) materials for non-compostable materials. Where non-compostable layers are required, design them so as to be easily separable from compostable layers, either before (preferred) or after composting.

2. Use only truly compostable materials that decompose completely (i.e., are assimilated and metabolized by micro-organisms) and safely (i.e., neither produce nor release toxic or other harmful substances).

## A.   Summary of All Packaging Purchased by Perseco

In its "Primary Packaging Classification Report," Perseco provides a summary of actual packaging purchased by U.S. McDonald's restaurants for a particular year. The 1989 data here have been adjusted to reflect the recent switch from sandwich polystyrene foam to paper-based wraps. It provides good information on customer-related packaging (including packaging used to serve food as well as utensils, napkins, carry-out bags and other ancillary items), but it does not provide

15                                                    (continues)

data on the amount of secondary packaging. The Perseco report also does not take into account food purchases and associated packaging. Lastly, the report is a purchasing report and provides no information on how such materials are managed after use, or how much packaging material leaves the restaurant with take-out meals. Data from this report are shown in Figure 1.6.

FIGURE 1.6:
SUMMARY OF PERSECO'S PRIMARY PACKAGING
CLASSIFICATION REPORT

Perseco Packaging Quantified
(Based on 1989)**

**Tables**

| PAPER-BASED PACKAGING: | EXAMPLES | Percentage |
|---|---|---|
| Tissue Products | Napkins, Towels | 17% |
| Paperboard | Fry Cartons | 31% |
| Molded Pulp | Carry-Out Trays | 5% |
| Paper | Bags, Wraps | 27% |
| | Paper Subtotal | 81% |

| PLASTIC-BASED PACKAGING: | EXAMPLES | Percentage |
|---|---|---|
| Foamed Polystyrene | Hot Cup, Breakfast Platters | 3% |
| Non-foamed polystyrene | Cutlery, lids, salad containers | 12% |
| | Polystyrene Subtotal | 15% |
| Polyethylene | Can Liners | 3% |
| Polypropylene | Straws | 1% |
| | Plastic Subtotal | 19% |
| | GRAND TOTAL | 100% |

149 lbs/day/restaurant

0.07 lbs per customer served

** Adjusted to reflect conversion from polystyrene foam clamshells to paper-based wraps.

16

## B.    On-Premise Waste Characterization Study (WCS)

To assist the work of the task force, Perseco and McDonald's initiated a restaurant waste characterization study (WCS), based on hand-sorting and classification of refuse from two restaurants deemed generally representative of the McDonald's system. The results from this study are summarized in Figure 1.7. This report delineates what is discarded on the premises of a typical McDonald's restaurant, whether in the restaurant itself or in the outdoor receptacles (including materials brought in by customers). The study does not address the materials used in the take-out and drive-thru portions of McDonald's business.

## C.    Packaging Materials Audit

For this study, a Perseco team physically separated and weighed all components of the packaging—primary, secondary and tertiary—used in a McDonald's restaurant. Such detailed information had not been available before. The audit provides data on secondary packaging utilized for all the packaging items used in the restaurant, including packaging associated with both Perseco items and McDonald's food products. It provides insight into the range of different packaging materials utilized in secondary packaging, which is almost all used behind the counter and is invisible to the customer.

A sample page from the full report is shown in Figure 1.8. To illustrate the usefulness of such a database, examine the data for a case of 12-ounce cold drink cups. Its corrugated shipping container weighs three pounds and contains 20 inner packs made from LDPE, each weighing seven grams, or 140 grams of LDPE all together. This type of information is an effective tool for identifying opportunities for source reduction and recycling and for measuring their effect on the amounts of solid waste generated, both at the micro level (individual packages) and the macro level (restaurant- or system-wide).

## D.    Distribution Center Audit

At the time the task force began its work, no information had been collected on the waste generated at the distribution centers that serve McDonald's, so a waste survey was initiated to gather the information. (The survey did not include distribution system wastes generated by McDonald's suppliers.) The results are summarized in Figure 1.9.

(continues)

FIGURE 1.7:
SUMMARY OF MCDONALD'S ON-PREMISE WASTE
CHARACTERIZATION STUDY[1]

| OVER-THE-COUNTER (OTC) | | BEHIND-THE-COUNTER (BTC) | |
|---|---|---|---|
| | % OF GRAND TOTAL | | % OF GRAND TOTAL |
| Uncoated Paper | 4% | Corrugated | 34% |
| Coated Paper | 7% | Putrescibles[2] | 34% |
| Polystyrene | 4% | LDPE | 2% |
| Non-McDonald's Waste | 4% | HDPE | 1% |
| Miscellaneous | 2% | Liquids[2] | 2% |
| | | Miscellaneous | 6% |
| | TOTALS | 21% | 79% |

GRAND TOTAL
238 lbs./day/restaurant
0.12 lbs per customer served

DEFINITIONS AND EXAMPLES    (examples are not an all-inclusive list):

| | |
|---|---|
| OVER-THE-COUNTER. | Waste in the customer sit-down area and from outside waste receptacles. |
| BEHIND-THE-COUNTER: | Waste behind the register counter, including kitchen and storage rooms. |
| POLYSTYRENE: | Hot cups and lids, cutlery, salad containers. |
| MISCELLANEOUS OTC: | Condiment packaging. |
| CORRUGATED: | Shipping boxes. |
| PUTRESCIBLES: | Food waste from customers, egg shells, coffee grounds, other food scraps. |
| LDPE: | Low-density polyethylene film wraps and plastic sleeves used as inner packaging in shipping containers. |
| HDPE: | High-density polyethylene plastic used mostly for jugs, e.g. syrup jugs. |
| LIQUIDS: | Excess, non-absorbed liquids measured during the waste audit. |
| MISC. BTC: | Durables, equipment, office paper, secondary packaging other than corrugated boxes. |

[1] Based on a two-restaurant, one-week-long waste audit performed 11/12 - 11/18/90 in Denver, CO and Sycamore, IL. Figures have been adjusted to reflect the conversion from sandwich foam to paper wraps.

[2] Most putrescibles and liquids are from behind the counter; the waste characterization study was not able to separately quantify customer organic waste, so all such waste is listed in the behind-the-counter category.

18

FIGURE 1.8:
EXCERPT FROM PERSECO'S PACKAGING MATERIALS AUDIT REPORT

PACKAGING MATERIAL AUDIT

| Item # | Description | Ctgry | Qty | Shipping Material | Wgt | Inner qty | Inner Wgt | Inner Material | Inner Packaging Dvdrs Qty | Dvdrs Wgt | Dvdrs Material | Portion Control Packaging Pkg Comp | Pkg Wgt | Pkg Material | Comments |
|---|---|---|---|---|---|---|---|---|---|---|---|---|---|---|---|
| 12919 | STANDARD (B) BAG | (P) | 1.420 | KRAFT | .248 | ( ) | | | ( ) | | | ( ) | | | |
| 13111 | STANDARD (C) BAG | (P) | .616 | KRAFT | .270 | ( ) | | | ( ) | | | ( ) | | | |
| 14102 | STANDARD (D) BAG | (P) | .622 | KRAFT | .453 | ( ) | | | ( ) | | | ( ) | | | |
| 14302 | STANDARD (E) BAG | (P) | .039 | KRAFT | .496 | ( ) | | | ( ) | | | ( ) | | | |
| 15500 | COLD CUP - (12 OZ) | (P) | 1.072 | CORRUGATED | 3.000 | (20) | 7.00 | LDPE | ( ) | | | ( ) | | | GRAPHICS ON INNER PACK |
| 15600 | COLD CUP - (16 OZ) | (P) | 2.058 | CORRUGATED | 3.387 | (24) | 6.50 | LDPE | ( ) | | | ( ) | | | GRAPHICS ON INNER PACK |
| 15700 | COLD CUP - (22 OZ) | (P) | 3.001 | CORRUGATED | 3.581 | (**) | 8.50 | LDPE | ( ) | | | ( ) | | | GRAPHICS ON INNER PACK |
| 16306 | FRENCH FRY BOX - (MED) | (P) | .496 | CORRUGATED | 2.039 | (4) | 62.50 | CHIPBOARD | ( ) | | | ( ) | | | |
| 16601 | APPLE PIE BOX | (P) | .340 | CORRUGATED | .500 | (1) | | | ( ) | | | ( ) | | | |
| 17000 | COFFEE STIRRER | (P) | .201 | CORRUGATED | .956 | (1) | 33.00 | LDPE | (1) | 65.50 | CORRUGATED | ( ) | | | WAX COATING INSIDE CORRUGATED |
| 19001 | SAUSAGE BISCUIT WRAP | (P) | .113 | CORRUGATED | .548 | (3) | 4.50 | LDPE | ( ) | | | ( ) | | | |
| 19300 | NAPKINS | (P) | 4.238 | CORRUGATED | 2.222 | (12) | 25.00 | | ( ) | | | ( ) | | | |
| 19900 | SAUSAGE/EGG BISCUIT WRAP | (P) | .091 | CORRUGATED | .548 | (3) | 4.50 | LDPE | ( ) | | | ( ) | | | |
| 22000 | PORTA-PACK INSULATED BOX | (P) | .002 | CORRUGATED | 3.011 | (4) | 68.50 | PS | ( ) | | | ( ) | | | WAX COATING OUTSIDE WT-(17.5) |
| 22301 | WRAPPED STRAWS | (P) | 1.189 | CORRUGATED | 2.320 | (16) | 9.50 | LDPE | ( ) | | | ( ) | | | |
| 23203 | HOT CUP - REGULAR | (P) | 1.053 | CORRUGATED | 2.240 | (28) | 4.50 | LDPE | ( ) | | | ( ) | | | GRAPHICS ON INNER PACK |
| 23305 | HOT CUP LID - (REG) | (P) | .349 | CORRUGATED | 2.042 | (28) | 5.50 | LDPE | ( ) | | | ( ) | | | |
| 25400 | WRAPPED KNIFE | (P) | .644 | CORRUGATED | 1.165 | ( ) | | | ( ) | | | (X) | .50 | LDPE | |
| 25702 | HOTCAKES W/SAUSAGE LID | (P) | .327 | CORRUGATED | 1.974 | (4) | 15.50 | LDPE | ( ) | | | ( ) | | | |
| 25802 | BREAKFAST BASE | (P) | 1.271 | CORRUGATED | 1.965 | (4) | 19.50 | LDPE | ( ) | | | ( ) | | | |
| 25904 | BIG BREAKFAST LID | (P) | .518 | CORRUGATED | 1.981 | (4) | 14.50 | LDPE | ( ) | | | ( ) | | | |
| 26103 | SUNDAE CUP | (P) | .461 | CORRUGATED | 3.694 | (24) | 6.00 | LDPE | ( ) | | | ( ) | | | |
| 26305 | SCRAMBLED EGG LID | (P) | .480 | CORRUGATED | .972 | (4) | 15.50 | LDPE | ( ) | | | ( ) | | | YELLOW TINT |
| 26403 | HOTCAKES LID | (P) | .334 | CORRUGATED | 1.985 | (4) | 16.00 | LDPE | ( ) | | | ( ) | | | |
| 26802 | MCNUGGET BOX - (6 PC) | (P) | .542 | CORRUGATED | .847 | (8) | 14.50 | LDPE | ( ) | | | ( ) | | | |
| 26900 | WRAPPED SPOON | (P) | .785 | CORRUGATED | 1.164 | ( ) | | | ( ) | | | (X) | .50 | LDPE | |
| 28402 | HAMBURGER WRAP | (P) | .294 | CORRUGATED | .659 | (5) | 4.00 | LDPE | ( ) | | | ( ) | | | |

(continues)

FIGURE 1.9:
SUMMARY OF RESULTS FROM McDONALD'S
DISTRUBUTION CENTER WASTE AUDIT

| | ITEMS/DESCRIPTION | % OF TOTAL WASTE | COMMENTS |
|---|---|---|---|
| 1. | Corrugated Scrap:<br>• Waxed<br>• Unwaxed<br>• Slip sheet | 37% | 26 of 34 locations recycle. Represents all corrugated scrap usage. |
| 2. | Pallets<br>• Wood<br>• Plastic | 35% | 20 of 34 locations operate under return/recycle/reuse program. |
| 3. | Stretchwrap Film | 10% | Locations compact and/or recycle. |
| 4. | Food<br>• Frozen<br>• Perishables<br>• Damaged | 10% | 18 of 34 locations either return or ship unused product to food bank locations. |
| 5. | Dry Products | 5% | 6 of 34 locations return unusable product or ship to food bank locations. |
| 6. | Office Waste Paper | 3% | 15 of 34 locations recycle. |
| | GRAND TOTAL (Total Waste Disposed) | 900 lbs./day/distribution center | |

## E.   <u>Supplier Materials Use and Waste</u>

At this time, a reliable and feasible means to monitor and track supplier waste has yet to be identified. The major difficulty is that many of McDonald's 600 suppliers are national companies that do business with others besides McDonald's. The task force did visit several suppliers and among the aspects of their operations examined were waste generation and management.

**Conclusion section**

## CONCLUSIONS

At the outset of this project, the task force worked to ensure that we had a firm basis from which to conduct research and draft recommendations. This foundation had several elements:

\*   The agreement establishing the task force maintained the independence of EDF and McDonald's while directing the task force to conduct detailed analyses of a wide range of reduction, reuse, recycling and composting options.

\*   The task force extensively discussed and agreed upon a set of criteria and guiding environmental principles for selecting and evaluating waste reduction options. These placed a strong emphasis on source reduction, which not only reduces solid waste, but also environmental impacts from raw materials extraction and manufacturing processes.

\*   The task force spent considerable effort understanding how the McDonald's system works and which aspects of McDonald's operations were most important in supporting waste reduction changes. The EDF members of the task force worked in restaurants and read a variety of reports on McDonald's past and present operations. The task force also spent a day in a restaurant discussing packaging and food delivery systems with McDonald's National Director of Operations Development (a task force member), and toured distribution and supply facilities.

\*   Detailed information on the composition of McDonald's packaging and solid waste was provided by four separate analyses, including the Perseco "Primary Packaging Classification Report," on-premise waste characterization studies conducted at two restaurants, a packaging materials audit conducted in a restaurant storeroom, a survey of McDonald's distribution centers, and a review of McDonald's suppliers' waste management practices. One significant finding was that about 80% of McDonald's on-premise waste (by weight) is generated behind the counter and therefore presents the greatest opportunities for waste reduction.

The groundwork laid in the steps above was critical in allowing us to ultimately develop the Waste Reduction Policy and Action Plan presented in the Executive Summary.

(continues)

# CHAPTER 2:
# SOURCE REDUCTION

## I.   <u>INTRODUCTION</u>

**Introduction**

       **Evidence**

Source reduction occupies the highest tier in the waste management hierarchy because of its benefits throughout the lifecycle of a product or package, from production through disposal. Source reduction means less material that must be manufactured, shipped, stored, and eventually discarded and managed as waste, whether through disposal or recycling. For this reason, the task force placed highest priority on source reduction in its evaluation of McDonald's materials use and waste management practices.

       **Claim**

       **Evidence**

The recent decision by McDonald's to replace polystyrene foam clamshells with a thin paper-based wrap illustrates the general preferability of source reduction over other waste management methods, including recycling. The substantial reduction in both volume of primary packaging—ranging from 70 to 90% for various menu items—and volume and weight of shipping (secondary) packaging—approaching 90%—achieved by the new wraps relative to foam is a significant benefit. Moreover, this benefit accrues whether the packaging is used by in-store or take-out/drive-thru customers and however the packaging is managed after discard.

       **Claim**

In contrast, while there is a greater technical ability to recycle foam than to recycle the new wraps, McDonald's experienced numerous problems that translated into actual recycling of much less than half of its total use of foam. Even from strictly a solid waste perspective, the net result is a far greater decrease through source reduction than could have been achieved through recycling. . . .

McDonald's Waste Reduction Action Plan commits the company to implementing source reduction at several different levels throughout the system. First, McDonald's has committed to incorporating explicit waste reduction criteria into its development of packaging. These criteria will follow the principle of the waste management hierarchy, which ranks source reduction as the most preferable alternative. Second, McDonald's has committed to establishing and maintaining a program to reduce the amount of materials used in its packaging.

**Forecast**

This chapter discusses McDonald's past and present source reduction activities, some complications and tradeoffs that arise in making source reduction decisions,

22

and new contributions of the task force. Among the most important of these contributions are a long-term commitment to eliminating the use of chlorine-bleached paper wherever feasible, and an ongoing pursuit of improvements in all packaging materials, including the new sandwich wraps. . . .

FIGURE 2.1:

## RELATIVE ENERGY AND ENVIRONMENTAL RELEASES OF FOUR SANDWICH PACKAGING OPTIONS

**Bar graph**

**Provides legend.**

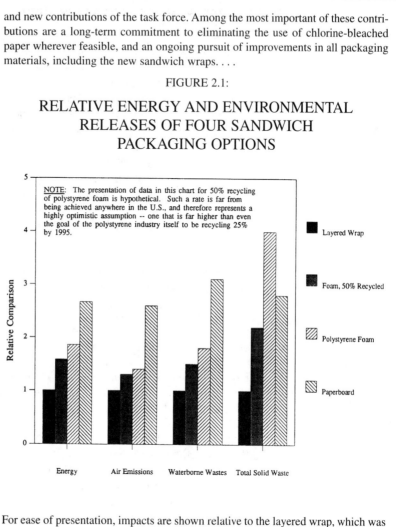

NOTE: The presentation of data in this chart for 50% recycling of polystyrene foam is hypothetical. Such a rate is far from being achieved anywhere in the U.S., and therefore represents a highly optimistic assumption -- one that is far higher than even the goal of the polystyrene industry itself to be recycling 25% by 1995.

Legend:
- Layered Wrap
- Foam, 50% Recycled
- Polystyrene Foam
- Paperboard

*Relative Comparison* (y-axis)

Energy    Air Emissions    Waterborne Wastes    Total Solid Waste

**Labels axes.**

For ease of presentation, impacts are shown relative to the layered wrap, which was assigned a value of 1.0.

Based on data from Franklin Associates' 1991 Report evaluating the layered wrap, and Franklin Associates' 1990 Report evaluating polystyrene foam and paperboard sandwich containers.

23

Source: Final Report by the Waste Reduction Task Force to the McDonald's Corporation–Environmental Defense Fund, April 1991. Reprinted with permission from 1991 Environmental Defense Fund and McDonald's Corporation.

**Suggestions for Discussion and Writing**

1. The executive summary is seven pages. Why do you think the writers felt it necessary to make it that long? Compare the information in the executive summary with the information in the excerpts from the first two chapters in the main report. How did the writers derive the information for the summary? Now compare the categories of information in the executive summary with the categories of information listed in the table of contents (TOC). Does the summary include all of the information in the TOC? Does it follow the sequence in which the information is listed in the TOC?

2. For Chapter 1 what claims do the writers make to persuade their primary audience, the executives at McDonald's, to accept their recommendations? Do you think the readers will accept these claims? Why or why not?

3. How does the introduction to Chapter 1 facilitate readers' reading processes and help readers comprehend the information in the chapter?

4. What types of visual text do the writers use? How do Figures 1.5, 1.7, and 1.9 in the report facilitate readers' fluency and comprehension?

5. Examine the language and syntax. What terms or sentences are inappropriate for the readers? Highlight them and explain why they are inappropriate in the margin.

6. What is the tone of the introduction to the executive summary? Serious? Humorous? Factual? Opinionated? Conciliatory? Blaming? Impersonal? Personal? Praising? Self-congratulatory? Whining? Demanding? Rationalizing? Courteous? Peremptory? Cavalier? Tactless? What adjectives do you notice? Underline them. Adjectives are seldom used in a scientific report. Why do you think the writers use them?

7. One of the major purposes of the report, from the point of view of McDonald's, is to persuade customers and potential customers to eat at McDonald's because it is environmentally friendly. What claims are made in the introductory section executive summary to persuade readers of this?

8. What strategies do the writers use to achieve interparagraph coherence in Chapter 2? What strategies do they use to indicate the relationship between Chapter 2 and the report as a whole?

## Research Report: *Advances in Lean Ground Beef Production*

This document reports the results of a research study. It is a final report for the sponsors of a project to develop lean beef.

The report will be read by the people who sponsored the research study. It will also be read by experts in the field who are interested in knowing the results for their own work. Furthermore, the report will be studied by product developers, who can use the results to develop marketable products and by public relations people who will use this report to advertise products.

The purpose of the ground beef report is to interest potential clients in using the process. The report is therefore written for generalists and novices, not just experts. Because generalists and novices as well as experts will read the ground beef report, the document is in the form of a small booklet rather than printed on the $8\frac{1}{2} \times 11$ inch paper on which research reports are usually printed. The smaller size makes it easier to read and gives the text an airy appearance. However, in content and tone, it does not differ from the conventions of a research report.

Both tables and graphs help readers, especially generalists, comprehend the statistical analyses. Headings and subheadings help readers predict what they will read.

As you read, consider how the report adheres to conventions and develops a line of reasoning that readers can follow.

However, the writers adhere to the conventions for maintaining a formal and impersonal tone; they use the names of their organizations and do not refer to themselves in first person plural.

As you read the report, consider the writers' claims and conclusions as they develop their arguments.

# *A*DVANCES IN LEAN GROUND BEEF PRODUCTION

Bulletin 606    March 1990
Alabama Agricultural Experiment Station
Lowell T. Frobish, Director
Auburn University
Auburn University, Alabama

# CONTENTS

*Page*

(continues)

FIRST PRINTING 4M, MARCH 1990
SECOND PRINTING 2M, DECEMBER 1990

*Information contained herein is available to all persons without
regard to race, color, sex, or national origin.*

# Advances In Lean Ground Beef Production
## DALE L. HUFFMAN and W. RUSSELL EGBERT[1,2]

## INTRODUCTION

OVER 3 BILLION pounds of ground beef products are consumed annually in the United States, which accounts for 44 percent of the total fresh beef cuts available for consumption (4). These products generally contain between 20 and 30 percent fat. A large segment of today's consumer population is health conscious and is concerned about dietary fat. These consumers avoid meat products with high fat content, such as ground beef. Current trends reflect a shift in consumers' consumption of fats, with a decrease in the intake of visible separable fats and an increase in consumption of low-fat animal products.

A consumer climate survey has indicated that health oriented and active life style consumers make up 50 percent of the population. These two groups of consumers are characterized by their low consumption of red meat. This survey indicated the U.S. population in general is concerned about weight control and caloric intake (5). As today's consumers continue to become more health conscious, their demand for lower fat ground beef products will rapidly expand.

It is important that the red meat industry develop low-fat ground beef products tailored to meet the needs of today's diet conscious consumers. The simple reduction of the fat in ground beef to 5-10 percent would be the most efficient method of developing low-fat ground beef products. However, ground beef with a fat content in this range is generally considered less palatable than ground beef with 20-30 percent fat. Therefore, a project was initiated to develop acceptable lean ground beef products. The approach combined present knowledge about the texture, juiciness, and flavor of currently produced ground beef products with changes in the technologies used to produce these products.

The objective of this project was to develop "lean" (90-95 percent) ground beef products with significantly reduced fat levels, which are as acceptable to the consumer in the same form as current ground beef items.

---

[1]Professor and Research Associate of Animal and Dairy Sciences.

[2]This study was funded in part by a grant from the Beef Industry Council of the National Live Stock and Meat Board, Chicago, Illinois, and the Alabama Cattlemen's Association, Montgomery, Alabama. The contribution of nonmeat ingredients by the Marine Colloids Division of FNIC Corporation, Philadelphia, Pennsylvania, and A.C. Legg Packing Company, Birmingham, Alabama, is appreciated. Cooperation of John Morrell and Company, Montgomery, Alabama, is also acknowledged with appreciation.

(continues)

# EXPERIMENTAL DEVELOPMENT

Development of a low fat ground beef product was approached in a series of logical steps (studies), each building on the results of the previous experiments. The first study explored the level of fat desired in ground beef patties based on consumer ratings. Study II was designed to determine consumers' ability to distinguish between ground beef patties with varying levels of fat. The third study was designed to determine the effect of cooking method on sensory properties of ground beef patties and the sensory property differences between ground beef products with differing fat levels. The objectives of Study IV were the same as for Study III, however a different method of cookery was used. The fifth study determined the effect of grind size on sensory traits of ground beef patties. The objective of Study VI was to determine the effect of various nonmeat ingredients on the sensory properties of ground beef patties. Study VII determined the effect of the addition of salt and hydrolyzed vegetable protein on sensory properties of lean ground beef patties. The final study was designed to confirm the findings of the earlier studies and to demonstrate that the overall project objective—the development of a lean ground beef product with sensory properties similar to those of a ground beef product containing 20 percent fat control—had been accomplished.

# DESIGN AND METHODOLOGY
## Overall Processing Procedure

Each of the "lean" ground beef products developed was compared to a control and was processed using manufacturing practices that yield high quality products. Fresh beef cap meat and 50/50 beef trimmings were each ground twice through a 1/2-inch (1.27-cm) grinder plate. Samples of both the ground cap meat and 50/50 trimmings were taken using the "grab" method. Samples were finely ground using a Kitchenaid mixer with grinder attachments and analyzed for fat content by ether extraction (2). The ground cap meat (lean component) and 50/50 trimmings (fat component) were vacuum packaged in approximately 10-pound (3.0-kg) packages, frozen, and held at –4°F (–20°C). Prior to manufacturing, the coarse ground lean and fat meat components were thawed at 41–44°F (5–7°C) for approximately 12 hours.

The low fat ground beef products were manufactured using the appropriate quantities of coarse ground lean and fat components as previously formulated. The appropriate amounts of lean and fat were combined and mixed with various combinations of the following: (1) lecithin (or other appropriate phospholipid emulsifying agents); (2) carrageenan (or other non-gel-forming food gums); and/or (3) beef extract and/or other beef flavor enhancers. After the meat and nonmeat ingredients had been mixed (approximately 1 minute), the products were finely ground. These finely ground products were then made into 4-ounce patties using a Hollymatic (Super 54) pattie machine. Ground beef patties were stored (2 days) at 38°F (3°C) until sensory evaluation and cooking loss analyses were completed.

## Cooking Methods

Ground beef patties were: (1) oven broiled at 350°F (177°C) for 8 minutes to a well-done state using a Blodgett forced air convection oven (G. S. Blodgett Company, Burlington, Vermont); (2) griddle broiled to a well-done state on a Model TG-72 Special McDonald's grill (Wolf Range Corporation) at a temperature of 330°F (165°C) for 3 1/2 minutes (2 minutes on the first side, 1 1/2 on the other); or (3) grill broiled to a well-done state on an Emberglo open hearth broiler (Model 310, Mid-Continent Metal Products Co., Chicago, Illinois) for 6 minutes (4 minutes on the first side and 2 minutes on the opposite side).

## Cooking Loss

Cooking yields were determined by the difference in weight for three patties from each treatment weighed prior to cooking and after equilibration to room temperature 68°F (20°C). Patties evaluated for cooking loss were blotted with paper towels after cooking.

## Proximate Analysis

Raw and cooked (from cooking loss determination) samples for proximate analysis were ground three times with a Kitchenaid mixer with grinder attachments and the samples were stored frozen at –4°F. Samples held at this temperature were used for determination of moisture, petroleum ether-extractable lipid, and protein content of the raw products. Moisture, lipid, and protein content of each product was determined with AOAC (2) approved methods.

## Sensory Evaluation

Cooked patties were cut into six wedges. These were held in a conventional oven at 104°F (40°C) until evaluated by a 9-member trained sensory panel (1) for juiciness (initial and sustained), tenderness, flavor, and overall acceptability on an 8-point hedonic scale (1 = extremely dry, extremely tough, extremely bland, extremely unacceptable and 8 = extremely juicy, extremely tender, extremely intense, and extremely acceptable, respectively). Texture was rated on a 7-point hedonic scale (1 = more sandy, 4 = typical of ground beef, and 7 = more mushy). Panel members were selected from students, faculty, and staff of the Department of Animal and Dairy Sciences. Panelists were served one wedge of each of the treatments in a random order. Unsalted crackers, apple juice, and water at room temperature were also served. Each treatment was evaluated once by each panelist on three separate occasions.

## Statistical Evaluation

The experimental data were statistically analyzed using the general linear model (12) where applicable. When a significant F-value (P<0.05) was found, Tukeys' mean separation procedure (13) was employed to determine differences between means.

(continues)

## STUDY 1: CONSUMER ACCEPTABILITY OF GROUND BEEF PRODUCTS WITH VARYING FAT LEVELS

### Design

Ground beef patties were formulated to contain five different levels of fat (5, 10, 15, 20, and 25 percent), using the cap meat and 50/50 trim as previously described. The patties were griddle broiled as previously described and evaluated by a 30-member untrained consumer-type panel. Panelists were instructed to evaluate the samples for overall acceptability on a 10-number descriptive analysis (1) scale (0 = dislike extremely and 10 = like extremely). The study was replicated three times and the data were analyzed using analysis of variance procedures as previously described.

## Results and Discussion

*Proximate Analysis*

Proximate analysis data for raw products are presented in table 1. These analyses confirmed that the products contained the desired fat level as formulated (5, 10, 15, 20, and 25 percent fat). As the level of fat in the raw ground beef products increased, the level of moisture decreased ($P<0.05$). This supports the findings of other researchers that an inverse relationship exists between fat and moisture content in ground beef (*14, 9*). Percent protein also decreased ($P<0.05$) as the fat level of the ground beef products increased. Generally, other researchers have reported that protein content of ground beef with differing fat levels did not differ (*11, 7*).

TABLE 1. PROXIMATE ANALYSIS OF RAW GROUND BEEF
WITH DIFFERING LEVELS OF FAT, STUDY I[1]

| Fat level | Analysis[2] | | |
|---|---|---|---|
| | Moisture | Fat | Protein |
| | *Pct.* | *Pct.* | *Pct.* |
| 5 | 72.34[b] | 5.81[f] | 21.60[b] |
| 10 | 69.05[c] | 10.04[c] | 20.93[b] |
| 15 | 65.55[d] | 14.16[d] | 18.93[c] |
| 20 | 61.89[c] | 19.35[c] | 18.15[c] |
| 25 | 58.12[f] | 22.22[b] | 16.17[d] |

[1]From Neale (*10*)
[2]Means within a column with different superscripts are significantly different ($P<0.05$).

Composition of the cooked ground beef products conform to the same trends as the raw products, table 2. Both moisture and protein contents decreased ($P<0.05$) as fat levels increased, with the exception of ground beef with fat levels of 20 and 25 percent which did not differ ($P>0.05$).

Table 2. PROXIMATE ANALYSIS OF COOKED GROUND BEEF
WITH DIFFERING LEVELS OF FAT, STUDY I[1]

| Fat level, pct. | Analysis[2] | | | |
|---|---|---|---|---|
| | Moisture | Fat | Protein | Cooking loss |
| | *Pct.* | *Pct.* | *Pct.* | *Pct.* |
| 5 | 61.58[b] | 8.96[f] | 29.20[b] | 29.66[dc] |
| 10 | 59.20[c] | 12.63[c] | 27.75[c] | 28.56[c] |
| 15 | 56.39[d] | 16.13[d] | 26.39[d] | 31.71[cd] |
| 20 | 54.15[c] | 18.59[c] | 25.67[c] | 33.74[c] |
| 25 | 50.68[f] | 22.16[b] | 25.28[f] | 39.13[b] |

[1]From Neale (*10*).
[2]Means within a column with different superscripts are significantly different (P<0.05).

### Cooking Loss

Percent cooking loss increased (P<0.05) as the fat level of the ground beef products increased, table 2. These results agree with other researchers that cooking yields are inversely related to the fat content of the product (*8,11,9,6*).

### Sensory Evaluation

Overall acceptability scores for the ground beef products with differing levels of fat are presented in table 3. Panelists found ground beef with 15 and 20 percent fat to be more desirable (P<0.05) than ground beef with 5 percent fat. Overall acceptability tended to decrease with decreasing fat levels. Ground beef with 25 percent fat was not different (P> 0. 05) from any of the other fat levels.

TABLE 3. OVERALL ACCEPTABILITY SCORES OF GROUND BEEF PATTIES
WITH DIFFERING LEVELS OF FAT, STUDY I

| Fat Level pct. | Overall acceptability scores[1,2,3] |
|---|---|
| 5 | 4.82[c] |
| 10 | 5.13[dc] |
| 15 | 5.77[cd] |
| 20 | 6.07[c] |
| 25 | 5.35[dc] |

[1]From Neale (*10*)
[2]Overall acceptability score on a 10 to 0 scale (10 = like extremely, 0 = dislike extremely).
[3]Means within different superscripts are significantly different (P<0.05).

(continues)

## Conclusions

Consumer panelists gave the highest (P<0.05) ratings for overall acceptability to ground beef patties formulated to 20 percent fat, followed by patties formulated to 15 percent fat. An increase or decrease in the fat content of ground beef patties from a fat content of 20 percent resulted in a decrease in overall acceptability of the products, figure 1. Based on consumer ratings, ground beef products formulated to 20 percent fat should be used as the control in the development of ground beef products with reduced fat levels.

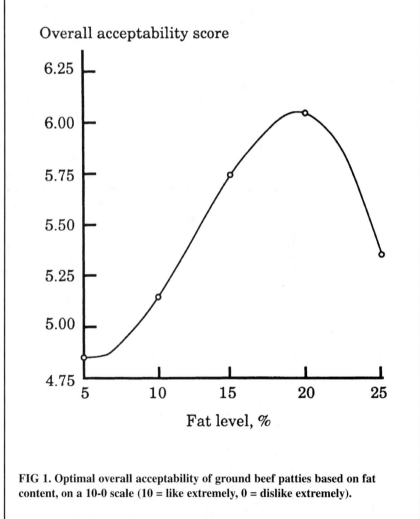

**FIG 1. Optimal overall acceptability of ground beef patties based on fat content, on a 10-0 scale (10 = like extremely, 0 = dislike extremely).**

# STUDY II: CONSUMERS' ABILITY TO DISTINGUISH DIFFERENCES IN GROUND BEEF PRODUCTS WITH VARYING LEVELS OF FAT

## Design

Ground beef products were formulated to contain 12.5, 15.0, 17.5, 20.0, and 22.5 percent fat using the raw materials and procedures as described previously. These products were evaluated by a consumer panel using a triangle test (*3*). Panelists were served three samples (two samples of one product and one of a second allotted at random) and asked to identify the different sample. All product combinations were evaluated by a 50-member consumer panel. Differences between the ground beef products were determined based on the number of correct responses (*3*).

## Results and Discussion

No difference (P>0.05) was found between ground beef with a fat level of 12.5 and 15.0 percent, 17.5 and 20.0 percent, and 12.5 and 22.5 percent, table 4. This was expected for the products that differed by only 2.5 percent fat, however not finding a significant difference between ground beef containing 12.5 and 22.5 percent was unexpected. For this reason, the 12.5 and 22.5 percent ground beef products were reformulated and tested a second time with a 50-member consumer panel. The results of the second consumer panel were similar to the first in that the sensory panel could not detect (P>0. 05) a difference between the samples (could not identify the different sample). However, panelists were able to detect differences of 5 and 7.5 percent in all other cases, table 4. Panelists were also able to detect a difference between ground beef with 15.0 percent fat compared to 17.5 percent and 20. 0 percent compared to 22.5 percent (a 2.5 percent difference).

TABLE 4. CONSUMERS' ABILITY TO DISTINGUISH BETWEEN GROUND BEEF
WITH DIFFERING LEVELS OF FAT, STUDY II

| Fat level pct. | Fat level[1], pct. | | | |
|---|---|---|---|---|
| | 15.0 | 17.5 | 20.0 | 22.5 |
| 12.5 . . . . . . . . . . . . . . . . . . . . . . . . . . . . . . . . . . . . . . . . . . . . . . . . . . . . . . NS | | * | ** | NS |
| 15.0 . . . . . . . . . . . . . . . . . . . . . . . . . . . . . . . . . . . . . . . . . . . . . . . . . . . . . | | ** | * | ** |
| 17.5 . . . . . . . . . . . . . . . . . . . . . . . . . . . . . . . . . . . . . . . . . . . . . . . . . . . . . | | | NS | ** |
| 20.0 . . . . . . . . . . . . . . . . . . . . . . . . . . . . . . . . . . . . . . . . . . . . . . . . . . . . . | | | | * |

[1]NS = no significant difference (P>0.05).* = significant difference (P<0.05). ** = highly significant difference (P<0.01).

(continues)

A summary of the ability of consumers to detect fat level differences is shown in table 5. Approximately 30 percent of the consumers were able to detect a fat level difference of 2.5 percent, approximately 50 percent of the consumers could detect a fat level difference of 5.0 percent, about 54 percent of the consumers could detect a difference of 7.5 percent, but only approximately 42 percent of the consumers could detect a difference in fat of 10 percent.

TABLE 5. PERCENTAGE OF CONSUMERS ABLE TO DISTINGUISH BETWEEN GROUND BEEF
PRODUCTS WITH A GIVEN DIFFERENCE IN FAT LEVEL, STUDY II

| Fat level differences, pct. | Percentage of consumers |
|---|---|
| 2.5 | 30.4 |
| 5.0 | 49.3 |
| 7.5 | 54.2 |
| 10.0 | 41.8 |

## Conclusions

From this study it was determined that consumers could consistently distinguish between ground beef products that differed by 5 to 7.5 percent fat. Further research in this area could provide useful information for explaining the observation that no difference ($P > 0.05$) was detected between ground beef products containing 12.5 and 22.5 percent fat (10 percent difference). Based on these results, it appears that less than half of consumers are able to detect fat level differences of ground beef patties in the range of 2.5 percent to 10.0 percent. . . .

## SUMMARY

The objective upon initiation of this project was to develop a lean ground beef product with a fat content of 10 percent or less that was as acceptable as the control (as determined by consumer panel to possess the most acceptable sensory characteristics) which contained 20 percent fat. Evidence from this study indicates that this objective has been accomplished. Two products were developed with sensory characteristics that do not differ from the control. These developed products along with the control possess more intense beef flavor and have a greater overall acceptability than the original ground beef product containing 10 percent fat (untreated).

The two products developed contain approximately 40 percent less fat than the control product (with a 20 percent fat content) on an "as eaten" basis. The caloric savings obtained from the consumption of these products compared to the control product are between 55 and 60 kcal/100 g serving or a caloric decrease of 22-23 percent on an "as eaten" basis. Based on a 100-g serving, calories from the lean ground beef are distributed as follows: 100 kcal from fat and 98 kcal from protein;

whereas in the control ground beef the calories are distributed in the following: 168 kcal from fat and 87 kcal from protein.

The development of lean ground beef products (10 percent fat content) which possess sensory properties comparable to ground beef products with 20 percent fat content was accomplished through the following:

(1) An increase in the particle size of the ground product through the use of a large-sized grinder plate.

(2) The addition of small quantities of salt and hydrolyzed vegetable protein as flavor enhancers.

(3) The addition of carrageenan as a product stabilizer.

These newly developed lean ground beef products are tailored to meet the needs of the diet conscious consumer. The products contain less fat and calories than traditional ground beef products. Upon introduction of these products to the retail market, consumers will have the opportunity to substitute the lean ground beef products for the higher-fat ground beef products currently available

# LITERATURE CITED

(*1*) AMSA. 1978. Guidelines for Cookery and Sensory Evaluation of Meat. American Meat Science Association, Chicago, Ill.

(*2*) AOAC. 1980. Official Methods of Analysis. 13th ed. Association of Official Analytical Chemists, Washington, D.C.

(*3*) ASTM. 1968. Manual on Sensory Testing Methods. American Society for Testing and Materials. Philadelphia, Penn.

(*4*) BREIDENSTEIN, B. C. AND WILLIAMS, J. C. 1986. The Consumer Climate for Red Meat: Special Issue. American Meat Institute, Washington D.C. and the National Live Stock and Meat Board. Chicago, Ill.

(*5*) BURKE MARKETING RESEARCH. 1987. The Consumer Climate for Meat Study. Prepared for the National Live Stock and Meat Board, Chicago, Ill., and the American Meat Institute, Washington, D.C.

(*6*) HOELSCHER, L. M., SAVELL, J. W, HARRIS, J. M., CROSS, H. R., AND RHEE, K. S. 1987 Effect of Initial Fat Level and Cooking Method on Cholesterol Content and Caloric Value of Ground Beef Patties. J. Food Sci. 52:883.

(*7*) HOLDEN, J. M., LANZA, E., AND WOLF, W. R. 1986. Nutrient Composition of Retail Ground Beef. J. Agric. Food Chem. 34:302.

(*8*) KENDALL, P.A., HARRISON, D. L., AND DAYTON, A. D. 1974. Quality Attributes of Ground Beef on the Retail Market. J. Food Sci. 39:610.

(*9*) KREGEL, K. K., PRUSA, K. J., AND HUGHES, K. V. 1986. Cholesterol Content and Sensory Analysis of Ground Beef as influenced by Fat Level, Heating, and Storage. J. Food Sci. 51:1162.

(*10*) NEALE, M.G. 1989. An Innovative Approach to Lean Ground Beef Production. M.S. Thesis, Auburn University.

(continues)

(*11*) ONO, K., BERRY, B. W., AND PAROCZAY, E. 1985. Contents and Retention of Nutrients in Extra Lean, Lean, and Regular Ground Beef. J. Food Sci. 50:701.

(*12*) SAS INSTITUTE INC. 1982. SAS User's Guide: Basic, 1982. Edition. SAS Institute Inc., Cary, N.C.

(*13*) TUKEY, J. W. 1953. The Problem of Multiple Comparisons, Princeton University, Princeton, N.J. Cited in: Principles and Procedures of Statistics. R. G. Steele and J. H. Torrie (Ed). McGraw-Hill Book Company, New York, N.Y.

(*14*) WOOLSEY, A. P. AND PAUL, P. C. 1969. External Fat Cover Influence on Raw and Cooked Beef. 1. Fat and Moisture Content. J. Food Sci. 34:554.

## Suggestions for Discussion and Writing

1. List the major categories in the ground beef report. Compare these to the conventional categories of a final report. In what ways does the ground beef report adhere to the conventions of a research report? In what ways does it differ?

2. What claims do the writers of the ground beef report make in the introduction to persuade readers that their research is important? Do you think these claims relate to your values and attitudes?

3. Read the Experimental Development section of the ground beef report. What implicit claims do the writers make? What evidence do they use to support their claims? What prior knowledge would readers need to comprehend this section?

4. Study the visual text and graphics in the report. Why do you think the writers use the various graphs and tables?

5. The ground beef report includes the objective for the research. Locate the objective and bracket it. Study the conclusions in the two reprinted sections. Indicate which conclusion relates to which part of the objective by writing the appropriate part of the objective in the margin next to the conclusion to which it's related.

## "The Plague of Athens," *The History of the Peloponnesian Wars*, Thucydides, Fifth Century, B.C.

Following the conventions of classical times, Thucydides placed his technical documents in their historical context. His description of the plague is a "mini report" within his larger report on the history of the Peloponnesian wars. According to Thucydides, because the areas surrounding Athens were continuously under attack by the Spartans, much of the rural population had moved to the city for safety. Furthermore, much of the army was concentrated in the city. Thus, Athens was extremely crowded when it was hit by the plague. Over a course of three years, the plague wiped out one-third of Athens's best troops, along with a large portion of the civilian population.

Thucydides explains that his purpose in writing the *History* is not to entertain but rather to provide "an exact knowledge of the past as an aid to the interpretation of the future." The document is considered to be the first history that reflects a scientific approach to data collection. In his preface, Thucydides criticizes the contemporary writers of his time who "took little pains" to investigate the truth and contrasts their methods with his own process of proceeding "upon the dearest of data, and arriving at conclusions as exact as can be expected in matters of such antiquity."

Both in structure and style, "The Plague of Athens" resembles contemporary reports. The author states his purpose and explains the sources of his data and the limitations of his study. Thucydides is factual, reporting only what he has experienced firsthand or heard from a "reliable source." His efforts at detailed observation are obvious throughout the document.

While the *History* itself is written for an audience of historians, Thucydides specifically writes this segment for physicians. And, in fact, this account has intrigued physicians over the centuries. Until recently, however, physicians have been unable to match the symptoms described with any known disease. But in 1985, Dr. D. A. Languir and his colleagues suggested that Thucydides may have been describing influenza complicated by Toxic Shock Syndrome (TSS). They named the illness "Thucydides Syndrome." Three months after publication of their diagnosis, a large outbreak of influenza occurred in Minnesota in which ten patients developed TSS. Their symptoms matched those in Thucydides' description.

In the following article, which appeared in the *Journal of the American Medical Association,* Dr. Bruce Dan describes the remarkable resemblance between Thucydides' description of the Athens plague and the Minnesota influenza outbreak.

# Toxic Shock Syndrome: Back to the Future

*All speculation as to its origin and its causes, if causes can be found adequate to produce so great a disturbance, I leave to other writers, whether lay or professional, for myself, I shall simply set down its nature, and explain the symptoms by which perhaps it may be recognized by the student, if it should ever break out again.*

Thucydides, 460–400 BC

In words that would transmit the heart-sinking fear of the contagion across 24 centuries, Thucydides dutifully chronicled the clinical features of the plague of Athens. The horrifying epidemic would not only kill tens of thousands of soldiers and civilians, it would alter the course of the Peloponnesian War, and change history.

Although the awful spectacle was depicted as "men dying like sheep,"[1] the pestilence was notable for far more than just its numerical devastation. It was a bewildering constellation of signs and symptoms that no physician or historian had ever recorded before. Its victims, previously in good health, were suddenly stricken with high fever, sneezing, coughing, and hoarseness. They rapidly developed nausea, vomiting, and diarrhea, with a puzzling combination of conjunctival and oropharyngeal hyperemia and erythroderma. So unbearable was their fever that many of the sick plunged themselves into cooling water to relieve their "agonies of unquenchable thirst." Despite the obvious dehydration of those struck down by the enigmatic malady, their bodies "did not waste away . . . but held out to a marvel against its ravages," suggesting the possibility of edema masking their hypovolemia. Those who survived the ordeal suffered amnesia, and many lost their fingers and toes to peripheral gangrene.

Strangely, physicians who attended the sick had the highest mortality rates during this pandemic of the Mediterranean basin, but after three years (430 to 427 BC) of annihilation, the disease vanished as quickly as it appeared.

There it would have ended: a curious medical footnote to history if it were not for the interest and, as it turns out, the prescience of a few medical scholars. In what must be one of the most incredible twists of timing, Langmuir et al[2] published in October 1985 a new hypothesis for the plague of Athens. The authors speculated that the entity that perplexed Thucydides was in fact the unlikely occurrence of influenza followed by an upper respiratory tract superinfection with *Staphylococcus aureus*—but not just any strain of *S aureus*. They presumed that the bacteria must have produced toxic shock syndrome toxin-1 (TSST-1)—the glory that was Greece was terminated by toxic shock syndrome.

Langmuir et al named the entity—influenza complicated by infection with a toxin-producing strain of *S aureus*—Thucydides syndrome, but they were careful to state that the plague of Athens might not be identical with modern toxic shock syndrome (TSS). Indeed, some authors have suggested Rift Valley fever as an alternative explanation for the epidemic.[3]

Nonetheless Langmuir and colleagues suggested, with apparent clairvoyance, "that Thucydides syndrome may reappear as a minor or even major manifestation of some future epidemic or pandemic of influenza." Within three months of the publication of their article, the prediction came true.

In this issue of THE JOURNAL, MacDonald et al[4] and Sperber and Francis[5] report the emergence (or reemergence) of Thucydides syndrome. The two articles detail ten cases of TSS complicating influenza-like illnesses during the winter of 1985–1986. Of the ten patients described (four male and six female), six died. All ten patients showed typical clinical features of TSS,[6] and in nine of the cases where cultures of respiratory secretions were obtained, *S aureus* was found. Of the eight staphylococcal isolates tested for enterotoxin production, seven produced TSST-1 (the characteristic marker for TSS-producing strains), and one produced enterotoxin B (also suggested as being associated with TSS).[7]

The pathogenesis of the syndrome seems to be the acquisition of influenza followed by the development of a staphylococcal tracheitis or pneumonia. (Influenza B infection was confirmed in three of the four cases in which throat cultures or acute and convalescent titers were obtained.) Although staphylococcal pneumonia has been a well-known consequence of influenza,[8] apparently no more than a superficial infectious mucositis of the laryngobronchial tree is required for the development of TSS in these patients.

In this manner it parallels the pathogenesis of menstrually associated TSS, which can present as a devastating illness with only a minimal inflammation of the vaginal mucosa. The lack of dramatic local inflammation is also a consistent finding in wound-associated and postoperative TSS.[9] As the authors point out, it now behooves physicians to be aware of this new entity and to consider quickly the diagnosis of TSS in a patient with influenza whose condition suddenly worsens.

Whether the plague of Athens was actually Thucydides syndrome we shall probably never know, but the entity does exist today. Its actual incidence will be determined only by active surveillance systems set into motion before influenza activity develops in communities.

We had previously commented that *S aureus* was a harbinger of things to come.[10] We did not know that it was also a reminder of things past.

Bruce B. Dan, MD

1. Finley J H Jr (trans-ed): *The Complete Writings of Thucydides: The Peloponnesian War.* New York: Random House, 1951, chap 7.

2. Langmuir AD, Worthen TD, Solomon J, et al: The Thucydides syndrome: A new hypothesis for the plague of Athens. *N Engl J Med* 1985;313:1027–1030.

3. Morens DM, Chu MC: The plague of Athens. *N Engl J Med* 1986;314:855.

4. MacDonald KL, Osterholm MT, Hedberg CW, et al: Toxic shock syndrome: A newly recognized complication of influenza and influenzalike illness. *JAMA* 1987;257:1053–1058.

5. Sperber SJ, Francis JB: Toxic shock syndrome during an influenza outbreak. *JAMA* 1987; 257:1086–1087.

6. Shands KN, Schmid GP, Dan BB, et al: Toxic shock syndrome in menstruating women: Association with tampon use and *Staphylococcus aureus* and clinical features in 52 cases. *N Engl J Med* 1980; 303:1463–1442.

7. Schlievert PM: Staphylococcal enterotoxin B and toxic shock syndrome toxin-1 are significantly associated with nonmenstrual TSS. *Lancet* 1986;1:1149–1150.

8. Finland M, Peterson OL, Strauss E: Staphylococcic pneumonia during an epidemic of influenza. *Arch Intern Med* 1942;70:183–205.

9. Reingold AL, Hargrett NT, Dan BB, et al: Nonmenstrual toxic shock syndrome: A review of 130 cases. *Ann Intern Med* 1982;96:871–874.

10. Shands KN, Dan BB, Schmid GP: Toxic shock syndrome: The emerging picture. *Ann Intern Med* 1981;94:264–266.

Source: Dan, Bruce. 1987. "Toxic Shock Syndrome: Back to the Future," *Journal of the American Medical Association.* 257, no. 8 (27 February): 1094–95.

As you read Thucydides' description, consider how closely it resembles contemporary reports.

---

Forecast of chapter.

Interpretation by Dr. Languir, et al.

Description of symptoms.

# CHAPTER VII

## SECOND YEAR OF THE WAR—THE PLAGUE OF ATHENS— POSITION AND POLICY OF PERICLES—FALL OF POTIDÆA

B.C. 430: *Outbreak of the plague—Description of symptoms—Demoralization a consequence of the plague—Peloponnesians ravage coast—Plague overtakes invaders—Pericles blamed—He defends himself—Fined first, then restored to full power—Peloponnesian envoys detained by Sitalces and put to death by Athenians—Ambraciots attack Amphilochian Argos—Fall of Potidæa.*

Such was the funeral that took place during this winter, with which the first year of the war came to an end. In the first days of summer the Lacedæmonians and their allies, with two-thirds of their forces as before, invaded Attica, under the command of Archidamus, son of Zeuxidamus, king of Lacedæmon, and sat down and laid waste the country. Not many days after their arrival in Attica the plague first began to show itself among the Athenians. It was said that it had broken out in many places previously in the neighbourhood of Lemnos and elsewhere; but a pestilence of such extent and mortality was nowhere remembered. Neither were the physicians at first of any service, ignorant as they were of the proper way to treat it, but they died themselves the most thickly, as they visited the sick most often; nor did any human art succeed any better. Supplications in the temples, divinations, and so forth were found equally futile, till the overwhelming nature of the disaster at last put a stop to them altogether.

It first began, it is said, in the parts of Ethiopia above Egypt, and thence descended into Egypt and Libya and into most of the king's country. Suddenly falling upon Athens, it first attacked the population in Piræus,—which was the occasion of their saying that the Peloponnesians had poisoned the reservoirs, there being as yet no wells there—and afterwards appeared in the upper city, when the deaths became much more frequent. All speculation as to its origin and its causes, if causes can be found adequate to produce so great a disturbance, I leave to other writers, whether lay or professional; for myself, I shall simply set down its nature, and explain the symptoms by which perhaps it may be recognized by the student, if it should ever break out again. This I can the better do, as I had the disease myself, and watched its operation in the case of others.

That year then is admitted to have been otherwise unprecedentedly free from sickness; and such few cases as occurred, all determined in this. As a rule, however, there was no ostensible cause; but people in good health were all of a sudden attacked by violent heats in the head, and redness and inflammation in the eyes, the inward parts, such as the throat or tongue, becoming bloody and emitting an unnatural and fetid breath. These symptoms were followed by sneezing and hoarseness, after which the pain soon reached the chest, and produced a hard cough. When it fixed in the stomach, it upset it; and discharges of bile of every kind named by physicians ensued, accompanied by very great distress. In most cases also an ineffectual retching followed, producing violent spasms, which in some cases ceased soon after, in others much later. Externally the body was not very hot to the touch, nor pale in its appearance, but reddish, livid, and breaking out into small pustules and ulcers. But internally it burned so that the patient could not bear to have on him clothing or linen even of the very lightest description; or indeed to be otherwise than stark naked. What they would have liked best would have been to throw themselves into cold water; as indeed was done by some of the neglected sick, who plunged into the rain-tanks in their agonies of unquenchable thirst, though it made no difference whether they drank little or much. Besides this, the miserable feeling of not being able to rest or sleep never ceased to torment them. The body meanwhile did not waste away so long as the distemper was at its height, but held out to a marvel against its ravages; so that when they succumbed, as in most cases, on the seventh or eighth day to the internal inflammation, they had still some strength in them. But if they passed this stage, and the disease descended further into the bowels, inducing a violent ulceration there accompanied by severe diarrhœa, this brought on a weakness which was generally fatal. For the disorder first settled in the head, ran its course from thence through the whole of the body, and even where it did not prove mortal, it still left its mark on the extremities; for it settled in the privy parts, the fingers and the toes, and many escaped with the loss of these, some too with that of their eyes. Others again were seized with an entire loss of memory on their first recovery, and did not know either themselves or their friends.

**Environmental effect.**

But while the nature of the distemper was such as to baffle all description, and its attacks almost too grievous for human nature to endure, it was still in the following circumstance that its difference from all ordinary disorders was most clearly shown. All the birds and beasts that prey upon human bodies, either abstained from touching them (though there were many lying unburied), or died after tasting them. In proof of this, it was noticed that birds of this kind actually disappeared; they were not about the bodies, or indeed to be seen at all. But of course the effects which I have mentioned could best be studied in a domestic animal like the dog.

**Psychological effect.**

Such then, if we pass over the varieties of particular cases, which were many and peculiar, were the general features of the distemper. Meanwhile the town enjoyed an immunity from all the ordinary disorders; or if any case occurred, it ended in this. Some died in neglect, others in the midst of every attention. No remedy was found that could be used as a specific; for what did good in one case, did harm in another. Strong and weak constitutions proved equally incapable of resistance, all alike being swept away, although dieted with the utmost precaution. By far the most terrible feature in the malady was the dejection which ensued when any one felt himself sickening, for the despair into which they instantly fell took away their power of resistance, and left them a much easier prey to the disorder; besides which, there was the awful spectacle of men dying like sheep, through having

(continues)

caught the infection in nursing each other. This caused the greatest mortality. On the one hand, if they were afraid to visit each other, they perished from neglect; indeed many houses were emptied of their inmates for want of a nurse: on the other, if they ventured to do so, death was the consequence. This was especially the case with such as made any pretensions to goodness: honour made them unsparing of themselves in their attendance in their friends' houses, where even the members of the family were at last worn out by the moans of the dying, and succumbed to the force of the disaster. Yet it was with those who had recovered from the disease that the sick and the dying found most compassion. These knew what it was from experience, and had now no fear for themselves; for the same man was never attacked twice—never at least fatally. And such persons not only received the congratulations of others, but themselves also, in the elation of the moment, half entertained the vain hope that they were for the future safe from any disease whatsoever.

**Demographic effect.**

An aggravation of the existing calamity was the influx from the country into the city, and this was especially felt by the new arrivals. As there were no houses to receive them, they had to be lodged at the hot season of the year in stifling cabins, where the mortality raged without restraint. The bodies of dying men lay one upon another, and half-dead creatures reeled about the streets and gathered round all the fountains in their longing for water. The sacred places also in which they had quartered themselves were full of corpses of persons that had died there, just as they were; for as the disaster passed all bounds, men, not knowing what was to become of them, became utterly careless of everything, whether sacred or profane. All the burial rites before in use were entirely upset, and they buried the bodies as best they could. Many from want of the proper appliances, through so many of their friends having died already, had recourse to the most shameless sepultures: sometimes getting the start of those who had raised a pile, they threw their own dead body upon the stranger's pyre and ignited it; sometimes they tossed the corpse which they were carrying on the top of another that was burning, and so went off.

**Ethical effect.**

Nor was this the only form of lawless extravagance which owed its origin to the plague. Men now coolly ventured on what they had formerly done in a corner, and not just as they pleased, seeing the rapid transitions produced by persons in prosperity suddenly dying and those who before had nothing succeeding to their property. So they resolved to spend quickly and enjoy themselves, regarding their lives and riches as alike things of a day. Perseverance in what men called honour was popular with none, it was so uncertain whether they would be spared to attain the object; but it was settled that present enjoyment, and all that contributed to it, was both honourable and useful. Fear of gods or law of man there was none to restrain them. As for the first, they judged it to be just the same whether they worshipped them or not, as they saw all alike perishing; and for the last, no one expected to live to be brought to trial for his offences, but each felt that a far severer sentence had been already passed upon them all and hung ever over their heads, and before this fell it was only reasonable to enjoy life a little.

**Literary digression.**

Such was the nature of the calamity, and heavily did it weigh on the Athenians; death raging within the city and devastation without. Among other things which they remembered in their distress was, very naturally, the following verse which the old men said had long ago been uttered:

'A Dorian war shall come and with it death.'

So a dispute arose as to whether dearth and not death had not been the word in the verse; but at the present juncture, it was of course decided in favour of the latter; for the people made their recollection fit in with their sufferings. I fancy, however, that if another Dorian war should ever afterwards come upon us, and a dearth should happen to accompany it, the verse will probably be read accordingly. The oracle also which had been given to the Lacedæmonians was now remembered by those who knew of it. When the God was asked whether they should go to war, he answered that if they put their might into it, victory would be theirs, and that he would himself be with them. With this oracle events were supposed to tally. For the plague broke out so soon as the Peloponnesians invaded Attica, and never entering Peloponnese (not at least to an extent worth noticing), committed its worst ravages at Athens, and next to Athens, at the most populous of the other towns. Such was the history of the plague.

Source: Chapter VII, The plague of Athens, In *The Complete Writings,* 1934. Reprinted with permission of Random House, Inc.

## Suggestions for Discussion and Writing

1. Locate the conventional sections of a report in Thucydides' description of the plague: background, statement of the problem, methods and procedures, and results and discussion. Place a heading at the beginning of each section.

2. Why do you think Thucydides reminds the reader that he reports only what he has experienced or heard? He says, for example, "If causes can be found adequate to produce so great a disturbance, I leave it to other writers" (300).

3. Select a passage that demonstrates the writer's efforts at detailed observation. What details does he include? Look at the paragraphs on pages 300 and 301. Languir and his associates have interpreted these sections according to modern medical diagnoses. What details does Thucydides include that enable them to make these diagnoses?

4. Thucydides says he is writing for physicians. What terms does he use that are a part of the medical discourse community and may not be understood by people outside the medical profession?

5. Find passages that discuss material that would not be found in a typical contemporary scientific report. Bracket them. Explain why these passages would be atypical today.

## *De Aquus* [Aqueducts of Rome], Frontinus, A.D. 97

*De Aquus* by Frontinus, water commissioner under Nerva Caesar, is our best source of information about the Roman aqueducts. At the time Frontinus assumed his position, forty percent of the water supply was lost or stolen, the aqueducts were contaminated, and the water department was plagued by graft and corruption. Frontinus's purpose in writing *De Aquus* was to provide information to serve as a baseline for repairing the aqueducts and reforming management. Using his information, engineers were able to double the water supply.

Frontinus's text is one of the earliest examples of an information report written for a diverse audience of readers. Both governmental leaders and engineers used the book. Long technical descriptions are interspersed with direct addresses to government leaders whose economic and political support for the continuing repairs was essential. Throughout the book, he addresses the emperor and gives him credit for improvements to the water system.

Like many documents of that time, Frontinus places the report in its historical and political context. And like Thucydides, he uses a personal tone. In the introduction, he indicates his own purpose for writing the document.

As you read the following selection, notice how Frontinus uses various strategies, such as a forecast, to facilitate readers' reading processes. Notice, too, how he supports his claim that technology serves society.

# SEXTUS JULIUS FRONTINUS
# TWO BOOKS ON
# THE AQUEDUCTS OF ROME

INASMUCH as every task assigned by the Emperor demands especial attention; and inasmuch as I am incited, not merely to diligence, but also to devotion, when any matter is entrusted to me, be it as a consequence of my natural sense of responsibility or of my fidelity; and inasmuch as Nerva Augustus (an emperor of whom I am at a loss to say whether he devotes more industry or love to the State) has laid upon me the duties of water commissioner, an office which concerns not merely the convenience but also the health and even the safety of the City, and which has always been administered by the most eminent men of our State; now therefore I deem it of the first and greatest importance to familiarize myself with the business I have undertaken, a policy which I have always made a principle in other affairs.

For I believe that there is no surer foundation for any business than this, and that it would be otherwise impossible to determine what ought to be done, what ought to be avoided; likewise that there is nothing so disgraceful for a decent man as to conduct an office delegated to him, according to the instructions of assistants. Yet precisely this is inevitable whenever a person inexperienced in the matter in hand has to have recourse to the practical knowledge of subordinates. For though the latter play a necessary rôle in the way of rendering assistance, yet they are, as it were, but the hands and tools of the directing head. Observing, therefore, the practice which I have followed in many offices, I have gathered in this sketch (into one systematic body, so to speak) such facts, hitherto scattered, as I have been able to get together, which bear on the general subject, and which might serve to guide me in my administration. Now in the case of other books which I have written after practical experience, I consulted the interests of my successors. The present treatise also may be found useful by my own successor, but it will serve especially for my own instruction and guidance, being prepared, as it is, at the beginning of my administration.

And lest I seem to have omitted anything requisite to a familiarity with the entire subject, I will first set down the names of the waters which enter the City of Rome; then I will tell by whom, under what consuls, and in what year after the founding of the City each one was brought in; then at what point and at what milestone each water was taken; how far each is carried in a subterranean channel, how far on substructures, how far on arches. Then I will give the elevation of each, [the plan] of the taps, and the distributions that are made from them; how much each aqueduct brings to points outside the City, what proportion to each quarter within the City; how many public reservoirs there are, and from these how much is delivered to public works, how much to ornamental fountains (*munera,* as the more polite call them), how much to the water-basins; how much is granted in the name of

(continues)

Caesar; how much for private uses by favour of the Emperor; what is the law with regard to the construction and maintenance of the aqueducts, what penalties enforce it, whether established by resolutions of the Senate or by edicts of the Emperors. . . .

## BOOK I

Since I have given in detail the builders of the several aqueducts, their dates, and, in addition, their sources, the lengths of their channels, and their elevations in sequence, it seems to me not out of keeping to add also some separate details, and to show how great is the supply which suffices not only for public and private uses and purposes, but also for the satisfaction of luxury; by how many reservoirs it is distributed and in what wards; how much water is delivered outside the City; how much in the City itself; how much of this latter amount is used for water-basins, how much for fountains, how much for public buildings, how much in the name of Caesar, how much for private consumption. But before I mention the names *quinaria, centenaria,* and those of the other ajutages by which water is gauged, I deem it appropriate to state what is their origin, what their capacities, and what each name means; and, after setting forth the rule according to which their proportions and capacities are computed, to show in what way I discovered their discrepancies, and what course I pursued in correcting them.

The ajutages to measure water are arranged according to the standard either of digits or of inches. Digits are the standard in Campania and in most parts of Italy; inches are the standard in . . . Now the digit, by common understanding, is $\frac{1}{16}$ part of a foot; the inch $\frac{1}{12}$ part. But precisely as there is a difference between the inch and the digit, just so the standard of the digit itself is not uniform. One is called square; another, round. The square digit is larger than the round digit by $\frac{3}{14}$ of its own size, while the round is smaller than the square by $\frac{3}{11}$ of its size, obviously because the corners are cut off.

Later on, an ajutage called a *quinaria* came into use in the City, to the exclusion of the former measures. This was based neither on the inch, nor on either of the digits, but was introduced, as some think, by Agrippa, or, as others believe, by plumbers at the instance of Vitruvius, the architect. Those who represent Agrippa as its inventor, declare it was so designated because five small ajutages or punctures, so to speak, of the old sort, through which water used to be distributed when the supply was scanty, were now united in one pipe. Those who refer it to Vitruvius and the plumbers, declare that it was so named from the fact that a flat sheet of lead 5 digits wide, made up into a round pipe, forms this ajutage. But this is indefinite, because the plate, when made up into a round pipe, will be extended on the exterior surface and contracted on the interior surface. The most probable explanation is that the *quinaria* received its name from have a diameter of $\frac{3}{4}$ of a digit, a standard which holds in the following ajutages also up to the 20-pipe, the diameter of each pipe increasing by the addition of $\frac{1}{4}$ of a digit. For example the 6-pipe is six quarters in diameter, a 7-pipe seven quarters, and so on by a uniform increase up to a 20-pipe. . . .

## BOOK II

This is the schedule of the amount of water as reckoned up to the time of the Emperor Nerva and this is the way in which it was distributed. But now, by the foresight of the most painstaking of sovereigns, whatever was unlawfully drawn by the water-men, or was wasted as the result of negligence, has been added to our supply; just as though new sources had been discovered. And in fact the supply has been almost doubled, and has been distributed with such careful allotment that wards which were previously supplied by only one aqueduct now receive the water of several. Take for example the Caelian and the Aventine Hills, to which Claudia alone used to run on the arches of Nero. The result was, that whenever any repairs caused interruptions, these densely inhabited hills suffered a drought. They are now supplied by several aqueducts, above all, by Marcia, which has been rebuilt on a substantial structure and carried from Spes Vetus to the Aventine. In all parts of the City also, the basins, new and old alike, have for the most part been connected with the different aqueducts by two pipes each, so that if accident should put either of the two out of commission, the other may serve and the service may not be interrupted.

The effect of this care displayed by the Emperor Nerva, most patriotic of rulers, is felt from day to day by the present queen and empress of the world; and will be felt still more in the improved health of the city, as a result of the increase in the number of the works, reservoirs, fountains, and water-basins. No less advantage accrues also to private consumers from the increase in number of the Emperor's private grants; those also who with fear drew water unlawfully, now free from care, draw their supply by grant from the sovereign. Not even the waste water is lost; the appearance of the City is clean and altered; the air is purer; and the causes of the unwholesome atmosphere, which gave the air of the City so bad a name with the ancients, are now removed. . . .

What shall we say of the fact that the painstaking interest which our Emperor evinces for his subjects does not rest satisfied with what I have already described, but that he deems he has contributed too little to our needs and gratification merely by such increase in the water supply, unless he should also increase its purity and its palatableness? It is worth while to examine in detail how, by correcting the defects of certain waters, he has enhanced the usefulness of all of them. For when has our City not had muddy and turbid water, whenever there have been only moderate rain-storms? And this is not because all the waters are thus affected at their sources, or because those which are taken from springs ought to be subject to such pollution. This is especially true of Marcia and Claudia and the rest, whose purity is perfect at their sources, and which would be not at all, or but very slightly, made turbid by rains, if well-basins should be built and covered over.

The two Anios are less limpid, for they are drawn from a river, and are often muddy even in good weather, because the Anio, although flowing from a lake whose waters are very pure, is nevertheless made turbid by carrying away portions of its loose crumbling banks, before it enters the conduits—a pollution to which it is subject not only in the rain-storms of winter and spring, but also in the showers of summer, at which time of the year a more refreshing purity of the water is demanded.

(continues)

One of the Anios, namely Old Anio, running at a lower level than most of the others, keeps this pollution to itself. But New Anio contaminated all the others, because, coming from a higher altitude and flowing very abundantly, it helps to make up the shortage of the others; but by the unskilfulness of the water-men, who diverted it into the other conduits oftener than there was any need of an augmented supply, especially Claudia, which, after flowing in its own conduit for many miles, finally at Rome, as a result of its mixture with Anio, lost till recently its own qualities. And so far was New Anio from being an advantage to the waters it supplemented that many of these were then called upon improperly through the heedlessness of those who allotted the waters. We have found even Marcia, so charming in its brilliancy and coldness, serving baths, fullers, and even purposes too vile to mention.

It was therefore determined to separate them all and then to allot their separate functions so that first of all Marcia should serve wholly for drinking purposes, and then that the others should each be assigned to suitable purposes according to their special qualities, as for example, that Old Anio, for several reasons (because the farther from its source it is drawn, the less wholesome a water is), should be used for watering the gardens, and for the meaner uses of the City itself.

But it was not sufficient for our ruler to have restored the volume and pleasant qualities of the other waters; he also recognized the possibility of remedying the defects of New Anio, for he gave orders to stop drawing directly from the river and to take from the lake lying above the Sublacensian Villa of Nero, at the point where the Anio is clearest; for inasmuch as the source of Anio is above Treba Augusta, it reaches this lake in a very cold and clear condition, be it because it runs between rocky hills and because there is but little cultivated land even around that hamlet, or because it drops its sediment in the deep lakes into which it is taken, being shaded also by the dense woods that surround it. These so excellent qualities of the water, which bids fair to equal Marcia in all points, and in quantity even to exceed it, are now to supersede its former unsightliness and impurity; and the inscription will proclaim as its new founder, "Imperator Caesar Nerva Trajanus Augustus."

We have further to indicate what is the law with regard to conducting and safeguarding the waters, the first of which treats of the limitation of private parties to the measure of their grants, and the second has reference to the upkeep of the conduits themselves. In this connection, in going back to ancient laws enacted with regard to individual aqueducts, I found certain points wherein the practice of our forefathers differed from ours. With them all water was delivered for the public use, and the law was as follows: "No private person shall conduct other water than that which flows from the basins to the ground" (for these are the words of the law); that is, water which overflows from the troughs; we call it "lapsed" water; and even this was not granted for any other use than for baths or fulling establishments; and it was subject to a tax, for a fee was fixed, to be paid into the public treasury. Some water also was conceded to the houses of the principal citizens, with the consent of the others.

To which authorities belonged the right to grant water or to sell it, is variously given even in the laws, for at times I find that the grant was made by the aediles, at other times by the censors; but it is apparent that as often as there were censors in the government these grants were sought chiefly from them. If there were none, then the aediles had the power referred to. It is plain from this how much more our forefathers cared for the general good than for private luxury, inasmuch as even the water which private parties conducted was made to subserve the public interest.

The care of the several aqueducts I find was regularly let out to contractors, and the obligation was imposed upon these of having a fixed number of slave workmen on the aqueducts outside the City, and another fixed number within the City; and of entering in the public records the names also of those whom they intended to employ in the service for each ward of the City. I find also that the duty of inspecting their work devolved at times on the aediles and censors, and at times on the quaestors, as may be seen from the resolution of the Senate which was passed in the consulate of Gaius Licinius and Quintus Fabius.

Source: Reprinted by permission of the Publishers of the Loeb Classical Library from Frontinus. *Strategems Aqueducts* (pp. 331–55) with English translation by Charles E. Bennett, Cambridge, Mass.: Harvard University Press, 1961.

## Suggestions for Discussion and Writing

1. What claims does Frontinus make concerning technology's purpose in serving society? What evidence does he use to support these claims?

2. Compare the information in the preface to the information contained in the executive summary of a modern report. What similarities and differences do you notice?

3. Frontinus includes in his discussion of ajutages in Book I the origin of the word "quinaria, a standard of measure." Why do you think he includes this discussion? Would such a discussion be included in a report written today? Why or why not? In his discussion, he includes explanations by other experts. Why does he refer to them?

4. Bracket a passage in which Frontinus uses a concrete example to support his observations in Book II.

5. Bracket a passage that demonstrates how Frontinus uses the cause/effect organization pattern typical of report narrative to organize his information.

6. Why does Frontinus take pains throughout this section to compliment the emperor on his efforts to improve the aqueducts? See if you can find similar types of passages in the McDonald's or ground beef reports. Bracket them if you find them. Why do you think the writers may have included or omitted such passages?

---

## *Relief Work in Mississippi Flood and Principles Of Flood Control,* **Herbert Hoover, 1927**

Herbert Hoover, thirty-first president of the United States, graduated as a mining engineer in Stanford's first class in 1895. Prior to becoming president, he served as Secretary of Commerce from 1921 to 1928. In 1927 President Calvin Coolidge called upon him to direct relief operations after an "unprecedented flood on the lower Mississippi." The flooded area stretched down the river 1,000 miles from Cairo, Illinois, to the Gulf of Mexico. The rescue effort Hoover directed was a monumental undertaking by any standard.

To keep the president informed of his progress in providing relief, Hoover wrote a series of periodic reports on his efforts. Periodic reports are conventionally organized chronologically, from present status to future plans. In the report printed in this chapter, Hoover begins by describing the present situation in relation to two aspects: rescue and the care of those displaced from their homes. He then discusses two aspects that will need to be addressed in the future: rehabilitation and flood control.

The report is factual; Hoover includes statistical data to support his claims. The style is similar to that of today; the tone is impersonal, sentences and paragraphs are short, and the list and headings are used to facilitate readers' processes and behavior.

Hoover recognized that, although the reports were being sent to the president, they would also be read by many governmental officials. The press, too, would read and publish them, thereby making the information available to the general public. He was correct. The reports were published verbatim in the *Herald Tribune* and the *New York Times*.

Hoover was also acutely aware of the economic, political, and social factors related to the event. Human suffering and financial losses had been great. The public needed to be reassured that measures would be taken to prevent a recurrence. Hoover therefore tried to maintain a positive tone in his reports, stressing the progress of the relief effort and underscoring the need for future flood control measures.

RELIEF WORK IN MISSISSIPPI FLOOD

AND PRINCIPLES OF FLOOD CONTROL—REPORT

TO PRESIDENT COOLIDGE

20 July 1927

The flood relief embraces four stages with four district organizations required—i.e., rescue, care of exiles, rehabilitation and flood control. The situation dates as follows:

The rescue organization has been mobilized, several hundred thousand people were brought out of the water since President Coolidge decided on the consolidation of all effort on the 21st of April. There has been a known loss of only three lives. The refugee organization** has been largely demobilized.

Of the total people flooded, 68,000 were dependent on public relief for food, shelter, etc., for from six to eight weeks. They were mostly placed in camps, but some fed in the upper stories. All but 20,000 of the

(continues)

people in the camps have been returned to their homes, although 150,000 are still being fed in their homes. Such health conditions have been maintained in camps that the total deaths were actually less than the births. The people generally were sent home clothed and in better health than when they were brought to the camps.

NEW CROPS PLANTED ON 2,000,000 ACRES

The county reconstruction committees*** are making substantial progress with rehabilitation in 101 counties considerably flooded; 2,000,000 acres of crops that were lost have been replanted with staple crops from seed furnished by the relief. We hope the acreage was planted early enough to secure a full crop. The area should, with the help being given, recover to self-support by the fall. On the other hand, about 1,000,000 acres of original crops, mostly in 20 counties, have remained under water so long as to prevent planting of staple products and the ultimate recovery of the proportion will be delayed by another year.

The inability to pay interest, State, and levy
taxes, and for some individuals among them to find food
for winter, may create a continuous problem. The people
involved are discouraged. In all counties the destitute
sufferers were furnished with a supply of food and fed on
leaving camps; and, if their homes were destroyed, they
were given their tents, they are being given household
furniture, implements and animals. The repairs of damaged
homes and building of new homes, for those who can not
provide for themselves, are now in progress. It can not
all be done in a day, but these services should all be
completed by October. Sanitary measures are being set up
in every county to put down malaria and other communicable
diseases. The financial condition is that with economy we
can complete all these programs—seed, feed, food, furniture, animals, house construction and sanitation. By the
first of November we estimate that we shall have spent on
the whole relief operation $13,400,000 from the Red Cross
fund, $7,000,000 on equipment and supplies from the
Federal Government, $3,000,000 free railway transportation
and provided $1,100,000 for county health cleanup units
from other sources. With economy we should have left

(continues)

$3,000,000 from the $16,500,000 subscribed to the Red
Cross with which to face continued necessities after
November 1.

FURTHER INVESTIGATION TO BE MADE IN AUTUMN

It is impossible at this date to determine what
the necessities will be after that date. That can only be
told when we have had a chance to examine the actual situation both as to the crop, amount of employment and the
volume of need in other directions. We will make an exhaustive investigation into the questions, during the fall
and will establish further cooperative measures in the situation. This has nothing to do with future flood control
policies.

Whatever else may be provided by the State
Legislation in relief from taxation certainly they should
have relief from the Federal Government for repairs to the
levees.

The greatest of all measures needed is prompt and effective flood control and quick legislation for that will restore confidence and from confidence will come a recovery in values and in business. It becomes clear that flood control must embrace the following principles:

(A) Higher and consequently wider levees and the extension of Federal responsibility for levees on some of the tributaries.

(B) A safety valve upon the levee system by the provision of a "spillway," or more properly called a "by-pass," to the Gulf to protect New Orleans and southern Louisiana—most probable using the Atchafalaya River for this purpose.

(C) Of further safety measures the engineers are examining the possible extension of the by-pass in the northward from the Atchafalaya towards the Arkansas, the possible erection of emergency flood basins and the possibilities of storage in the tributaries.

(continues)

CONTROL OF RIVER SAID TO BE POSSIBLE

There is no question that the Mississippi River can be controlled if a bold and proper engineering plan is developed. It is not possible for the country to contemplate the constant jeopardy which now exists to 1,500,000 of its citizens or the stupendous losses which the lack of adequate control periodically brings about. Furthermore, flood control means the secure development of some 20,000,000 acres of land capable of supporting five to ten million of Americans. The cost of such work if spread over 10 years would be an inconsiderable burden upon the country. It is not incompatible with national economy to prevent $10 of economic loss by the expenditure of $1 Federal outlay. In the face of their great losses and their present destitution I do not see how people along the river can contribute much more than the maintenance of the central works after they have been constructed.

     *—The agencies brought into rescue action comprised, the Engineer Corps personnel and equipment, the Coast Guard personnel and equipment, the lighthouse personnel and equipment, the Missippi River Commission personnel and equipment, the Missouri Pacific Railway personnel and equipment, the Illinois Central personnel and equipment, the Standard Oil Company personnel and equipment, the

American Legion, the Red Cross, the various local organizations.

      \*\*—The agencies brought in to assist in the
exile stage comprised the Red Cross, with its general
staff and 90 local committees, Army Quartermaster
Department equipment, Public Health Service, personnel and
supplies, State health service, personnel and supplies,
National Guard.

      \*\*\*—The county reconstruction committees repre-
sented the authority of the following organizations: The
Red Cross, State reconstruction commissions, Public Health
Service, Department of Agriculture, emergency loan corporations, county authorities.

Source: Report by Herbert Hoover to President Calvin Coolidge, "Relief Work in Mississippi Flood and Principles of Flood Control." Reprinted with permission by Herbert Hoover Library. Herbert Hoover Papers, Public Statements File, #763

## Suggestions for Discussion and Writing

1. List the strategies Hoover uses to help readers skim through the report to locate information.

2. What claims does Hoover make in his argument for flood control? What evidence does he use to support his claims?

3. In what ways does Hoover maintain a positive tone in his report? Provide examples.

4. Years after the flood, in his memoirs, Hoover describes his experience in the relief effort. Read the excerpt from his memoirs printed opposite. Compare the tone, point of view, and focus of the description of the flood in the memoirs with his description in the report to the president. Hoover, as the objective scientific observer writing the progress report, becomes Hoover, the sympathetic participant, as he writes his memoirs.

5. In 1993, sixty-six years after the great flood of the lower Mississippi, the great flood of the upper Mississippi occurred, causing havoc to

farmlands and cities alike and calling into question the system of levees that had been built as a result of the 1923 disaster. Government experts were called upon to oversee the 1993 flood relief efforts, just as Hoover had been assigned to do two-thirds of a century earlier. The excerpt in Figure 6.3 (page 321) comes from a report to the Illinois legislature by the Illinois Department of Transportation, which was responsible for overseeing flood relief efforts in that state. Compare the tone, point of view, and focus of the introductory description with the descriptions in Hoover's memoir (below) and in Hoover's report to the President (pages 311–317). Which of the documents do you think is the most humanity-centered? Why?

---

## CHAPTER 18

✳✦✳✳✦✳✳✦✳✳✦✳✳✦✳

## AN INTERLUDE—RELIEF IN THE MISSISSIPPI FLOOD OF 1927

The cause of the unprecedented flood on the lower Mississippi River in 1927 was the coincidence of floods on the Ohio, the Missouri, and the upper Mississippi which brought down more water than the lower Mississippi could carry tranquilly to the sea between its thousand miles of levees. The levees broke in scores of places. The area ultimately flooded was, in places, as much as 150 miles wide and stretched down the river 1,000 miles, from Cairo to the Gulf. . . .

For rescue work we took over some forty river steamers and attached to each of them a flotilla of small boats under the direction of Coast Guardsmen. As the motorboats we could assemble proved insufficient, the sawmills up and down the river made me 1,000 rough boats in ten days. I rented 1,000 outboard motors from the manufacturers, which we were to return. (But after it was all over we could find only 120 motors. Undoubtedly every fisherman in the territory motorized his transportation.) We established great towns of tents on the high ground. We built wooden platforms for the tents, laid sewers, put in electric lights, and installed huge kitchens and feeding halls. And each tent-town had a hospital.

As the flood receded we rehabilitated the people on their farms and homes, providing tents to the needy and building material, tools, seed, animals, furniture, and what not to start them going again. We established sanitary measures to put down malaria, typhoid, pellagra and generally prevention of contagious disease, all of which we continued after the flood.

As at this time we all believed in self-help, I financed the operation by three actions. We put on a Red Cross drive by radio from the flood area, and raised $15,000,000. I secured $1,000,000 from the Rockefeller Foundation to finance the after-flood campaign of sanitation to be matched by equal contributions from the counties. We organized a nonprofit organization through the United States Chamber of Commerce to provide $10,000,000 of loans at low rates, for rehabilitation, every cent of which was paid back. But those were days when citizens expected to take care of one another in time of disaster and it had not occurred to them that the Federal Government should do it.

The railways furnished for my immediate staff and myself a free, special train of Pullmans and a dining car. There we lived when not on our "mother" steamers. We directed the train to points as near as possible to any special emergency and often went round and round east and west of the flood area from Cairo to New Orleans. We usually traveled only by day, as many sections of the tracks were under shallow water. Wherever we stopped the railway officials connected our train with the telephone. The three months spent in this residence were not particularly comfortable, for the ordinary heat of the season was lifted by the superheaters of railway yards and iron Pullmans.

But there was a lighter side to this job. . . .

We had many experiences . . . Our engineers had learned to time the advance of the flood accurately and to determine within a few feet the height that the water would reach at any point. That was a simple problem in hydraulics. In the routine work in advance of the flood we ran our train into a . . . town and, assembling the Mayor, the town council, and the principal residents, informed them that the town would go under water about a certain date, and that this would rise to certain levels in relation to the depot platform but would probably not reach to the second floor or the roofs of most houses and stores. Therefore they should move all possessions upstairs to correspond with the levels we gave them and should build a reserve of rough boats so as to move all the people to the high ground some ten miles distant if the waters came up suddenly.

Further, we instructed them how to build a camp with the lumber, tents, and other materials which we would send, and be prepared for a two months' stay. We advised the Mayor to organize a police force to stay behind and guard the town. Except for the intelligent Mayor, the audience was incredulous. Finally a gentleman arose in the back of the room and said to me:

"You are a Wall Streeter. You intend to rob us. I am a surveyor in this parish. My father was a surveyor before me. There never has been a flood here, there never will be a flood here!"

That was what the inhabitants wanted to hear. They followed this false prophet— all but the Mayor. He educated himself by visiting some of our camps farther north and did his job in building the camp for the 15,000 inhabitants. Also he followed

(continues)

directions and built a cement wall around the electric light plant in the town, which he connected with the camp.

A few weeks later, I was rung up on our train in the middle of the night by the Mayor.

"The water is rushing over the town!"

"Yes, we told you it would. Are your people moving? Have you had your boats built? Have all the furniture and goods been taken upstairs or onto the roofs?"

"Not any boats, not any goods moved. That surveyor persuaded the people not to do anything in spite of everything I could say. But the camp is all ready."

I knew that, for our engineers had seen to it.

I told the Mayor that we would run our train down at once and would arrive early in the morning; that we would bring a trainload of boats; that he should get all the people on to the roofs. The railway station was the highest point in town. We ran into it just at daylight with the water up to the the axles of the train. On the platform a mob of terror-stricken people were standing thick as bristles up to their ankles in steadily rising water. As our engineer feared that the mob would climb onto the train, he ran us past the station. From the rear platform I called out that the water would probably rise more but to be patient, for we had a trainload of flatcars with boats following us. I called for the Mayor, who soon appeared in rubber boots, grinning broadly. I asked him:

"Why so cheerful, Mayor"

"Well, we had only two motorboats in town. A few minutes ago I saw that surveyor grab one and go due east. He said that he was leaving for good!". . . .

I received a lasting impression from this experience. I had organized relief among many peoples in Europe. One of our difficulties there had been to find sufficient intelligence, organizational ability and leadership in the many villages and towns to carry on the local work. But in this organization among Americans the merest suggestion sparked efficient and devoted organization—indeed often in advance of specific request. The reasons for this reach to the very base of our American system of life. In this there also lies a special tribute to the peoples of these states.

Source: "An Interlude: Relief in the Mississippi Flood of 1927." *The Memoirs of Herbert Hoover, The Cabinet and the Presidency, 1920–1933.* New York: Macmillan, 1952. Courtesy of the Herbert Hoover Presidential Library Association.

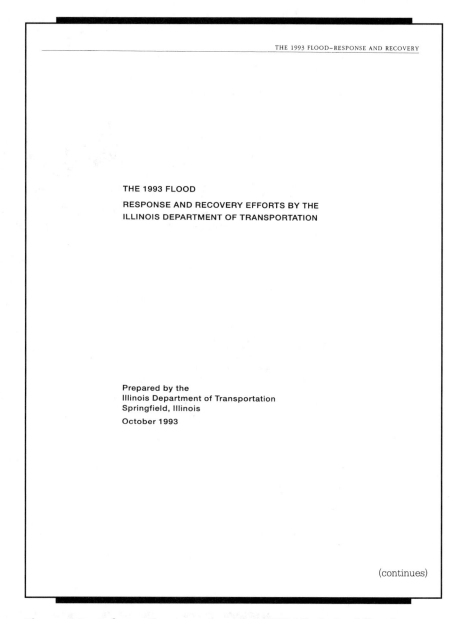

THE 1993 FLOOD

RESPONSE AND RECOVERY EFFORTS BY THE
ILLINOIS DEPARTMENT OF TRANSPORTATION

Prepared by the
Illinois Department of Transportation
Springfield, Illinois

October 1993

(continues)

**Figure 6.3    Information report on the 1993 Mississippi flood.** Source: Illinois Department of Transportation.

**322** The Documents

## INTRODUCTION

The great flood of '93 swept into Illinois in late June in the northern part of the state. More than three months later, parts of central and southern Illinois are still submerged and threatened by new flooding and heavy rains. The devastation of farmland and communities, as well as the disruption of people's lives and businesses, has been unprecedented.

In Illinois alone, 872,000 acres of farmland were flooded, destroying $425 million worth of corn, soybean and other crops. The flood impacted some 82 communities. Of these, 59 communities were actually flooded or sustained significant damage. A number of Illinois residents have lost their homes and, many more, the use of their homes. Still others have been forced to accept unemployment when their employers suspended work due to the flood or the commute over lengthy detours became too cumbersome and expensive.

The flood closed nearly 300 miles of roads and 12 bridges on the state highway system, and all four ferry routes crossing the Mississippi and Illinois rivers. Approximately 80 miles of state highways are still flooded. Another 900 miles of local roads and streets were inundated with flooding. Many of them are still not passable.

At one point in July, portions of the Joe Page Bridge approach at Hardin carrying Illinois 100 across the Illinois River were 16 feet under water. The Central Illinois Expressway at the U.S. 36/Illinois 336 interchange was 12 feet under water. A break in the Len Small levee near Miller City in Alexander County formed a new channel for the Mississippi River eroding some of the land to a depth of 80 feet.

Responding to this record-breaking flood has placed extraordinary demands on government agencies at every level, including the Illinois Department of Transportation (IDOT). This report summarizes IDOT's involvement in the flood response and recovery efforts.

**Figure 6.3** *(continued)*

## FLOOD RESPONSE

Through August 31, IDOT spent $6.8 million on operations and identified an estimated $10.7 million in repairs. More than 610 IDOT employees worked on

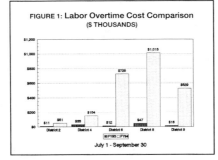

FIGURE 1: **Labor Overtime Cost Comparison**
($ THOUSANDS)

July 1 - September 30

direct flood-related duties. As of August 31, they have contributed nearly 77,000 regular work hours and 83,600 hours of overtime on a wide range of emergency measures. The work included: shoring up levees, building temporary roads and ferry landings, hauling sand and rock and other materials needed to hold back the flood waters, evacuating communities, airlifting hospital patients and stranded residents and motorists, directing motorists to alternate routes and controlling sightseer traffic, transporting bottled water and other supplies, keeping watch on the river levels to ensure the safety of the public, cleaning up debris, and repairing roads and bridges to reopen them.

Overtime pay for eligible employees during the first quarter of state fiscal year 1994 was $2.5 million compared to $118,500 for the first quarter of fiscal year 1993. That is a 2,000 percent increase. Figure 1 provides a comparison of cost by each district affected by the flood. Those statistics do not reflect the uncompensated

overtime hours accumulated by many employees who are neither entitled to overtime pay nor time off.

Many of the 400 orange IDOT trucks, 75 endloaders and other pieces of equipment used in the flood, have accumulated years of wear in only a few months. Figure 2 summarizes the utilization of dump trucks in the five flooded highway districts. Expenditures for aggregate rock, sand and other materials during the first quarter of fiscal year 1993 also increased significantly.

Since the flood began in late June, IDOT has:

- **Hauled 146,500 tons of rock and 64,000 tons of sand.** The combined tonnage filled about 22,000 truck loads that would form a 70-mile convoy. The rock was used to construct temporary road-ways on top of existing flooded pavement to bring the road above the flood level. Some of the aggregate was also used to reinforce levees. The sand was used for sandbags to build temporary levees to keep the water from flooding highways, public works facilities and other critical structures. Many of the sandbags also were used to raise the height of existing levees.

- **Hauled or stacked 7 million sandbags** to various locations. These sandbags, laid end-to-end, would form a 2,000-mile chain that would stretch from Springfield to Seattle, Washington.

- **Supplied 42 heavy duty pumps** that were put into service in several communities and areas of the state to prevent damage to public works facilities, to expedite drainage of flooded areas, and to keep roads open. Large rented pumps, consuming almost 60,000 gallons of diesel fuel per month, are being used to help keep open segments of Illinois 146 in southern Illinois, a key access route to the Cape Girardeau Bridge.

(continues)

**Figure 6.3** *(continued)*

- **Provided air transportation support** for a number of flood relief missions. State helicopters were stationed in Quincy for 25 days to provide emergency medical evacuation service after both bridges were closed. In addition, IDOT's Division of Aeronautics provided flight support missions for the Illinois Emergency Management Agency (IEMA) and supplied the Illinois National Guard with special rescue nets. For the most recent rescue operation, IDOT provided helicopter service to evacuate critically ill residents out of Bath, a town near the Illinois River that was flooded from a seeping underground aquifer in the region. The IDOT helicopters and fixed wing aircraft put in 194 hours of flood relief duty at a cost of more than $146,000.

- **Assisted with the evacuation** of several communities threatened with flooding by providing trucks and personnel.

- **Provided traffic patrol** at locations where needed. In several cases, IDOT employees were stationed at closed roads on a 24-hour basis to make sure motorists did not attempt to ford rising waters, putting themselves in danger. This was necessary at the McArthur Bridge across the Mississippi River between Gulfport, Illinois and Burlington, Iowa. A 24-hour patrol also was set up at both ends of a closed section of Illinois 84 at Albany in Whiteside County to prevent motorists, including sightseers, from taking unnecessary risks and to minimize wave action against sandbag levees around homes in the area. In addition, District 6 employees provided 24-hour road blocks at the Hamilton-Keokuk Bridge, the Hannibal Bridge and the Louisiana Bridge.

- **Provided flood risk assessments** using elevation readings and other critical data to assist IEMA with carrying out the state's comprehensive flood response strategy. The information from IDOT's Division of Water Resources was

used to identify potential problem areas in advance so that work crews, equipment and supplies could be directed for timely response measures.

Water Resources engineers also worked with the U.S. Corps of Engineers in assessing the flood protection provided by levees along the Mississippi and Illinois rivers. This was particularly critical with some levees in the St. Louis metro-east area because of the large population that was at risk.

Some of the major emergency projects undertaken by IDOT are highlighted as follows:

**In District 9, IDOT**

- Constructed a 1.5 mile crushed-rock roadway over the existing pavement of Illinois 3 in Alexander County to keep open an important truck access route to the Cape Girardeau Bridge across the Mississippi River in southern Illinois. The road was raised an average of two feet by placing 45,000 tons of rock on top of the existing surface. Cost: $716,000.

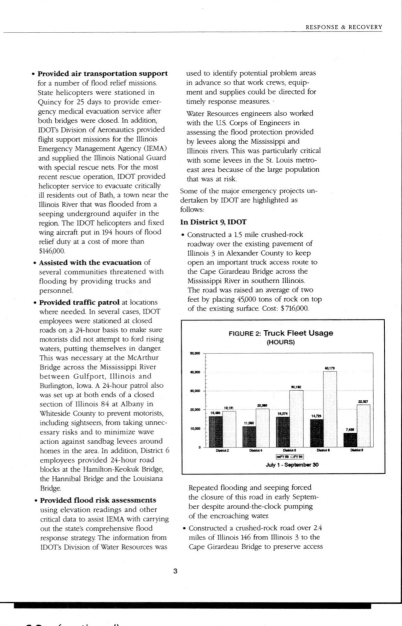

FIGURE 2: **Truck Fleet Usage** (HOURS)

July 1 - September 30

Repeated flooding and seeping forced the closure of this road in early September despite around-the-clock pumping of the encroaching water.

- Constructed a crushed-rock road over 2.4 miles of Illinois 146 from Illinois 3 to the Cape Girardeau Bridge to preserve access

3

**Figure 6.3** *(continued)*

## CHAPTER SUGGESTIONS FOR DISCUSSION AND WRITING

1. The introduction to this section of the textbook suggests that a model final report contains the following sections in the following sequence: (1) problem/background/rationale, (2) short- and long-range objectives, (3) methods/procedures, (4) results, (5) conclusions, and (6) recommendations. The reports included in this section, however, seldom follow this exact model. Sometimes the sequence of sections differs, sometimes a section is omitted (there are no recommendations in the lean ground beef report), and sometimes a section is only one paragraph (objectives in the lean ground beef report). Other times a section comprises several chapters. Sometimes each chapter contains a single section, but other times each chapter contains all of the sections (McDonald's, lean ground beef report). Why do you think the writers made these modifications?

2. Compare the design and methodology section of the lean ground beef production report with Section VI, Understanding McDonald's Packaging and Waste, in Chapter 1 of the McDonald's/EDF report. The ground beef report was written for experts and generalists (people who have some knowledge of the field but are not experts). The McDonald's/EDF report was written for novices as well as generalists. What differences in content, language, numerical data, and syntax do you notice, if any?

# CHAPTER 7

# Environmental Impact Statements

## INTRODUCTION

The Environmental Impact Statement (EIS) is perhaps the most controversial of all documents. It had a political birth in 1969 when it was introduced as part of the National Environmental Policy Act (NEPA), which was passed by the U.S. Congress to protect the environment. It has been involved in political controversies ever since. An EIS has the power to bring immense governmental projects to a screeching halt. In 1993 opponents of NAFTA (North American Free Trade Agreement) used the EIS as a means to delay the agreement, claiming that an EIS needed to be written before Congress could act on it. Environmental impact statements have resulted in delaying or even permanently halting a wide variety of construction projects, causing private contractors and the federal government to lose hundreds of thousands of dollars but guaranteeing citizens a safer environment and preserving historical and archaeological sites. An EIS finding indicating that an African slave burial ground was located on the site where a federal courthouse was to be built in Manhattan has not only forced the government to find another site for the courthouse but will cost the government several million dollars. However, it will preserve the site as a historic landmark and will guarantee that the black community oversees the scientific study of the remains from the gravesite (see Part III).

An environmental impact statement is a special type of report. It must accompany all proposals of projects that involve federal monies and which may affect the environment. Because our environment encompasses our historical and cultural environments as well as our physical environment, these statements often involve a wide variety of specialists, from archaeologists to biologists to psychologists. Usually these specialists write their own section of the statement, and an editor puts the individual sections together into a single document.

# CONTEXT

## Purpose

The purpose of an EIS is to provide decision makers responsible for determining whether to accept a proposal with information concerning a project's potential impact on the environment. Thus, according to NEPA, the EIS serves "as an action-forcing device to insure that the policies and goals defined in the Act (NEPA) are infused into the ongoing programs and actions of the Federal Government. . . . An environmental impact statement is more than a disclosure document. It shall be used by Federal officials in conjunction with other relevant material to plan actions and make decisions."

## Content

According to the legislation that specifies the information to be included, an EIS should contain a complete and unbiased discussion of all potential impacts on the environment that could result from implementing a project. In addition, if a project could impact the environment negatively, then the statement should suggest alternative plans to lessen or reduce the potential problems. For example, if an EIS indicates that a construction project may disturb a historic cemetery, as in the case of the Foley Square project, discussed in Part III, the EIS may suggest the following alternatives: (1) digging up the cemetery for further anthropological and historic studies, (2) relocating the cemetery, or (3) leaving the cemetery untouched.

## The EIS Review Process

The legislation was designed to make certain that decision makers knew as much as possible about a project's impact on the environment, including all potential negative effects, before determining whether to approve a proposal. Thus, the act required that anyone affected by a project have an opportunity to have input into the decision. The EIS process, according to NEPA, ensures that "environmental information is available to public officials and citizens before decisions are made and actions taken . . . The NEPA process is intended to help public officials make decisions that are based on understanding of environmental consequences, and take actions that protect, restore, and enhance the environment."

This legislation has resulted in a procedure in which a preliminary EIS is written and sent to various interested agencies for their input. The public also has access to the reports, which may be published in local papers or left at such public facilities as local libraries. Interested citizens can then offer their opinions on the project through public hearings, which are held by the writer's organization. The report is then revised to take account of the comments. The comments along with the writers' responses are published in the final draft of the report.

## Audience

The primary audience for an EIS is the decision makers who will read the project proposal, but government officials, representatives of private organizations, and interested citizens serve as intermediary readers. Thus, the audience is extremely varied, ranging from experts to novices. These readers also represent a broad range of interests, some that may be sympathetic to the EIS, and some that may be hostile.

## Conventions

In addition to specifying the audience and purpose of an EIS, the legislation also provides guidelines for the format, content, organization, and style. It suggests that the report should be fewer than 300 pages. However, many EIS reports are far longer when their appendices are included. It also indicates that the reports should be "written in plain language" and use appropriate graphics so that decision makers and the public can readily understand them. It further recommends that agencies hire writers and editors to write, review, and edit the statements from scientific and technical experts. The following format is suggested.

1. Cover sheet.

2. Summary.

3. Table of contents.

4. Purpose of and need for action.

5. Alternatives, including proposed action.

**§ 1502.11 Cover sheet.**

The cover sheet shall not exceed one page. It shall include:

(a) A list of the responsible agencies including the lead agency and any cooperating agencies.

(b) The title of the proposed action that is the subject of the statement (and if appropriate the titles of related cooperating agency actions), together with the State(s) and county(ies) (or other jurisdiction if applicable) where the action is located.

(c) The name, address, and telephone number of the person at the agency who can supply further information.

(d) A designation of the statement as a draft, final, or draft or final supplement.

(e) A one paragraph abstract of the statement.

(f) The date by which comments must be received (computed in cooperation with EPA under § 1506.10).

The information required by this section may be entered on Standard Form 424 (in items 4, 6, 7, 10, and 18).

**§ 1502.12 Summary.**

Each environmental impact statement shall contain a summary which adequately and accurately summarizes the statement. The summary shall stress the major conclusions, areas of controversy (including issues raised by agencies and the public), and the issues to be resolved (including the choice among alternatives). The summary will normally not exceed 15 pages.

**§ 1502.13 Purpose and need.**

The statement shall briefly specify the underlying purpose and need to which the agency is responding in proposing the alternatives including the proposed action.

**§ 1502.14 Alternatives including the proposed action.**

This section is the heart of the environmental impact statement. Based on the information and analysis presented in the sections on the Affected Environment (§ 1502.15) and the Environmental Consequences (§ 1502.16), it should present the environmental impacts of the proposal and the alternatives in comparative form, thus sharply defining the issues and providing a clear basis for choice among options by the decisionmaker and the public. In this section agencies shall:

(a) Rigorously explore and objectively evaluate all reasonable alternatives, and for alternatives which were eliminated from detailed study, briefly discuss the reasons for their having been eliminated.

(b) Devote substantial treatment to each alternative considered in detail including the proposed action so that reviewers may evaluate their comparative merits.

(c) Include reasonable alternatives not within the jurisdiction of the lead agency.

(d) Include the alternative of no action.

(e) Identify the agency's preferred alternative or alternatives, if one or more exists, in the draft statement and identify such alternative in the final statement unless another law prohibits the expression of such preference.

(f) Include appropriate mitigation measures not already included in the proposed action or alternatives.

**§ 1502.15 Affected environment.**

The environmental impact statement shall succinctly describe the environment of the area(s) to be affected or created by the alternatives under consideration. The descriptions shall be no longer than is necessary to understand the effect of the alternatives. Data and analyses in a statement shall be commensurate with the importance of the impact, with less important material summarized, consolidated, or simply referenced.

(continues)

Agencies shall avoid useless bulk in statements and shall concentrate effort and attention on important issues. Verbose descriptions of the affected environment are themselves no measure of the adequacy of an environmental impact statement.

### § 1502.16 Environmental consequences.

This section forms the scientific and analytic basis for the comparisons under § 1502.14. It shall consolidate the discussions of those elements required by sections 102(2)(C) (i), (ii), (iv), and (v) of NEPA which are within the scope of the statement and as much of section 102(2)(C)(iii) as is necessary to support the comparisons. The discussion will include the environmental impacts of the alternatives including the proposed action, any adverse environmental effects which cannot be avoided should the proposal be implemented, the relationship between short-term uses of man's environment and the maintenance and enhancement of long-term productivity, and any irreversible or irretrievable commitments of resources which would be involved in the proposal should it be implemented. This section should not duplicate discussions in § 1502.14. It shall include discussions of:

(a) Direct effects and their significance (§ 1508.8).

(b) Indirect effects and their significance (§ 1508.8).

(c) Possible conflicts between the proposed action and the objectives of Federal, regional, State, and local (and in the case of a reservation, Indian tribe) land use plans, policies and controls for the area concerned. (See § 1506.2(d).)

(d) The environmental effects of alternatives including the proposed action. The comparisons under § 1502.14 will be based on this discussion.

(e) Energy requirements and conservation potential of various alternatives and mitigation measures.

(f) Natural or depletable resource requirements and conservation potential of various alternatives and mitigation measures.

(g) Urban quality, historic and cultural resources, and the design of the built environment, including the reuse and conservation potential of various alternatives and mitigation measures.

(h) Means to mitigate adverse environmental impacts (if not fully covered under § 1502.14(f)).

[43 FR 55994, Nov. 29, 1978; 44 FR 873, Jan. 3, 1979]

### § 1502.17 List of preparers.

The environmental impact statement shall list the names, together with their qualifications (expertise, experience, professional disciplines), of the persons who were primarily responsible for preparing the environmental impact statement or significant background papers, including basic components of the statement (§§ 1502.6 and 1502.8). Where possible the persons who are responsible for a particular analysis, including analyses in background papers, shall be identified. Normally the list will not exceed two pages.

### § 1502.18 Appendix

If an agency prepares an appendix to an environmental impact statement the appendix shall:

(a) Consist of material prepared in connection with an environmental impact statement (as distinct from material which is not so prepared and which is incorporated by reference (§ 1502.21)).

(b) Normally consist of material which substantiates any analysis fundamental to the impact statement.

(c) Normally be analytic and relevant to the decision to be made.

(d) Be circulated with the environmental impact statement or be readily available on request.

Source: *Regulations for Implementing the Procedural Provisions of the National Environmental Policy Act.* From the Council on Environmental Quality, Executive Office of the President. July 1986.

# THE DOCUMENT

## Context

Prior to determining a site for the Superconducting Super Collider, the Department of Energy (DOE) prepared a general EIS. Once DOE officials determined the exact site, a supplemental EIS needed to be written that was concerned with the specific site. DOE contracted with the Argonne National Laboratory in Illinois to study the proposed site and to write the EIS.

### *The National Context*

The document was written in a context in which economic, political, and social factors played a large part in determining whether the proposal for the project would finally be approved. Many scientists believed the United States needed to construct the complex site to keep ahead as a world scientific power. However, many citizens were not sure the country could afford the project and did not want their tax dollars spent on it.

### *The Local Context*

Many of the residents where the Super Collider would be built read the EIS and responded to it. Many of these residents didn't understand what a Super Collider was or whether it could affect their health or safety. However, they knew that the construction of the project should improve the area's economy.

Located in Ellis County, Texas, the projected site for the Super Collider lies within a ninety-mile radius of the Dallas/Ft. Worth area. The area is mainly rural, and many of the 75,000 people living in the county are former urbanites who moved to the area to escape pollution, crowded conditions, and traffic. These residents were concerned with the ways the project would affect their lifestyle.

Although much of the Dallas area had grown rapidly, growth in Ellis County had not been as great. The Super Collider was expected to help the area catch up by creating more than 9,400 new jobs. Regional earnings were expected to increase, as was demand for goods. However, with the influx of new workers to staff the project, it was expected that the counties would need to increase the number of teachers and law enforcement officials and to boost fire protection and health services. Residents wanted to know how these services would be provided and who would pay for them. They did not want to pay higher taxes to cover the increased costs.

The construction of the Super Collider would cause the dislocation of approximately 500 people, who would have to move to make way for the project. It would also affect several hundred others in the surrounding area who had looked out on a rural landscape. They would now view the walls and roads of the project. There would also be an increase in traffic and noise during project construction. These people were very concerned with the site location and construction.

## The EIS

Excerpts from the EIS supplement statement for the Super Collider follow. The actual document is 312 pages, not including the ancillary documents. As you can see by skimming through the excerpts included here and by scanning the table of contents, the report includes the ancillary documents required by NEPA: cover sheet, summary, table of contents, and appendices. It also provides readers with an abstract and a list of tables. The list of preparers can give you some idea of the number of people involved in writing the document and the range of specialties included.

By examining the table of contents, you can see that the document is organized in a modified cause/effect pattern. Chapters 2 and 3 of the report relate to the cause (the proposed project and the environment involved), and Chapters 4 and 5 relate to the effects of the project. The chapters themselves follow a modified analytical pattern in which each topic is discussed as a separate chunk. For example, both Chapters 3 and 4 are divided into the major environmental categories involved, such as earth, water, biotic, and land resources; climate and meteorology; and baseline noise and vibration. In Chapter 4, each of the subcategories follows a problem/solution organizational pattern. The writer presents the effect on the environment (the problem) and then discusses measures for mitigating the effect (solving the problem).

The following excerpts cover items d, e, f, and g from NEPA §1502.16 (see page 330). The excerpts from Chapter 1 provide a general overview of an EIS and its review process as well as an overview of this specific EIS. The excerpts from Chapter 2 discuss the proposed actions, the excerpts from Chapter 3 examine the affected environment, and the excerpts from Chapter 4 study the consequences of the project.

The descriptions in Chapters 3 and 4 are very detailed so that experts in the field can evaluate the information. The subsections on mitigative measures in Chapter 4 are implicitly persuasive. Writers are trying to persuade readers that these measures will solve the negative environmental consequences described in the respective sections.

As you examine the excerpts from this statement, consider how you would respond if you were a resident in the area where the project is to be located.

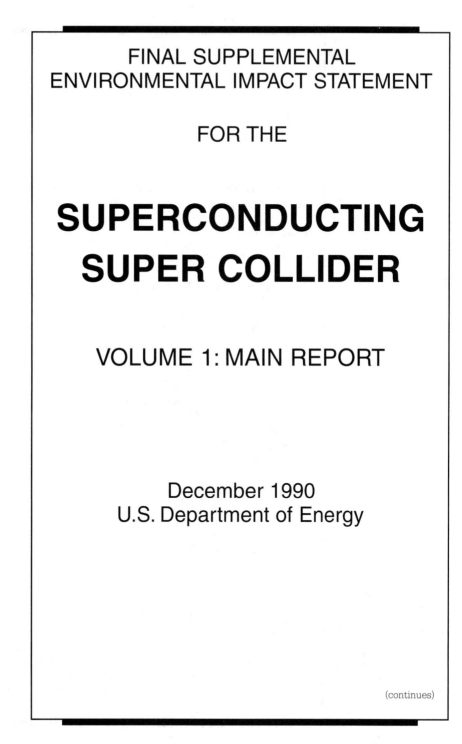

FINAL SUPPLEMENTAL
ENVIRONMENTAL IMPACT STATEMENT

FOR THE

# SUPERCONDUCTING
# SUPER COLLIDER

## VOLUME 1: MAIN REPORT

December 1990
U.S. Department of Energy

(continues)

December 1990

## COVER SHEET, VOLUME 1

**LEAD AGENCY**

U.S. Department of Energy (DOE)

**TITLE**

Final Supplemental Environmental Impact Statement (SEIS) for the Superconducting Super Collider

**CONTACT**

For further information, contact:

1.  Joseph R. Cipriano, Manager
    SSC Project Office
    U.S. Department of Energy
    1801 North Hampton Avenue
    DeSoto, TX 75115

2.  Carol Borgstrom, Director
    Office of NEPA Oversight
    Office of the Assistant Secretary for Environment, Safety and Health
    U.S. Department of Energy (EH-25)
    1000 Independence Avenue, S.W.
    Washington, DC 20585
    (202) 586-4600

3.  William Dennison
    Acting Assistant General Counsel for Environment
    U.S. Department of Energy (GC-11)
    1000 Independence Avenue, S.W.
    Washington, DC 20585
    (202) 586-6947

**The abstract fulfills guideline 1502.11(e). This is mainly written for readers of computer data bases in which the report will be entered, rather than those who will read a hard copy of the report.**

**ABSTRACT**

The proposed action evaluated in the SEIS is the construction and operation of the Superconducting Super Collider (SSC), the largest scientific instrument ever built, in Ellis County, Texas. The SSC would be a laboratory facility designed to investigate the basic structure of matter. It would be a particle accelerator capable of accelerating each of two counter-rotating beams of protons to an energy of 20 trillion electron volts. The two proton beams would then be made to collide, and the results of these collisions (at energies up to 40 trillion electron volts) would be studied by scientists.

On November 10, 1988, the Secretary of Energy identified the Texas site as the preferred alternative for the location of the SSC. The DOE published a final EIS in December 1988, and a Record of Decision was signed that documented DOE's decision to proceed with the SSC and to formally select the site in Ellis County. In the EIS and the Record of Decision, the DOE committed to prepare a supplemental EIS prior to construction in order to analyze more fully impacts based on a site-specific design and to assess alternative measures to mitigate potentially adverse impacts. Public hearings were held in the vicinity of the site during September 1990. This final SEIS reflects comments received during those hearings and in written letters.

**Page numbering before main text uses small Roman numerals.**

iv

(continues)

CONTENTS, VOLUME 1

(continues)

**Lists Appendices. Glossary is part of Appendix.**

**Includes responses to comments as required legally.**

TABLES

**Table of Contents for tables.**

(continues)

1-1

# 1  SUMMARY

## 1.1  OVERVIEW

The U.S. Department of Energy (DOE)* has proposed that the United States build the Superconducting Super Collider (SSC), a state-of-the-art laboratory facility for the study of high-energy physics. The proposed SSC would be the largest scientific instrument ever built.

On November 10, 1988, the Secretary of Energy identified the Texas site as the preferred alternative for the location of the SSC. The DOE published a final environmental impact statement (FEIS) in December 1988, and a Record of Decision (ROD) was signed that documented DOE's decision to proceed with the SSC and to formally select the site in Ellis County. In the EIS and the ROD, the DOE committed to prepare a supplemental EIS (SEIS) prior to construction in order to analyze more fully impacts based on a site-specific design and to assess alternative measures to mitigate potentially adverse impacts.

This SEIS has been prepared to take into account design modifications that have been made to the SSC since the ROD was published. These modifications have been made both to accommodate technical improvements to the SSC and to adapt the SSC conceptual design to the Texas site. Wherever possible, and particularly where no significant changes have occurred in the SSC design since the ROD, this SEIS relies on the analyses and assessments presented in the December 1988 EIS. This chapter summarizes the information in this SEIS, emphasizing any changes that have occurred since publication of the EIS.

The basic purpose of the SSC is to gain a better understanding of the fundamental structure of matter. This machine will be capable of accelerating two beams of subatomic particles (protons) to an energy of 20 trillion electron volts (TeV). The two beams will then be made to collide, and the results of these collisions (at 40 TeV) will be studied by scientists. The SSC could create particle collisions at energies 20 times higher than can be achieved at existing accelerators. This means that the SSC could probe the properties of matter at distances 20 times smaller than can now be done with existing and planned particle accelerators. The SSC will enable the United States to maintain its world leadership in the field of high-energy physics.

The SSC is expected to result in other benefits as well. Besides providing scientific data, the SSC could be a source of spin-off technology with applications in

_____
*Appendix A contains a glossary and a list of acronyms and abbreviations.

---

*Sidebar annotations (left margin):*

The term "Summary" is misleading. This section is not an executive summary but a summary of the project.

Provides background of document.

Explains that once a site was selected, a supplemental EIS was written to look more closely at the environmental impact of the study on that specific site and to decide on ways to solve adverse effects.

States purpose of document.

Gives forecast for chapter.

Gives background of project.

Explains benefits of project.

Uses jargon and acronyms extensively.

Tells reader how to learn what jargon and acronyms are.

Pages are numbered at top of page so readers can quickly and easily locate information. Chapter and page are both listed.

1-2

other fields. Within the past 10 years, the technology developed for high-energy physics has made new products possible, such as equipment used for medical diagnostics and therapy, improved computer components and new superconducting magnet materials.

Projecting to the future, discoveries resulting from SSC may lead to benefits that are currently impossible to envision. Looking back in time, one sees that research in subatomic physics over the last 80 years was essential to the development of technology, including portions of computers, that constitutes a significant portion of our current gross national product. On a broader scale, the wonder and excitement resulting from discoveries made possible by the SSC may provide inspiration for young people to enter careers in science and engineering. This atmosphere could contribute to maintaining America's economic competitiveness in an increasingly technological world.

Uses a numerical system of outlining to indicate to readers the hierarchy of the information.

## 1.2 PROPOSED ACTION AND MODIFICATION OF ORIGINAL PROPOSED ACTION AND ALTERNATIVES

The proposed action assessed in this SEIS is to construct and operate the SSC at the selected site in Texas. This SEIS relies heavily on the analyses and assessments presented for the Texas site in the final EIS (FEIS). As was anticipated in the FEIS (Vol. I, Section 3.1.1), some design details have been modified to accommodate environmental and technical aspects of the Texas site. These modifications are discussed in general terms below.

Defines words readers may not know in text, so readers can read fluently.

Describes the difference between the FEIS and the SEIS.

The proposed layout, or *footprint,* of the SSC, which identifies land areas above and below ground, was developed by responding to environmental and operating requirements and by adapting the required SSC configuration to the specific geological and topographical features of the Texas site. Variances in these requirements from those set forth in the site proposal from the state of Texas (Texas National Research Laboratory Commission 1987) and in the initial conceptual design (SSC Central Design Group 1986a) (which together provided the basis for the EIS assessment) have arisen from developments in the design of the accelerator and from the evolution of the requirements for the experimental program since the Invitation for Site Proposals (ISP) was issued by the DOE in 1987. In addition, analysis of geotechnical data has led to the currently proposed slight shift and counterclockwise rotation of the facility on the site compared with the placement described in the Texas site proposal. . . .

Composites were prepared by overlaying maps of floodplains, land use, land parcelization,* slopes, and soils for use in the preliminary siting of buildings. . . .

Uses footnotes to define community-specific terms.

---

*The term *land parcelization* is used in this document to refer to the distribution of lands into parcels by ownership and use. The principal consideration here was to avoid (to the extent possible) siting SSC surface facilities in such a way as to divide a large, single-owner land parcel into two or more smaller, divided parcels and thus lower the value and utility of the land.

(continues)

Uses headings and subheadings so readers can quickly and easily locate information as well as accurately predict what they will read in the following chunk.

Headings and subheadings are in boldfaced type.

Headings are in all capital letters so readers can easily see them.

### 1.3  PURPOSE AND NEED FOR THE PROPOSED ACTION

The purpose and need for the SSC have not changed from those described in the FEIS (Vol. 1, Chapter 2). In January 1990, a High-Energy Physics Advisory Panel (HEPAP) Subpanel on SSC Physics was convened to review the technical changes discussed in Section 1.2 of this SEIS and to investigate the potential usefulness of an SSC with somewhat reduced energy. In a letter transmitting the HEPAP subpanel's 1990 report to the DOE, HEPAP Chairman Francis E. Low (1990) stated:

> Timely completion of the Superconducting Super Collider (SSC) remains the highest priority of the national High Energy Physics program. The physics research to be done with the SSC is essential to the improvement of present human understanding of the fundamental forces of nature and the underlying constituents of the physical universe in which we live. The SSC is certain to be a focus of worldwide scientific attention for decades to come.

The subpanel's report (DOE 1990) contains the following statements:

> The very spirit of physics is to explore the unknown. This makes it impossible for us to predict precisely what we will discover in the future. Based on our present knowledge, however, we are confident that the SSC will explore a region in which major new discoveries will be made. The SSC specifications were established with this goal in mind, and experience gained since 1983 has strengthened our conviction of the importance of constructing a proton-proton collider with the beam energy of 20 TeV and luminosity of $10^{33}$ cm$^{-2}$ see $^{-1}$, as originally proposed.

and

> . . . Based on recent experimental and theoretical findings, the SSC Laboratory judges these technical changes (higher injection energy and a larger magnet aperture) to be required for reliable operation. We have reviewed these technical changes and the reasons for making them, and we conclude that implementing them will ensure confidence in reliable and timely operation of the SSC.

The HEPAP report thus reiterates and reemphasizes the purpose and need for the proposed action as described in Chapter 2 (Vol. I) of the FEIS and supports the modifications described in this SEIS.

### 1.4  ENVIRONMENTAL CONSEQUENCES

Explains table so readers know what they are looking at.

Table 1.1 summarizes the potential environmental impacts associated with construction and operation of the SSC. The assessed impacts of the conceptual design as presented in the FEIS are provided for comparison with the assessed impacts of the site-specific conceptual design. The differences in impacts can be attributed to (1) the use of additional site-specific data not available for the assessments conducted for the EIS; (2) the application of more refined or sophisticated technical approaches, where appropriate; and (3) changes in the location of areas of surface disturbance.

### 1.5  FEDERAL PERMITS, LICENSES, AND OTHER ENTITLEMENTS

The DOE has examined the federal permits, licenses, and other entitlements that may be necessary to construct and operate the SSC in Texas. Various federal environmental statutes impose environmental protection and compliance requirements upon DOE, including compliance with applicable state and local regulations.

Chapter 5 discusses federal statutes that may apply to construction and operation of the SSC, including the Clean Water Act, the Clean Air Act, the Safe Drinking Water Act, the Solid Waste Disposal Act, the National Historic Preservation Act, the Endangered Species Act, the Farmland Protection Policy Act, and the Uniform Relocation and Real Property Acquisition Policies Act.

### 1.6  CHANGES IN THE SUPPLEMENTAL EIS FROM THE EIS

The SEIS includes site-specific analyses relevant to an exact location (i.e., a *footprint*) for the SSC project facilities. Specific design requirements developed since publication of the EIS and information gained from additional geotechnical test borings taken at the Texas site have resulted in a more exact location for the collider ring, service areas, east and west campus areas, utility lines, access roads, and other project features. . . .

**Includes references for readers who wish to read primary source.**

### 1.8  REFERENCES FOR SECTION 1

DOE, 1988, *Final Environmental Impact Statement, Superconducting Super Collider,* U.S. Department of Energy Report DOE/EIS-0138, Washington, D.C., Dec.

DOE, 1990, *Report of the HEPAP Subpanel on SSC Physics,* U.S. Department of Energy Report DOE/ER-0434, Washington, D.C., Jan.

Low, F.E., 1990, Chairman, High-Energy Physics Advisory Panel, Cambridge, Mass., letter to J. F. Decker, Office of Energy Research, U.S. Department of Energy, Washington, D.C., Jan. 12.

SSC Central Design Group 1986a, *SSC Conceptual Design,* Report SSC-SR-2020, March.

(continues)

Uses table to summarize data. Table is numbered and has a title. Table helps reader to perceive relation- ships among data. Columns are labeled. Space between columns, rows allows readers to see each column, row easily. Uses hanging indentation so readers easily see chunks in each row.

TABLE 1.1 **Comparison of Potential Environmental Impacts Associated with Constructing and Operating the SSC—EIS versus SEIS**

| Impact | SEIS | EIS |
|---|---|---|
| Earth Resources | | |
| Loss of oil and gas wells | None | None |
| Loss of metallic resources | None | None |
| Loss of quarries | None | None |
| Water Resources | | |
| Surface water use | Use of small increment of excess supply | Use of small increment of excess supply |
| Groundwater use | Use is projected to be about 14% of total use in Ellis County in 1986 | Increase groundwater withdrawal by 915 to 954 acre-ft/yr |
| Floodplains | Some surface facil- ities at 4 of the 18 service areas would be subject to flooding during the 100- year or greater flood | Some surface facilities at four external beam access areas and two service areas could be affected by flood- plains |
| Air Resources | | |
| Air quality (PM$_{10}$ NAAQS)[a] | Impacts are similar to EIS assessment | Small, incremental addi- tion to regional air emissions; PM$_{10}$ fugitive dust impact from con- struction activity is of a temporary and inter- mittent nature and will vary from location to location |
| Noise | | |
| Number of people living in areas with intermittent 70-75-dBA levels during construction | 0 (with mitigation) | 25 |
| Number of people living in areas with intermittent 60-70-dBA levels during construction | 40 (with mitigation) | 314 |

SSC Central Design Group, 1986b, *Conventional Facilities,* Attachment C to SSC Conceptual Design, Report SSC-SR-2020C, March.

SSCL, 1990, *Superconducting Super Collider Site-Specific Conceptual Design,* Superconducting Super Collider Laboratory Report SSC-SR-1056, Dallas, July.

Texas National Research Laboratory Commission, 1987, *Texas Response to Invitation for Site Proposals,* Austin, Sept.

Texas National Research Laboratory Commission, 1990, letter from M. Smith to M. Lazaro, Argonne National Laboratory, Argonne, Ill., May 17.

2-1

## 2  PROPOSED ACTION AND ALTERNATIVES

**Forecast.**

This chapter describes the proposed action and alternatives. Where appropriate, planned impact mitigation measures are presented. Site-adapted conceptual design details and design changes developed since the FEIS was published are included in the description of the proposed SSC facility. The most significant design change is an increase in the energy level of the HEB from 1 to 2 TeV. This change was deemed necessary on the basis of recently developed computer simulations indicating that, at 1 TeV, an unacceptably large fraction of the protons injected into the main storage rings would be lost even before the acceleration cycle began. Increasing the HEB energy level to 2 TeV significantly reduces this fractional loss. The impacts of doubling the HEB energy level are assessed throughout this SEIS. A 1-TeV HEB is no longer technically acceptable and is not considered a reasonable option.

The various technical and procedural alternatives for the SSC that were considered and evaluated for the FEIS (Vol. I, Section 3.3) are not analyzed in this document. The no-action alternative is the continuation of current conditions and trends that would take place in Texas if the SSC were not constructed. Impacts of the no-action alternatives are presented in Section 2.3 and subsections of Chapter 4.

### 2.1  PROPOSED ACTION

**Purpose of project.**

The proposed action is to construct and operate the SSC (a 20-TeV particle accelerator with its supporting systems and facilities), which would serve as a U.S. national laboratory for high-energy physics experiments. The five phases of the proposed project are:

- *Siting.* The siting phase consisted of DOE issuing its *Invitation for Site Proposals* (ISP), evaluating those proposals to develop the *best qualified list*

(continues)

(BQL), of seven sites, identifying the preferred site in the EIS, and selecting a site in the *Record of Decision* (ROD).

- *Preconstruction.* The preconstruction phase consists of activities at the selected site to confirm geotechnical conditions, to validate site engineering parameters, to acquire the necessary land parcels, and to perform the assessments or surveys necessary to verify site data for site-specific design.

- *Construction.* The construction phase would include continued design, as well as physical establishment of the tunnel, fabrication of technical components (including magnets, detectors, and support systems), construction of surface facilities and campus areas, and creation of infrastructure connections (roads and utility corridors).

- *Operations.* This primary and long-term phase would involve use of the SSC facilities for physics experiments. The operating life of the SSC is expected to be 25–35 years.

- *Decommissioning.* The decommissioning phase would involve removal, closure, decontamination, and other activities designed to remove the SSC, including its support facilities, from service. Additional NEPA review will be required before decision making on decommissioning can start.

Siting and preconstruction, as well as conceptual descriptions of constructive and operating activities, were covered in the EIS. This SEIS discusses additional site-specific and conceptual design developments since the FEIS was published. It also addresses issues raised during the public comment period for the DEIS that were not addressed in detail in the FEIS.

### 2.1.1 Description of the Proposed SSC

#### 2.1.1.1 Project Overview

The general features of the SSC accelerator and associated facilities have not changed since publication of the FEIS in December 1988; however, some design details have been altered, including those intended to adapt the plans to the features at the Texas site. Changes that have occurred since EIS publication are summarized in the following sections and, when possible, compared with the features assessed in the EIS.

The principal feature of the proposed SSC is the collider ring, a 54-mi-long oval tunnel. (The EIS assessment was for a 53-mi-long tunnel.) Approximately 10,000 superconducting magnets in the form of two rings, one atop the other, would focus and guide two proton beams around the tunnel. Within the magnets, the two proton beams (one in each magnet ring) would be accelerated in opposite directions to an energy of 20 TeV and made to collide with a combined energy of 40 TeV. Special facilities intermittently spaced around the collider ring would provide the power supplies to energize the magnets and the cryogenic system to keep the superconducting magnets cooled to a temperature near absolute zero.

Other prominent features of the proposed SSC design are the experimental areas, the injector facilities, and the campus areas. The experimental areas would contain the detectors used to record particle collision products, support buildings, and support facilities. The injector facilities would consist of four separate cascading accelerators in which the proton beams first would be formed and then accelerated to the required energy for injection into the ring magnets in the collider tunnel. The campus areas would include the main laboratory and administration building, the auditorium, warehouses, support facilities, and a number of shop buildings. . . .

**Graphic has caption and is numbered according to chapter and graphic number in chapter.**

**Uses a graphic so readers can *see* area being described.**

**Provides compass key.**

**Graphic is labeled.**

**FIGURE 2.1 Texas Proposed Footprint**

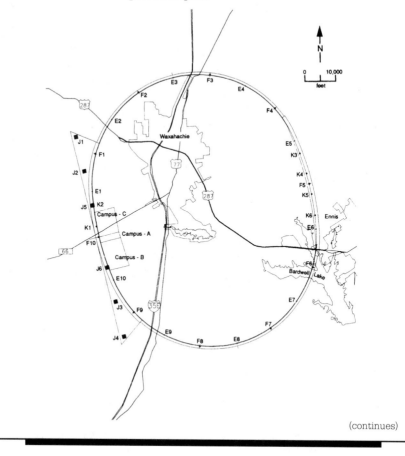

(continues)

3-1

## 3   AFFECTED ENVIRONMENT

### 3.1   EARTH RESOURCES

The following discussion summarizes material originally presented in the DEIS (Vol. IV, Appendix 5c) and provides additional information on faults near the site (Section 3.1.3).

#### 3.1.1   Physiography and Topography

The SSC site is in the Western Gulf Coast section of the Coastal Plains physiographic province. The area is characterized by submature to mature erosion of southeast-dipping strata. The eroded surface contains low, west-facing escarpments separated by flat to rolling prairies. The prairies follow the dip direction of the underlying rocks, generally sloping gently to the southeast.

Much of the site has a relatively flat to slightly rolling prairie surface, grading to rolling prairie at a few incised drainages. Elevations in the area range from 840 ft mean sea level (msl) at the crest of the White Rock escarpment to 350 ft msl where Waxahachie and Onion creeks depart to the southeast. The largest drainage, Waxahachie Creek, is incised 80–120 ft below the prairie surface.

The site is traversed by the tributaries and main stems of Red Oak, Waxahachie, Onion, and Chambers creeks, all of which flow southeast to join the Trinity River. Except for Waxahachie and Onion creeks, most of the streams flow intermittently where they cross the tunnel footprint and campuses.

#### 3.1.2   Stratigraphy

Table 3.1 provides a summary description of the major sedimentary units within and surrounding the site. The bedrock formations of the site are dominated by massive beds of chalk and shale, all of Cretaceous age. The Cretaceous units are part of a 1,750- to 4,400-ft-thick wedge of sediment that strikes north-northeast and dips southeast. The alluvium overlies the Cretaceous sediments and is composed of unconsolidated accumulations of gravel, sand, silt, and clay deposited as terraces along stream channels and in the floodplain. Units that will be affected by construction are, from oldest to youngest, the Eagle Ford shale, the Austin chalk, the Taylor marl, and alluvium. . . .

4-1

## 4 ENVIRONMENTAL CONSEQUENCES . . .

### 4.7 HUMAN HEALTH EFFECTS . . .

#### 4.7.1 Radiation Effects . . .

##### 4.7.1.1 Technical Approach and Methodology

**Notice 4th level heading.**

**Very technical. Most members of the general public will not understand. Explains calculations, methods used to determine the amount of radioactivity that would be released by the SSC.**

The Clean Air Act Assessment Package–1988 (CAP-88) (Beres 1989) was used to estimate the radiological impacts from normal operations of the SSC on the maximally exposed off-site individual and the general population. CAP-88 uses the AIRDOS-EPA code (Moore et al. 1979) to calculate environmental concentrations resulting from radionuclide emissions into the air. The results of the AIRDOS-EPA analyses are estimates of air and ground surface radionuclide concentrations; intake rates via inhalation of air; and ingestion of radioactivity via meat, milk, and fresh vegetables. The code is described in detail in Moore et al. (1979) and summarized in EPA (1989).

DARTAB is the code incorporated into the CAP-88 package to estimate the health effects resulting form airborne emissions of radionuclides (Begovich et al. 1981). DARTAB takes as input the environmental concentrations of radionuclides as calculated by AIRDOS-EPA and provides tabulations of predicted impacts of radioactive airborne effluents. DARTAB is described in Begovich et al. (1981) and RSIC (1987). RSIC (1987, 1990) provides a description of the input and output as well as auxilliary programs for the Clean Air Act Code (CAAC), CAP-88's predecessor, and for CAP-88.

The radiological impacts are also assessed for transportation of the low-level radioactive wastes (LLW) from SSC to the disposal sites. Such impacts include incident-free operations and accident conditions. For the FEIS, the RADTRAN III Code (Madsen et al. 1986) was used to calculate radiological risks from transport of LLW, given the assumption that the LLW disposal site was DOE's Hanford site. In this SEIS, the impacts are also assessed for the alternative LLW disposal site in Texas. For this assessment, the RADTRAN III Code is once again used for the calculation. For comparison, the transportation risks for the Hanford site were recalculated, but updated data and parameters were used.

**Explains how the impact on a person is determined. Too technical for the general public to understand.**

All radiological impacts are assessed collectively for the affected general population and individually for the hypothetical maximally exposed person, during transportation of radioactive materials involving either on-site or off-site activities. The radiological impacts are expressed as the 50-year committed effective dose equivalent ($CEDE_{50}$) for the exposed individuals and for the exposed population as a whole.

(continues)

**4.7.1.2   Source Terms and Assumptions for Impact Projections**

Relates to amount of
radon to which people
would be exposed.

The radiological hazards (i.e., radon and radon progeny source terms) given for the SSC site in the EIS were reviewed and determined to be accurate. However, because of the rather random distribution of radon-producing natural material, the expected absence of appreciable amounts of radon (well below the accepted working level) will be verified through an in-situ air sampling program during and following construction. The *working level* concept is used because of the complex decay schemes and behaviors of radon and its daughters. As a matter of practicality, the working level for occupationally exposed persons, as defined in 10 CFR 20, is taken as 100 pCi/L of Rn-222 at equilibrium. DOE Order 5480.11 lists a concentration guide for radon gas in air under control conditions of 80 pCi/L. As indicated in the DEIS (Appendix 10, Table 10.1.3–12), the SSC site is expected to have tunnel airspace radon concentrations of 6.2 pCi/L, with ventilation during periods of tunnel occupancy. Thus, radon is not expected to be a problem in the tunnels or excavated areas. . . .

**4.8   SOCIOECONOMICS AND INFRASTRUCTURE**

The information in this section is condensed from a supporting technical document (Robert D. Niehaus, Inc. 1990).

**4.8.1   Technical Approach and Methodology**

Describes how
calculated statistics
were done and on what
they were based.

- *Economic Activity.* The methodology for projecting SSC-related direct and secondary economic effects is essentially unchanged from that used in the DEIS (Vol. IV, Appendix 14, Section 14.1.2.3.B.1). Updated information on project spending, earnings, and employment was derived from cost estimates current as of May 25, 1990. In-migrant work-force impacts for the eight-county region of influence (ROI) were derived as the mid-range of a high and a low scenario, which are also discussed in the DEIS (Vol. IV, Appendix 14, Section 14.1.2.3.B.1).

- *Demographics and Housing.* Total population impacts include in-migrant workers and their families. The average household size of these workers (at the time they in-migrate) and the likely age and sex breakdowns were based on the observed composition of state-to-state migrants in the United States between 1980 and 1985. A cohort-component method was used to estimate children born to the in-migrants and to estimate deaths projected among this impact population. Where the impact population would reside within the region was projected for SSC direct workers by using a procedure based on entropy maximization that simultaneously considers the relative attractiveness of each local area and the travel time from each area to the main campus of the SSC. Housing unit requirements of the in-migrant population were estimated on the basis of documented housing tendencies associated with construction and operation of large-scale facilities.

- *Public Services.* Potential impacts to local public services from increased demand by SSC-induced in-migrants were determined for the region's key public services: public education, law enforcement, fire protection, and health care services. Impacts were determined for the ROI, the eight constituent ROI counties, 15 selected independent school districts (ISDs), and selected communities in the project area. . . .

### 4.8.3 Economic Activity

The regional economy is expected to experience beneficial increases in employment, income, and sales, beginning in the early construction years, as a result of SSC construction and operation in Ellis County (Table 4.27). By construction year 4, as many as 9,400 additional jobs would be created, including opportunities for direct employment at the SSC site and for secondary jobs supplying goods and services required by the project and satisfying the additional consumer demand created by direct project workers. As major construction efforts are completed, the number of workers required for the project would decline through construction year 7 and then rebound the following year because of the influx of workers required to complete several components of the project. SSC-related employment would reach its lowest level in year 10, providing approximately 4,200 direct and secondary jobs. During full operation, employment would stabilize and provide about 5,700 direct and secondary jobs in the region. Regional earnings also would escalate as a result of these job increases; sales demand would increase as a result of project-related expenditures for goods and services and additional consumer purchases made by SSC workers.

**Provides table (on following page) so readers of various occupations, counties can easily locate specific information related to them.**

(continues)

**TABLE 4.27   SSC-Related Changes in Regional Economic Activity[a]**

| Economic Attribute | \| Construction Year \| | | | | | | | | | \| Preoperation Year \| | | | Full Operation |
|---|---|---|---|---|---|---|---|---|---|---|---|---|---|
| | 1 | 2 | 3 | 4 | 5 | 6 | 7 | 8 | 9 | 10 | 11 | 12 | |
| SSC-Related Jobs (number) | 3,069 | 6,348 | 8,713 | 9,390 | 8,011 | 6,991 | 6,327 | 7,093 | 5,132 | 4,215 | 5,047 | 5,533 | 5,706 |
| Direct Jobs[b] | 1,550 | 2,746 | 3,740 | 3,893 | 3,513 | 3,296 | 3,126 | 3,733 | 2,641 | 2,427 | 2,893 | 3,158 | 3,248 |
| Construction | 824 | 1,675 | 2,279 | 2,262 | 1,867 | 1,417 | 971 | 653 | 416 | 0 | 0 | 0 | 0 |
| Crafts | 199 | 680 | 925 | 903 | 631 | 422 | 241 | 119 | 54 | 0 | 0 | 0 | 0 |
| Technical | 246 | 483 | 721 | 903 | 707 | 583 | 525 | 394 | 261 | 0 | 0 | 0 | 0 |
| Management, clerical | 379 | 511 | 633 | 627 | 529 | 411 | 206 | 140 | 101 | 0 | 0 | 0 | 0 |
| Operation and preoperation | 726 | 1,071 | 1,461 | 1,632 | 1,646 | 1,879 | 2,155 | 3,081 | 2,225 | 2,427 | 2,893 | 3,158 | 3,248 |
| Professional | 268 | 388 | 517 | 525 | 511 | 562 | 648 | 981 | 656 | 732 | 909 | 1,010 | 1,044 |
| Technical | 78 | 112 | 150 | 152 | 148 | 163 | 188 | 284 | 190 | 212 | 263 | 292 | 302 |
| Clerical and other | 360 | 521 | 694 | 705 | 686 | 754 | 869 | 1,316 | 880 | 983 | 1,220 | 1,356 | 1,401 |
| Visiting scientists | 20 | 50 | 100 | 250 | 300 | 400 | 450 | 500 | 500 | 500 | 500 | 500 | 500 |
| Secondary Jobs[c] | 1,519 | 3,602 | 4,973 | 5,497 | 4,499 | 3,695 | 3,201 | 3,360 | 2,491 | 1,788 | 2,154 | 2,375 | 2,458 |
| Manufacturing | 214 | 615 | 815 | 847 | 628 | 468 | 343 | 307 | 207 | 150 | 178 | 196 | 202 |
| Transportation, utilities | 117 | 266 | 358 | 438 | 404 | 383 | 397 | 456 | 363 | 282 | 342 | 380 | 395 |
| Trade | 433 | 980 | 1,392 | 1,538 | 1,248 | 1,008 | 870 | 921 | 679 | 471 | 565 | 621 | 642 |
| Services | 569 | 1,327 | 1,836 | 2,042 | 1,693 | 1,392 | 1,197 | 1,255 | 930 | 655 | 793 | 874 | 904 |
| Other | 185 | 415 | 572 | 631 | 526 | 443 | 394 | 421 | 311 | 230 | 276 | 304 | 314 |
| Location of Secondary Jobs | | | | | | | | | | | | | |
| Dallas County | 817 | 1,964 | 2,709 | 2,994 | 2,438 | 1,996 | 1,729 | 1,813 | 1,346 | 967 | 1,165 | 1,285 | 1,330 |
| Ellis County | 101 | 174 | 246 | 273 | 255 | 224 | 195 | 211 | 150 | 107 | 129 | 140 | 144 |
| Hill County | 12 | 30 | 41 | 46 | 37 | 30 | 26 | 28 | 21 | 15 | 18 | 20 | 20 |
| Johnson County | 49 | 110 | 153 | 169 | 142 | 118 | 102 | 108 | 79 | 57 | 68 | 75 | 78 |
| Kaufman County | 22 | 54 | 74 | 82 | 66 | 54 | 46 | 48 | 36 | 26 | 31 | 35 | 36 |
| Navarro County | 18 | 44 | 60 | 67 | 54 | 44 | 38 | 40 | 30 | 21 | 26 | 28 | 29 |
| Rockwall County | 10 | 25 | 34 | 37 | 30 | 24 | 21 | 22 | 16 | 12 | 14 | 16 | 16 |
| Tarrant County | 490 | 1,202 | 1,656 | 1,829 | 1,477 | 1,204 | 1,043 | 1,091 | 813 | 584 | 703 | 777 | 804 |
| Earnings (millions 1990$)[d] | 81.4 | 172.3 | 238.7 | 261.7 | 222.1 | 188.9 | 168.3 | 180.9 | 131.9 | 98.8 | 118.1 | 130.0 | 134.8 |
| Direct | 41.9 | 77.0 | 107.0 | 115.8 | 103.4 | 91.9 | 84.2 | 92.8 | 66.3 | 51.9 | 61.6 | 67.6 | 70.2 |
| Secondary | 39.5 | 95.3 | 131.7 | 145.8 | 118.7 | 97.0 | 84.0 | 88.1 | 65.6 | 46.9 | 56.6 | 62.4 | 64.6 |
| Sales Demand (millions 1990$) | 122.5 | 303.1 | 421.4 | 469.9 | 385.3 | 319.1 | 278.7 | 294.0 | 220.7 | 161.8 | 197.2 | 219.3 | 228.8 |
| Direct | 63.4 | 154.3 | 214.5 | 239.8 | 197.9 | 165.4 | 146.3 | 155.7 | 117.2 | 86.9 | 105.9 | 117.7 | 122.8 |
| Project purchases[e] | 31.8 | 96.3 | 133.9 | 152.5 | 120.0 | 96.3 | 82.9 | 85.9 | 67.2 | 47.8 | 59.5 | 66.8 | 70.0 |
| Consumer demand[f] | 31.6 | 57.9 | 80.6 | 87.2 | 77.9 | 69.2 | 63.4 | 69.9 | 49.9 | 39.1 | 46.4 | 50.9 | 52.9 |
| Secondary | 59.1 | 148.8 | 207.0 | 230.1 | 187.5 | 153.7 | 132.4 | 138.3 | 103.5 | 74.9 | 91.3 | 101.6 | 106.0 |

[a]Some totals may not be exact because of rounding.

[b]All direct jobs are located in Ellis County.

[c]Includes indirect and induced employment effects.

[d]Earnings from direct and secondary jobs.

[e]Purchases from regional sources used for construction and operation of the SSC.

[f]Demand by direct SSC workers.

Source: Robert D. Niehaus 1990, Table 3-4.

6-1

## 6 PREPARERS

The individuals who prepared the SEIS for the SSC project are identified below. The overall effort for the DOE was led by G.J. Scango, Acting Director of the SSC Project Engineering and Review Division within DOE's Office of the Superconducting Super Collider. The DOE Environmental Project Manager is T.A. Baillieul of DOE's Chicago Operations Office.

The authors of the document, listed by technical area, are as follows:

**Program Manager**

E.D. Pentecost, Argonne National Laboratory (ANL), Ph.D., Ecology; 16 years of experience in ecological assessments and 13 years of experience in project management

**Project Manager**

M.A. Lazaro, ANL, M.S., Atmospheric Sciences, and M.S., Nuclear Engineering, P.E.; 17 years of experience in atmospheric and environmental science research and assessment, 10 years of experience in project management, and 5 years of experience in radiological assessment

**On-Site Project Coordinator**

T.H. Filley, ANL, Ph.D., Hydrogeology; 6 years of experience in groundwater research and environmental assessment

**Earth Resource Assessment**

Lead: R.H. Pearl, ANL, M.A., Geology; 31 years of related experience
M.L. Werner, The Earth Technology Corporation, Ph.D., Geology; 17 years of experience in geological research and assessment

**Surface Water and Groundwater Resources Assessment**

Lead: S.C.L. Yin, ANL, M.S., Hydrology; 17 years of experience in surface water assessment
R.L. Bateman, The Earth Technology Corporation, Ph.D., Hydrology

**Air Resources Assessment**

Lead: M.A. Lazaro, ANL, M.S., Atmospheric Science, and M.S., Nuclear Engineering, P.E.; 17 years of experience in atmospheric and environmental science research and assessment, 10 years of experience in project management, and 5 years of experience in radiological assessment. . . .

This was a collaborative effort among many people in many different professional communities. While each wrote the sections related to their specific community, managers oversaw the entire project so it was written as a single entity when it was finally produced

Source: Final Supplemental Environmental Impact Statement for the Superconducting Super Collider, Volume II. December 1990, U.S. Department of Energy.

## CHAPTER SUGGESTIONS FOR DISCUSSION AND WRITING

1. Compare the excerpts with the guidelines on pages 329–350. Do you believe that this SEIS conforms to the guidelines? Cite instances where it does and does not.

2. Consider the section on radiation effects (4.7.1.1). In the first paragraph find terms or data that you do not think the general public will understand, and highlight them. Do the writers provide definitions of these terms or explanations of the data to help readers comprehend the information? If not, write a definition or explanation to help the general public comprehend the writers' message for the first paragraph of this section.

3. What are the environmental consequences to the local residents in terms of socioeconomics and infrastructure? Revise one of the sections so that local residents understand the consequences more easily.

4. What is the tone of the sections on human health effects and on socioeconomics and the infrastructure? Is the tone personal? Impersonal? Factual? Opinionated? Biased? Unbiased? What are the personal consequences of the Super Collider on the local residents? Do you believe these sections adhere to NEPA statements concerning the style and purpose of an EIS? Why or why not?

5. Read the letters in Figures 7.1 and 7.2. They were written in response to the report. What is the purpose of each? Are they persuasive or informative? What is the point of view and focus of each letter? What is the style? Personal? Impersonal? Formal? Informal? Businesslike? Subjective? Objective? Hostile? Skeptical? Friendly? Compare the point of view, focus, and style with the point of view, focus, and style of the EIS. Discuss whether you believe the EIS should have more closely reflected the point of view, focus, and style of the people affected by the project.

**Letter 1**

Melvin H. Hunter
2229 Mayfair Drive
Ennis, Texas, 75119

September 17, 1990

Mr. Thomas A. Baillieul
U.S. Department of Energy, EMD
9800 South Cass Avenue
Argonne, Illinois 60439

Dear Mr. Bailliieul:

I, Melvin Hunter, am a member of the Ennis Chamber of Commerce. I am excited about the Superconducting Super Collider and the impact it will have on our city.

However, I am very concerned with having safe access to the East Campus, where the experimental halls will be housed. I know your team has studied the issue and is considering making improvements to Ebenezer Road from FM 879 to FM 878 and improvements to the condemned bridges at Bone Branch and Cotton Creek. These improvements are necessary for public safety, environmental protection, project access and future economic development.

I feel very strongly that these improvements are critical to this project and commend you for your consideration of these improvements. I hope you will continue to give these improvements strong consideration. The business community is behind you.

Sincerely,

Melvin H. Hunter

**Letter 2**

I have attended 3 meetings, including one with Congressman Barton. A lot of the same questions were asked at each meeting and different answers were given each time. The only answer they gave that they agreed on, was to try to convince us there was no danger. The people living in this disposal area don't believe this. We have reports that say there is a danger. Some of us have small children that will have to grow up on this land. What about their future health and their unborn families. Are they going to be normal, healthy babies? I'm sure they thought there would be no ill effects from agent Orange and some of their other projects. Can you give us a better guarantee than this? As the old saying goes, "It is too late to pray when the devil comes!"

Matthew Bryant
Rt. 3 Box 222A
Waxahachie, TX 75165

**Figure 7.1   Letters written by local residents in response to the Superconducting Super Collider.**   Source: Final Supplemental Environmental Impact Statement for the Superconducting Super Collider, Volume II. December 1990, U.S. Department of Energy.

# PART III
# CASE STUDIES

In Part I you learned about the effect of context on writers' decisions in writing documents and on readers' perceptions in reading documents. You also studied strategies used by writers to ensure that readers understand and accept a message. In Part II you studied the conventions governing the four major types of documents. Now you have an opportunity to apply what you learned to the documents involved in the case studies in this section. Part III contains three case studies that involve a variety of documents, all of which are written in relation to a specific project or event. You will be able to follow these events as they unfold, and you will have an opportunity to learn how various written documents are used during different periods of an event's or project's history.

You will also have an inside look at the ways organizations operate and the ways people communicate within them. You will learn about organizational communities and how their conventions often govern the lines of communication among their personnel. Finally, you will discover how miscommunication and misunderstanding can occur as well as how, through careful planning, you can persuade your readers to understand, accept, and carry out your messages.

CHAPTER

8

# The Three Mile Island Nuclear Accident

## INTRODUCTION

On March 28, 1979, an accident occurred at the Three Mile Island (TMI) nuclear plant in Pennsylvania. The level of water used to cool the reactor core fell, creating the possibility of a core meltdown. Eventually engineers were able to control the problem and avert a disaster. Although there were no casualties, the cost of the accident was tremendous, and the long-term consequences continue to be felt. Nuclear power as a potentially major source of energy in the United States was effectively halted after the incident; no new nuclear power plants have been initiated since the event.

To understand what happened, you need some basic knowledge of how a nuclear reactor works in a utility. Read the following narrative, contained in the *Report of the President's Commission on the Accident at Three Mile Island* to acquire familiarity with the process involved in producing nuclear energy. The writers of the document realized that most of their readers (the President of the United States, Congressional leaders, persons involved with nuclear utilities, persons living near TMI, and interested citizens) would have difficulty understanding what happened at TMI because they would be unfamiliar with the process of generating electricity from nuclear fission. Therefore, they began the text with a description of the process. This section of the evaluation report provides an excellent example of a description of a complex process that is written for a wide range of readers, including novices (the general public). The writers use a variety of strategies to ensure readers' comprehension. Technical terms, such as *poisons*, are defined and cues are provided to help readers recognize the significance of a fact. For example, in paragraph 2, the writer explains the importance of free neutrons. The language is simple, with few polysyllabic words or words with Latin roots. Sentences are straightforward; the normal pattern of subject, predicate, and object is seldom broken.

Writers also recognize that many readers are biased against nuclear energy because they believe it is unsafe. They attempt to persuade readers to view nuclear energy positively by making claims that are related to these attitudes.

Each TMI plant is powered by its nuclear reactor. A reactor's function in a commercial power plant is essentially simple—to heat water. The hot water, in turn, produces steam, which drives a turbine that turns a generator to produce electricity. Nuclear reactors are a product of high technology. In recent years, nuclear facilities of generating capacity much larger than those of earlier years—including TMI-1 and TMI-2—have gone into service.4/

A nuclear reactor generates heat as a result of nuclear fission, the splitting apart of an atomic nucleus, most often that of the heavy atom uranium. Each atom has a central core called a nucleus. The nuclei of atoms typically contain two types of particles tightly bound together: protons, which carry a positive charge, and neutrons, which have no charge. When a free neutron strikes the nucleus of a uranium atom, the nucleus splits apart. This splitting—or fission—produces two smaller radioactive atoms, energy, and free neutrons. Most of the energy is immediately converted to heat. The neutrons can strike other uranium nuclei, producing a chain reaction and continuing the fission process. Not all free neutrons split atomic nuclei. Some, for example, are captured by atomic nuclei. This is important, because some elements, such as boron or cadmium, are strong absorbers of neutrons and are used to control the rate of fission, or to shut off a chain reaction almost instantaneously.5/

Uranium fuels all nuclear reactors used commercially to generate electricity in the United States. At TMI-2, the reactor core holds some 100 tons of uranium. The uranium, in the form of uranium oxide, is molded into cylindrical pellets, each about an inch tall and less than half-an-inch wide. The pellets are stacked one atop another inside fuel rods. These thin tubes, each about 12 feet long, are made of Zircaloy-4, a zirconium alloy. This alloy shell—called the "cladding"—transfers heat well and allows most neutrons to pass through.6/

TMI-2's reactor contained 36,816 fuel rods—208 in each of its 177 fuel assemblies. A fuel assembly contains not only fuel rods, but space for

(continues)

cooling water to flow between the rods and tubes that may contain control rods or instruments to measure such things as the temperature inside the core. TMI-2's reactor has 52 tubes with instruments and 69 with control rods.7/

Control rods contain materials that are called "poisons" by the nuclear industry because they are strong absorbers of neutrons and shut off chain reactions. The absorbing materials in TMI-2's control rods are 80 percent silver, 15 percent indium, and 5 percent cadmium. When the control rods are all inserted in the core, fission is effectively blocked, as atomic nuclei absorb neutrons so that they cannot split other nuclei. A chain reaction is initiated by withdrawing the control rods. By varying the number of and the length to which the control rods are withdrawn, operators can control how much power a plant produces. The control rods are held up by magnetic clamps. In an emergency, the magnetic field is broken and the control rods, responding to gravity, drop immediately into the core to halt fission. This is called a "scram."

The nuclear reactors used in commercial power plants possess several important safety features. They are designed so that it is impossible for them to explode like an atomic bomb. The primary danger from nuclear power stations is the potential for the release of radioactive materials produced in the reactor core as the result of fission. These materials are normally contained within the fuel rods.

Damage to the fuel rods can release radioactive material into the reactor's cooling water and this radioactive material might be released to the environment if the other barriers—the reactor coolant system and containment building barriers—are also breached.

A nuclear plant has three basic safety barriers, each designed to prevent the release of radiation. The first line of protection is the fuel rods themselves, which trap and hold radioactive materials produced in the uranium fuel pellets. The second barrier consists of the reactor vessel and the closed reactor coolant system loop. The TMI-2 reactor vessel, which holds the reactor core and its control rods, is a 40-foot high steel tank with walls 8-1/2 inches thick. This tank, in turn, is surrounded by two, separated concrete-and-steel shields, with a total thickness of up to 9-1/2 feet, which absorb radiation and neutrons emitted from the reactor core. Finally, all this is set inside the containment building, a 193-foot high, reinforced-concrete structure with walls 4 feet thick.9/

To supply the steam that runs the turbine, both plants at TMI rely on a type of steam supply system called a pressurized water reactor. This simply means that the water heated by the reactor is kept under high pressure, normally 2,155 pounds per square inch in the TMI-2 plant.

In normal operations, it is important in a pressurized water reactor that the water that is heated in the core remain below "saturation"—that is, the temperature and pressure combination at which water boils and turns to steam. In an accident, steam formation itself is not a danger, because it too can help cool the fuel rods, although not as effectively as the coolant water. But problems can occur if so much of the core's coolant water boils away that the core becomes uncovered.

An uncovered core may lead to two problems. First, temperature may rise to a point, roughly 2,200°F, where a reaction of water and the cladding could begin to damage the fuel rods and also produce hydrogen. The other is that the temperature might rise above the melting point of the uranium fuel, which is about 5,200°F. Either poses a potential danger. Damage to the zirconium cladding releases some radioactive materials trapped inside the fuel rods into the core's cooling water. A melting of the fuel itself could release far more radioactive materials. If a significant portion of the fuel should melt, the molten fuel could melt through the reactor vessel itself and release large quantities of radioactive material into the containment building. What might happen following such an event is very complicated and depends on a number of variables such as the specific characteristics of the materials on which a particular containment building is constructed.10/

The essential elements of the TMI-2 system during normal operation include:

- The reactor, with its fuel rods and control rods.
- Water, which is heated by the fission process going on inside the fuel rods to ultimately produce steam to run the turbine. This water, by removing heat, also keeps the fuel rods from becoming overheated.
- Two steam generators, through which the heated water passes and gives up its heat to convert cooler water in another closed system to steam.
- A steam turbine that drives a generator to produce electricity.
- Pumps to circulate water through the various systems.
- A pressurizer, a large tank that maintains the reactor water at a pressure high enough to prevent boiling. At TMI-2, the pressurizer tank usually holds 800 cubic feet of water and 700 cubic feet of steam above it. The steam pressure is controlled by heating or cooling the water in the pressurizer. The steam pressure, in turn, is used to control the pressure of the water cooling the reactor.

(continues)

Normally, water to the TMI-2 reactor flows through a closed system of pipes called the "reactor coolant system" or "primary loop." The water is pushed through the reactor by four reactor coolant pumps, each powered by a 9,000 horsepower electric motor. In the reactor, the water picks up heat as it flows around each fuel rod. Then it travels through 36-inch diameter, stainless steel pipes shaped like and called "candy canes," and into the steam generators.

In the steam generators, a transfer of heat takes place. The very hot water from the reactor coolant system travels down through the steam generators in a series of corrosion-resistant tubes. Meanwhile, water from another closed system—the feedwater system or "secondary loop"—is forced into the steam generator.

The feedwater in the steam generators flows around the tubes that contain the hot water from the reactor coolant system. Some of this heat is transferred to the cooler feedwater, which boils and becomes steam. Just as it would be in a coal- or oil-fired generating plant, the steam is carried from the two steam generators to turn the steam turbine, which runs the electricity-producing generator.

The water from the reactor coolant system, which has now lost some of its heat, is pumped back to the reactor to pass around the fuel rods, pick up more heat, and begin its cycle again.

The water from the feedwater system, which has turned to steam to drive the turbine, passes through devices called condensers. Here, the steam is condensed back to water, and is forced back to the steam generators again.

The condenser water is cooled in the cooling towers. The water that cools the condensers is also in a closed system or loop. It cools the condensers, picks up heat, and is pumped to the cooling towers, where it cascades along a series of steps. As it does, it releases its heat to the outside air, creating the white vapor plumes that drift skyward from the towers. Then the water is pumped back to the condensers to begin its cooling process over again.

Neither the water that cools the condensers, nor the vapor plumes that rise from the cooling towers, nor any of the water that runs through the feedwater system is radioactive under normal conditions. The water that runs through the reactor coolant system is radioactive, of course, since it has been exposed to the radioactive materials in the core.

The turbine, the electric generator it powers, and most of the feed-water system piping are outside the containment building in other structures. The steam generators, however, which must be fed by water from both the reactor coolant and feedwater systems, are inside the containment building with the reactor and the pressurizer tank.

A nuclear power facility is designed with many ways to protect against system failure. Each of its major systems has an automatic backup system to replace it in the event of a failure. For example, in a loss-of-coolant accident (LOCA)—that is, an accident in which there is a loss of the reactor's cooling water—the Emergency Core Cooling System (ECCS) automatically uses existing plant equipment to ensure that cooling water covers the core.

In a LOCA, such as occurred at TMI-2, a vital part of the ECCS is the High Pressure Injection (HPI) pumps, which can pour about 1,000 gallons a minute into the core to replace cooling water being lost through a stuck-open valve, broken pipe, or other type of leak. But the ECCS can be effective only if plant operators allow it to keep running and functioning as designed. At Three Mile Island, they did not.

Source: Excerpt from the *Report of the President's Commission on the Accident at Three Mile Island,* 1979.

## CONTEXT

Now that you have a basic understanding of nuclear energy, let's examine the events leading up to the TMI accident. The nuclear reactor at TMI had been manufactured by the Babcock and Wilcox (B & W) Company, which had also built reactors at several other utilities, including the Davis-Besse plant at Toledo, Ohio. Prior to the TMI accident, a problem had occurred at Davis-Besse similar to the one that began the chain of events that eventually led to the loss-of-coolant accident (LOCA) at TMI. If the operators at TMI had been aware of the previous problem at the Davis-Besse plant, and if they had known the new procedures for controlling the kind of situation that had occurred at Davis-Besse, they might have been able to correct the problem at TMI early in the chain of events and thus have prevented it from progressing. In fact, several B & W engineers had written memos that recommended informing TMI operators, but their recommendations were never acted on. The organizational structure of B & W and a failure in communications between the various B & W divisions involved with the reactor prevented the TMI operators from receiving the information concerning the new procedures. The Nuclear Regulatory Commission's report on the accident suggests that a "breakdown in communications" served as a precursor to the event.

This chapter discusses the oral and written communications related to the TMI accident that occurred internally at Babcock and Wilcox. As you read, you will see how the organizational context informed the drafting of the documents as well as readers' interpretation of the documents. You will also see how writers' failure to consider various conventions related to the organizational context caused miscommunication and misunderstanding.

## Organizational Structure of Babcock and Wilcox

Because the structure within an organizational community can often determine the lines of communication and influence the way in which messages are perceived, it is important to understand the organizational structure at B & W. B & W's Nuclear Power Generation Division (NPGD) was responsible for the manufacture of nuclear reactors. The division was divided into two major departments: Engineering and Nuclear Services. Each of these departments was further subdivided (see Figure 8.1). Although these divisions were not separated geographically, they represented different subcommunities within the company. Engineering was responsible for the design, manufacture, and operation of the reactor. Nuclear services was responsible for making certain that the reactor met specifications, that personnel were trained to operate it properly, and that procedures were written to provide operators with instructions for proper operation. Because these two subcommunities had different responsibilities, their points of view, focuses, and biases in relation to a project differed. Writers needed to recognize those differences when communicating with members of the other subcommunity.

If a writer does not consider the point of view of another subcommunity in writing to a reader who is a member of that subcommunity, it is likely that the reader will misinterpret the message.

## Chronology of Events

As the supplier of nuclear reactors, B & W had a responsibility to provide its customers, the utilities, with updated information about the equipment as the company acquired new knowledge in what was and continues to be a new field. Thus, when the Davis-Besse plant had a problem in September 1977, eighteen months prior to the TMI accident, Joseph Kelly, an engineer in the B & W plant design section of the engineering department, was sent

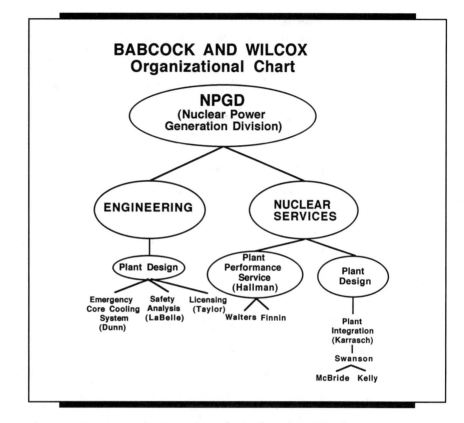

**Figure 8.1    Organization chart for Babcock & Wilcox.**

to investigate the problem. On returning to B & W, he briefed thirty employees on his findings. Kelly had discovered that the procedures the operators had been trained to use were inappropriate and that different procedures needed to be implemented. Among those attending the briefing was Bert Dunn, the manager of B & W's Emergency Core Cooling System (ECCS). After the meeting, Dunn discussed the problem with Kelly. In their discussion, he indicated that operators at other B & W plants needed to know what to do under circumstances similar to those that occurred at Davis-Besse. When Davis-Besse experienced the problem again on October 23, a month after the first occurrence, Kelly became sufficiently concerned to meet with John Lind, the lead instructor in the training division. Lind assured Kelly that operators were learning the appropriate procedures. However, the federal investigation led by John Kemeny and his committee later determined that Lind never incorporated the new procedures into the training.

### Memo One

Both Kelly and Dunn believed that this information should be disseminated to the operators at the various utilities. When Kelly finally realized that the new procedures were not being incorporated into the training, he wrote a memorandum (Figure 8.2) recommending that B & W's customers be

---

THE BABCOCK & WILCOX COMPANY
POWER GENERATION GROUP

TO:        Distribution                              _____EXHIBIT_____
                                                      FOR IDENTIFICATION
FROM:      J.J. Kelly, Plant Integration             _____R. ZERKIN

CUST:      Generic                                    File No.
                                                      Or. Ref.   VIII 3

SUBJ:      Customer Guidance On High Pressure        Date: November 1, 1977
           Injection Operation

                              DISTRIBUTION

              B. A. Karrasch                D. W. LaBelle
              E. W. Swanson                 N. S. Elliott
              R. J. Finnin                  D. F. Hallman
              B. M. Dunn

**Background**   Two recent events at the Toledo site have pointed out that perhaps we are not giving our customers enough guidance on the operation of the high pressure injection system.  On September 24, 1977, after depressurizing due to a stuck open electromatic relief valve, high pressure injection was automatically initiated.  The operator stopped HPI when pressurizer level began to recover, without regard to primary pressure. As a result, the transient continued on with boiling in the RCS, etc.

**Problem**      In a similar occurrence on October 23, 1977, the operator bypassed high pressure injection to prevent initiation, even though reactor coolant system pressure went below the actuation point.

**Solution**     Since these are accidents which require the continuous operation of the high pressure injection system, I wonder what guidance, if any, we should be giving to our customers on when they can safely shut the system down following an accident?  I recommend the following guidelines be sent:

**Recommendation**
              a)   Do not bypass or otherwise prevent the actuation of high/low pressure injection under <u>any</u> conditions except a normal, controlled plant shutdown.

              b)   Once high/low pressure injection is initiated, do not stop it unless:  Tave is stable or decreasing <u>and</u> pressurizer level is increasing <u>and</u> primary pressure is at least 1600 PSIG and increasing.

**Purpose of memo.**   I would appreciate your thoughts on this subject.

---

**Figure 8.2   Memorandum requesting action.** Source: Excerpt from the *Report of the President's Commission on the Accident at Three Mile Island*, 1979.

notified of the procedures in case a problem occurred similar to the one at Davis-Besse. The memo was sent to seven people at B & W who held positions of responsibility in both the engineering and nuclear services departments.

## Memo Two

The only response Kelly received was the memo in Figure 8.3, which appears to defend the actions of the Toledo operators and to question Kelly's recommended procedures. The memo was written by James Walters, an engineer in the Nuclear Services Department, whose name did not even appear in the distribution list, but whose supervisor, Donald Hallman, did. Hallman had not considered the memo important and had asked Walters to respond to it. Because Walters' supervisor did not consider the memo important, neither did Walters who didn't even bother to have his response

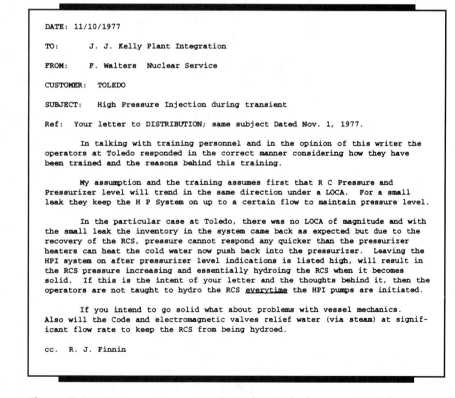

Claim

Evidence

Request

DATE: 11/10/1977

TO:       J. J. Kelly Plant Integration

FROM:     F. Walters  Nuclear Service

CUSTOMER:  TOLEDO

SUBJECT:   High Pressure Injection during transient

Ref: Your letter to DISTRIBUTION; same subject Dated Nov. 1, 1977.

In talking with training personnel and in the opinion of this writer the operators at Toledo responded in the correct manner considering how they have been trained and the reasons behind this training.

My assumption and the training assumes first that R C Pressure and Pressurizer level will trend in the same direction under a LOCA. For a small leak they keep the H P System on up to a certain flow to maintain pressure level.

In the particular case at Toledo, there was no LOCA of magnitude and with the small leak the inventory in the system came back as expected but due to the recovery of the RCS, pressure cannot respond any quicker than the pressurizer heaters can heat the cold water now push back into the pressurizer. Leaving the HPI system on after pressurizer level indications is listed high, will result in the RCS pressure increasing and essentially hydroing the RCS when it becomes solid. If this is the intent of your letter and the thoughts behind it, then the operators are not taught to hydro the RCS everytime the HPI pumps are initiated.

If you intend to go solid what about problems with vessel mechanics. Also will the Code and electromagnetic valves relief water (via steam) at significant flow rate to keep the RCS from being hydroed.

cc. R. J. Finnin

**Figure 8.3   A response memo that also includes a request for information. The original memo was handwritten.** Source: Excerpt from the *Report of the President's Commission on the Accident at Three Mile Island*, 1979.

typed or to proofread it. Walters implicitly argues against Kelly's recommendation, claiming that the Toledo operators acted properly and that if they had acted as Kelly suggested, the plant could have "gone solid," a condition Walters apparently considered to be as critical as a LOCA, although it is not.

Why didn't the other readers respond to Kelly's memo? Both Mathes (1989) and Herndl, Fennell, and Miller (1991) have commented on the weaknesses of his memo. Mathes suggests that Kelly did not take into account the organization's decision-making structure when he determined his audience. Kelly sent the memo to a list of people rather than to a single individual who would be responsible for making a decision. In addition, although Kelly was simply an engineer in the nuclear services department (he was not very high up on the organizational ladder), he sent the memo to a number of managers. According to organizational conventions, managers place more importance on messages from their peers or those above them in the organizational hierarchy than they do on messages from subordinates. Dunn, who was higher up in the hierarchy, probably should have signed the message instead of Kelly to ensure that managers read and responded to the memo.

Kelly's position in the hierarchy is also part of the reason that the memo has a tentative rather than an authoritative tone. Again, according to organizational conventions, people in a subordinate position cannot *tell* their superiors how to act; they can simply *suggest,* which is what Kelly does. Rather than make an outright recommendation that readers take action, Kelly simply requests that they express their opinion in regard to his recommendation. And throughout the memo, Kelly uses "hedge" words, words that indicate tentativeness and qualify the writer's statements so that his comments don't appear dogmatic. Herndl et al. suggest that Kelly assumes this deferential tone for political expediency. Because the readers are members of different divisions, he has no authority to tell them what to do. Thus, he is hoping that by referring to their own knowledge, he will persuade them to follow his suggestion.

Herndl et al. also suggest that readers did not respond because, as members of a different organizational community, their focus was on a different aspect of the project. As plant design engineers Kelly and Dunn's focus was on new information, such as that related to the new procedures, whereas Hellman and Walters, as members of Performance Services, focused their attention on prior organizational commitments such as the training already given to operators. A difference in focus probably also caused Kelly to fail to follow up on Walters' questions relating to going solid; Kelly correctly perceived a LOCA rather than going solid as a major problem.

### Memo Three

Probably all of these causes played a part in Kelly's failure to elicit a response. However, Dunn continued to worry about the problem, and three months later he wrote what has since been characterized as a $2.5 billion memorandum (Figure 8.4). Dunn's purpose was to "kick . . . tail," to get the people responsible for retraining the operators to incorporate the information learned at Davis-Besse. Unlike Kelly, he sent the memorandum to a single decision maker, James Taylor, who was the manager of licensing, the division responsible for safety concerns. His memo was also more authoritative than Kelly's. Dunn makes a recommendation, requesting action, and warns that the potential for a catastrophic accident exists if the recommendation is not acted on.

Dunn's decision to send the memo to Taylor was a good one. His decision to write a memo was a poor one. Rather than writing a memo, Dunn should have filled out a PSC (Preliminary Safety Concern) form. According to B & W procedures, all matters relating to safety concerns were to be reported on a PSC form that would go directly to the manager of licensing, who would then submit them to the appropriate engineers for analysis and evaluation. Because Taylor received a memo rather than the form, he concluded it related to a procedural rather than a safety concern and forwarded it to the nuclear services department instead of to the engineers.

### Memo Four

Dunn's memo was successful in getting nuclear services personnel to respond. They contacted him orally, agreeing that a potential problem existed. However, like Walters, they questioned the solution. After several discussions with them, Dunn revised his recommendation and sent the revision in another memo to Taylor (See Figure 8.5). Like his first, it is authoritative and makes a recommendation.

At this point Dunn assumed his recommendation would be sent to B & W's customers. In fact, those instructions were not sent until seven days after the TMI accident, 13 months after Dunn sent his memo.

### Memo Five

Walters, still questioning the solution reached between Dunn and his own division, deferred sending the recommendation. Eventually he wrote a memo to the manager of the plant integration division, Bruce Karrasch, requesting an evaluation of the problem (Figure 8.6). However, he did not make the mistake that Kelly made. He had his supervisor, Donald Hallman, sign the memo so that the memo would come from someone on the same

THE BABCOCK & WILCOX COMPANY
POWER GENERATION GROUP

TO:        Jim Taylor, Manager, Licensing             _____EXHIBIT_____
                                                      FOR IDENTIFICATION
FROM:      Bert M. Dunn, Manager                      _____R. ZERKIN
           ECCS Analysis (2138)

CUST.                                                 File No.
                                                     or Ref.

SUBJ:      Operator Interruption of High             Date:  February 9, 1978
           Pressure Injection

**Purpose**

This memo addresses a serious concern within ECCS Analysis about the
potential for operator action to terminate high pressure injection
following the initial stage of a LOCA.  Successful ECCS operation during
small breaks depends on the accumulated reactor coolant system inventory
as well as the ECCS injection rate.  As such, it is mandatory that full
injection flow be maintained from the point of emergency safety features
actuation system (ESFAS) actuation until the high pressure injection

**Background**

rate can fully compensate for the reactor heat load.  As the injection
rate depends on the reactor coolant system pressure, the time at which a
compensating match-up occurs is variable and cannot be specified as a
fixed number.  It is quite possible, for example, that the high pressure
injection may successfully match up with all heat sources at time t and
that due to system pressurization be inadequate at some later time t2.

**Problem**

The direct concern here rose out of the recent incident at Toledo.
During the accident the operator terminated high pressure injection due
to an apparent system recovery indicated by high level within the
pressurizer.  This action would have been acceptable only after the
primary system had been in a subcooled state.  Analysis of the data from
the transient currently indicates that the system was in a two-phase
state and as such did not contain sufficient capacity to allow high
pressure injection termination.  This became evident at some 20 to 30
minutes following termination of injection when the pressurizer level
again collapsed and injection had to be reinitiated.  During the 20 to
30 minutes of noninjection flow they were continuously losing important
fluid inventory even though the pressurizer indicated high level.  I
believe it fortunate that Toledo was at an extremely low power and
extremely low burnup.  Had this event occurred in a reactor at full

**Effect**

power with other than insignificant burnup it is quite possible, perhaps
probable, that core uncovery and possible fuel damage would have resulted.

**Solution**

The incident points out that we have not supplied sufficient information
to reactor operators in the area of recovery from LOCA.  The following
rule is based on an attempt to allow termination of high pressure
injection only at a time when the reactor coolant system is in a

**Figure 8.4    A recommendation memo.**    Source: Excerpt from the *Report of
the President's Commission on the Accident at Three Mile Island*, 1979.

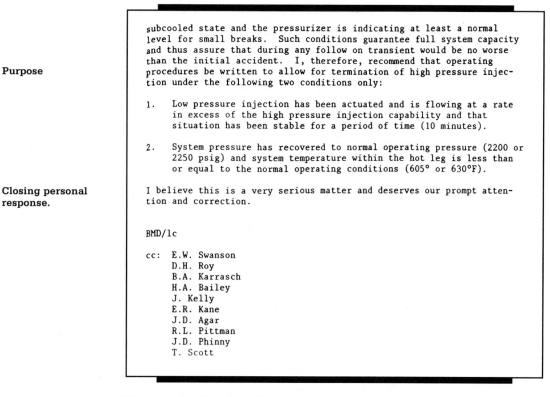

**Purpose**

subcooled state and the pressurizer is indicating at least a normal level for small breaks.  Such conditions guarantee full system capacity and thus assure that during any follow on transient would be no worse than the initial accident.  I, therefore, recommend that operating procedures be written to allow for termination of high pressure injection under the following two conditions only:

1.   Low pressure injection has been actuated and is flowing at a rate in excess of the high pressure injection capability and that situation has been stable for a period of time (10 minutes).

2.   System pressure has recovered to normal operating pressure (2200 or 2250 psig) and system temperature within the hot leg is less than or equal to the normal operating conditions (605° or 630°F).

**Closing personal response.**

I believe this is a very serious matter and deserves our prompt attention and correction.


BMD/lc

cc:  E.W. Swanson
     D.H. Roy
     B.A. Karrasch
     H.A. Bailey
     J. Kelly
     E.R. Kane
     J.D. Agar
     R.L. Pittman
     J.D. Phinny
     T. Scott

**Figure 8.4**   *(continued)*

level in the organizational hierarchy as the reader. In addition, he iterated the catastrophic warning that appeared in Dunn's memo. Despite these attentions to conventions, Karrasch never responded. In fact, when Karrasch received the memo, he skimmed it quickly, and then sent it to a subordinate to answer. According to his testimony after the accident, Karrasch admitted that he had not recognized the significance of the memo. "It seemed to me that it was a routine matter . . . and I sent it on [to be] answered in a rather routine manner" (135).

Not long after Karrasch received the memo, TMI experienced a situation similar to that at Davis-Besse. Unlike Davis-Besse, however, which had been at low power, TMI was at high power, and the problem escalated into what remains the worst nuclear accident in this country.

THE BABCOCK & WILCOX COMPANY
POWER GENERATION GROUP

TO:      Jim Taylor, Manager, Licensing              _____EXHIBIT_____
                                                      FOR IDENTIFICATION
FROM:    Bert M. Dunn, Manager,                       _____R. ZERKIN
         ECCS Analysis (2138)

CUST:                                                 File No.
                                                      or Ref.

SUBJ:    Operator Interruption of High       Date:   February 16, 1978
         Pressure Injection

In review of my earlier memo on this subject, dated February 9, 1978,
Field Service has recommended the following procedure for terminating
high pressure injection following a LOCA.

1.   Low pressure injection has been actuated and is flowing at a rate
     in excess of the high pressure injection capability and that
     situation has been stable for a period of time (10 minutes).  Same
     as previously stated.

2.   At X minutes following the initiation of high pressure injection,
     termination is allowed provided the hot leg temperature indication
     plus appropriate instrument error is more than 50°F below the
     saturation temperature corresponding to the reactor coolant system
     pressure less instrument error.  X is a time lag to prevent the
     termination of the high pressure injection immediately following
     its initiation.  It requires further work to define its specific
     value, but it is probable that 10 minutes will be adequate.  The
     need for the delay is that normal operating conditions are within
     the above criteria and thus it is conceivable that the high pressure
     injection would be terminated during the initial phase of a small
     LOCA.

I find that this scheme is acceptable from the standpoint of preventing
adverse long range problems and is easier to implement.  Therefore, I
wish to modify the procedure requested in my first memo to the one
identified here.

cc:  E.W. Swanson
     D.H. Roy
     B.A. Karrasch
     H.A. Bailey
     J. Kelly
     E.R. Kane
     J.D. Agar
     R.L. Pittman
     J. D. Phinny
     T. Scott
     R. Davis

**Figure 8.5    A memo containing a revision to a previous memo.**

Source: Excerpt from the *Report of the President's Commission on the Accident at Three Mile Island*, 1979.

BABCOCK & WILCOX COMPANY
POWER GENERATION GROUP

TO:  B.A. Karrasch, Manager, Plant Integration  cc: E.R. Kane
                          J.D. Phinney
FROM:  D.F. Hallman, Manager,          B.W. Street
     Plant Performance Services Section (2149)  B.M. Dunn
                          J.F. Walters
CUST:                  File No.
                  or Ref.

SUBJ:  Operator Interruption of High   Date: August 3, 1978
     Pressure Injection (HPI)

References: (1) B. M. Dunn to J. Taylor, same subject, February 9, 1978
       (2) B. M. Dunn to J. Taylor, same subject, February 16, 1978

References 1 and 2 (attached) recommend a change in B&W's philosophy for
HPI system use during low-pressure transients. Basically, they recommend
leaving the HPI pumps on, once HPI has been initiated, until it can be
determined that the hot leg temperature is more than 50°F below Tsat for
the RCS pressure.

Nuclear Service believes this mode can cause the RCS (including the
pressurizer) to go solid. The pressurizer reliefs will lift, with a
water surge through the discharge piping into the quench tank.

We believe the following incidents should be evaluated:

1. If the pressurizer goes solid with one or more HPI pumps continuing
  to operate, would there be a pressure spike before the reliefs open
  which could cause damage to the RCS?

2. What damage would the water surge through the relief valve discharge piping and quench tank cause?

To date, Nuclear Service has not notified our operating plants to change
HPI policy consistent with References 1 and 2 because of our above-
stated questions. Yet, the references suggest the possibility of
uncovering the core if present HPI policy is continued.

We request that Integration resolve the issue of how the HPI system
should be used. We are available to help as needed.

        /s/_____ _____
               D. F. Hallman

DFH/fch
Attachments            _____EXHIBIT_____
                FOR IDENTIFICATION
                _____R. ZERKIN

                        227

Go solid is a condition in which the entire system, including the pressurizer, is filled with water that could damage the system but not the core.

**Figure 8.6 A memo requesting information.** Source: Excerpt from the
*Report of the President's Commission on the Accident at Three Mile Island,* 1979.

## CHAPTER SUGGESTIONS FOR DISCUSSION AND WRITING

1. How does the discussion on the nuclear reactor on pages 359–363 help you understand the memos? What information do you learn from the description of the reactor that helps you understand the memos? Bracket the background information that you do not need and which, therefore, the writers could have omitted.

2. In his memo in Figure 8.2, Kelly wants his readers to inform B & W's customers of a change in procedures. While organizational conventions influence the way in which readers read Kelly's memo, readers' perceptions of the message are also influenced by the focus of the text, the organizational pattern and sequence of information, and the tone. Why do you think the writer's focus, organizational pattern, sequence of information, and tone may be inappropriate for his readers? Revise the memo so that readers might be persuaded to act immediately.

3. What claims does Kelly make to argue his point in Figure 8.2? What evidence does he use to support his claims? The response to his memo indicates his argument is not effective. What other claims or evidence do you think he might have used to persuade his readers?

4. Kelly uses a number of "hedge" words in his memo in Figure 8.2, words indicating that the claim may not *always* be true. Highlight the hedge words. In this case, hedge words may be inappropriate, but in many cases they are appropriate. Can you think of some situations in which someone may need to use hedge words? Provide some examples.

5. It has been said that argumentation is a process of negotiation between reader and writer. To develop an effective argument, a writer must establish goals that both writer and reader share. List some claims you think writers in the engineering division might make that the readers in nuclear services will accept.

6. Karrasch never responds to Hallman's memo because he doesn't think it is important. What aspects of the memo do you think give the

impression that it is unimportant? Revise the document so that Karrasch might consider it important.

7. Examine the memos of Kelly and Walters in terms of their language and syntax. Both men use technical terms and acronyms (a word in which each letter stands for another word, such as *NASA*, for the National Air and Space Administration). Explain how these terms may cause readers difficulty in understanding the documents. In the Kelly memo, the description of the Toledo incident is very short, and there is no explanation of such statements as, "The operator stopped HPI when pressurizer level began to recover without regard to primary pressure." Why do you think the writer omitted most explanations? Should he have included them? Why or why not?

8. Examine the organizational pattern and sequence of information in Dunn's February 9 memo. He does not indicate that the problem could lead to a possible meltdown until the end of the second paragraph.

   Dunn does not state his recommendation until after he has provided the reader with background on the situation and a description of the problem. If writers believe readers will be hostile to their suggestions, they usually lead up to a recommendation, as Dunn does, rather than present their recommendations at the beginning of their message.

   Revise the memo so that you discuss a potential meltdown in the first paragraph, and present your recommendations prior to providing the background. Which letter, your revision or the original, do you think would be more apt to catch the reader's attention and persuade him to do what you want? Why?

9. What similarities do you perceive between Kelly's and Dunn's failure to communicate their messages concerning a potential disaster at a nuclear power plant and Koncza's failure to communicate his message concerning a potential flood in the Chicago freight tunnels?

10. The Kemeny report was written for people with comparatively little knowledge of nuclear reactors. What kinds of strategies do the writers of the Kemeny report use to help novice readers understand how a nuclear reactor works?

## Notes

Herndl, Carl G., Barbara A. Fennell, and Carolyn R. Miller. 1991. Understanding failures in organizational discourse: The accident at Three Mile Island and the shuttle Challenger disaster. In *Text and the professions*, edited by Charles Bazerman and James Paradis. Madison, Wisc.: University of Wisconsin Press.

Mathes, J. C. 1989. Written communication: The industrial context. In *Worlds of writing*, edited by Carolyn Matalene. New York: Random House.

# The Challenger Disaster

## INTRODUCTION

On January 28, 1986, the space shuttle Challenger lifted off the launch pad at Kennedy Space Center. Seventy-three seconds into the launch an external fuel tank exploded, igniting a gaseous ball of flame that broke up the shuttle and killed all seven crew members. In addition to causing the loss of life of seven of America's "best and brightest," the disaster set the space program back several years, giving Europeans the opportunity to fill the gap and provide a commercial shuttle service for private American companies left without a means to orbit their satellites. The space program has not yet caught up.

In the aftermath of the disaster, a presidential commission formed to investigate the cause of the shuttle's failure. The commission eventually determined that the explosion was caused by an O-ring's failure to seal a joint.

The O-rings are used to seal four joints on the Solid Rocket Boosters (SRBs), which are jettisoned from the main rocket shortly after launch. These boosters are recaptured from the ocean, remanufactured, and reused. Prior to remanufacturing, the engines are examined for defects and degeneration caused by the flight. From the beginning, it was observed that the O-rings appeared charred, a condition that should not have occurred. Because the charring did not appear to cause any problem, however, and because a second O-ring served as a backup, managers did not halt future missions. But in 1985 a series of tests indicated that the backup O-ring might not reseal when temperatures dropped below 50 degrees Fahrenheit. Managers at the Marshall Space Flight Center never fully grasped the significance of this fact, and flights continued. When the Challenger blasted off on its fatal mission, the outside temperature was 36 degrees.

You may want some additional technical knowledge of the shuttle and the purpose of the O-rings on the SRBs to help you understand the documents you will read in this chapter. The following excerpt from the *Report of the Presidential Commission on the Space Shuttle Challenger Accident* provides a detailed description of the function of the O-rings.

---

Chapter I

# Introduction

The Space Shuttle concept had its genesis in the 1960s, when the Apollo lunar landing spacecraft was in full development but had not yet flown. From the earliest days of the space program, it seemed logical that the goal of frequent, economical access to space might best be served by a reusable launch system. In February, 1967, the President's Science Advisory Committee lent weight to the idea of a reusable spacecraft by recommending that studies be made "of more economical ferrying systems, presumably involving partial or total recovery and use."

In September, 1969, two months after the initial lunar landing, a Space Task Group chaired by the Vice President offered a choice of three long-range plans:

- A $8–$10 billion per year program involving a manned Mars expedition, a space station in lunar orbit and a 50-person Earth-orbiting station serviced by a reusable ferry, or Space Shuttle.
- An intermediate program, costing less than $8 billion annually, that would include the Mars mission.
- A relatively modest $4–$5.7 billion a year program that would embrace an Earth-orbiting space station and the Space Shuttle as its link to Earth.[1]

In March, 1970, President Nixon made it clear that, while he favored a continuing active space program, funding on the order of Apollo was not in the cards. He opted for the shuttle-tended space base as a long-range goal but deferred going ahead with the space station pending development of the shuttle vehicle. Thus the reusable Space Shuttle, earlier considered only the transport element of a broad, multi-objective space plan, became the focus of NASA's near-term future.

# The Space Shuttle Design

The embryo Shuttle program faced a number of evolutionary design changes before it would become a system in being. The first design was based on a "fly back" concept in which two stages, each manned, would fly back to a horizontal, airplane-like landing. The first stage was a huge, winged, rocket-powered vehicle that would carry the smaller second stage piggyback; the carrier would provide the thrust for liftoff and flight through the atmosphere, then release its passenger—the orbiting vehicle—and return to Earth. The Orbiter, containing the crew and payload, would continue into space under its own rocket power, complete its mission and then fly back to Earth.

The second-stage craft, conceived prior to 1970 as a space station ferry, was a vehicle considerably larger than the later Space Shuttle Orbiter. It carried its rocket propellants internally, had a flight deck sufficiently large to seat 12 space station-bound passengers and a cargo bay big enough to accommodate space station modules. The Orbiter's size put enormous weightlifting and thrust-generating demands on the first-stage design.

This two-stage, fully reusable design represented the optimum Space Shuttle in terms of "routine, economical access to space," the catch-phrase that was becoming the primary guideline for development of Earth-to-orbit systems.

For the launch system, NASA examined a number of possibilities. One was a winged but unmanned recoverable liquid-fuel vehicle based on the eminently successful Saturn 5 rocket from the Apollo Program. Other plans envisioned simpler but also recoverable liquid-fuel systems, expendable solid rockets and the reusable Solid Rocket Booster. NASA had been using solid-fuel vehicles for launching some small unmanned spacecraft, but solids as boosters for manned flight was a technology new to the agency. Mercury, Gemini and Apollo astronauts had all been rocketed into space by liquid-fuel systems. Nonetheless, the recoverable Solid Rocket Booster won the nod, even though the liquid rocket offered potentially lower operating costs. . . .

# Solid Rocket Boosters

The two solid-propellant rocket boosters are almost as long as the External Tank and attached to each side of it. They contribute about 80 percent of the total thrust at liftoff; the rest comes from the Orbiter's three main engines. Roughly two minutes after liftoff and 24 miles downrange, the solid rockets have exhausted their fuel. Explosives separate the boosters from the External Tank. Small rocket motors move them away from the External Tank and the Orbiter, which continue toward orbit under thrust of the Shuttle's main engines.

The Solid Rocket Booster is made up of several subassemblies: the nose cone, Solid Rocket Motor and the nozzle assembly. Marshall is responsible for the Solid Rocket Booster; Morton Thiokol, Inc., Wasatch Division, Brigham City, Utah, is the contractor for the Solid Rocket Motors. Each Solid Rocket Motor case is made of 11 individual cylindrical weld free steel sections about 12 feet in diameter. When assembled, they form a tube almost 116 feet long. The 11 sections are the forward dome section, six cylindrical sections, the aft External Tank-attach ring section, two stiffener sections, and the aft dome section.

The 11 sections of the motor case are joined by tang-and-clevis joints held together by 177 steel pins around the circumference of each joint.

After the sections have been machined to fine tolerances and fitted, they are partly assembled at the factory into four casting segments. Those four cylindrical segments are the parts of the motor case into which the propellant is poured (or cast). They are shipped by rail in separate pieces to Kennedy.

Joints assembled before the booster is shipped are known as factory joints. Joints between the four casting segments are called field joints; they are connected at Kennedy when the booster segments are stacked for final assembly. . . .

(continues)

**O-Ring Compression at Launch (Static)**

As noted previously, diameters measured just prior to assembly do not permit determination of conditions at launch because, among other things, the propellant slowly relaxes. For STS 51-L [Challenger Shuttle], the difference in the true diameters of the surfaces of tang and clevis measured at the factory was 0.008 inches. Thus, the average gap at the O-rings between the tang and clevis was 0.004 inches. The minimum gap could be somewhat less, and possibly metal-to-metal contact (zero gap) could exist at some locations.

During the investigation, measurements were made on segments that had been refurbished and reused. The data indicate that segment circumferences at the sealing surfaces change with repeated use. This expectation was not unique to this joint.

Recent analysis has shown and tests tend to confirm that O-ring sealing performance is significantly improved when actuating pressure can get behind the entire face of the O-ring on the upstream side of the groove within which the O-ring sits (Figure 18). If the groove is too narrow or if the initial squeeze is so great as to compress the O-ring to the extent that it fills the entire groove and contacts all groove surfaces, pressure actuation of the seal could be inhibited. This latter condition is relieved as the joint gap opens and the O-ring attempts to return to its uncompressed shape. However, if the temperature is low, resiliency is severely reduced and the O-ring is very slow in returning towards its original shape. Thus, it may remain compressed in the groove, contact all three surfaces of that groove, and inhibit pressure actuation of the seal. In addition, as the gap opens between the O-ring and tang surface allowing pressure bypass, O-ring actuation is further inhibited.

Two sub-scale dynamic test fixtures were designed and built that simulated the initial static gap, gap opening rate, maximum gap opening and ignition transient pressures. These fixtures were tested over a temperature range with varying initial static gap openings. A summary of results with initial gap openings of 0.020 and 0.004 inches is provided in Figure 19. The results indicate that with a 0.020-inch maximum initial gap, sealing can be achieved in most instances at temperatures as low as 25 degrees Fahrenheit, while with the 0.004-inch initial gap, sealing is not achieved at 25 degrees Fahrenheit and is marginal even in the 40 and 50 degree Fahrenheit temperature range. For the 0.004-inch initial gap condition, sealing without any gas blow-by, did not occur consistently until the temperature was raised to 55 degrees Fahrenheit. To evaluate the sensitivity to initial gap opening, four tests were conducted at 25 degrees Fahrenheit with an initial gap of 0.010 inch. In contrast to the tests at a 0.004 inch gap, these tests resulted in sealing with some minimal O-ring blow-by observed during the sealing process.

These tests indicate the sensitivity of the O-ring seals to temperature and O-ring squeeze in a joint with the gap opening characteristics of the Solid Rocket Motors.

It should be noted that the test fixture placed the O-rings at a specific initial gap and squeeze condition uniformly around the circumference. It is not certain what the effect of differences in circumferential gaps might be in full size joints. Such effects could not be simulated in the sub-scale test results reported above.

**Joint Temperature**

Analyses were conducted to establish STS 51-L [Challenger Shuttle] joint temperatures at launch. Some differences existed among the six 51-L field joints. The joints on the right Solid Rocket Motor had larger circumferential gradients than those on the left motor at launch. It is possible that the aft field joint of the right Solid Rocket Booster was at the lowest temperature at launch, although all joints had calculated local temperatures as low as 28 ± 5 degrees Fahrenheit. Estimated transient temperature for several circumferential locations on the joints are shown for the right Solid Rocket Motor aft field joint and the left motor aft field joint in Figures 20 and 21. These data are representative of other joints on the respective Solid Rocket Motors.

The investigation has shown that the low launch temperatures had two effects that could potentially affect the seal performance: (1) O-ring resilience degradation, the effects of which are explained above; and (2) the potential for ice in the joints. O-ring hardness is also a function of temperature and may have been another factor in joint performance.

Consistent results from numerous O-ring tests have shown a resiliency degradation with reduced temperatures. Figure 23 provides O-ring recovery from 0.040 inches of initial compression versus time. This shows how quickly an O-ring will move back towards its uncompressed shape at temperatures ranging from 10 to 75 degrees Fahrenheit. When these data are compared with the gap openings versus time from Figure 17, it can be seen that the O-rings will not track or recover to the gap opening by 600 milliseconds (gap full open) at low to moderate temperatures. These data show the importance of timely O-ring pressure actuation to achieve proper sealing.

It is possible that water got into some, if not all STS 51-L field joints. Subsequent to the Challenger accident, it was learned that water had been observed in the STS-9 [another shuttle flight] joints during restacking operations following exposure to less rain than that experienced by STS 51-L. It was reported that water had drained from the STS-9 joint when the pins were removed and that approximately 0.5 inch of water was present in the clevis well. While on the pad for 38 days, STS 51-L was exposed to approximately seven inches of rain. Analyses and tests conducted show that water will freeze under the environmental conditions experienced prior to the 51-L launch and could unseat the secondary O-ring. To determine the effects of unseating, tests were conducted on the sub-scale dynamic test fixture at Thiokol to further evaluate seal performance. For these tests, water was frozen downstream of the secondary O-ring. With ice present, there were conditions under which the O-ring failed to seal. . . .

# Findings

1.  A combustion gas leak through the right Solid Rocket Motor aft field joint initiated at or shortly after ignition eventually weakened

**Field Joint Distress**

| Flight | Joint | SRB (right or left) | Angular location | Joint Temp (°F) | Previous of Segments (2) | Types of Distress |
|--------|-------|---------------------|------------------|-----------------|--------------------------|-------------------|
| STS-2 | AFT | RH | 090 | 70 | none/none | Erosion |
| 41-B | FWD | LH | 351 | 57 | 1/none | Erosion |
| 41-C | AFT | LH | N/A | 63 | 1/1 | O-ring heat |
| 41-D | FWD | RH | 275/110 | 70 | 2/none | Erosion |
| 51-C | FWD | LH | 163 | 53 | 1/none | Erosion |
| 51-C(3) | MID | RH | 354 | 53 | 1/1 | Erosion |
| 61-A | MID | LH | 36-66 | 75 | none/none | Blow-by |
| 61-A | AFT | LH | 338/018 | 75 | none/none | Blow-by |
| 61-C | AFT | LH | 154 | 58 | 1/none | Erosion |
| 51-L | AFT | RH | 307 | 28 | 1/2 | Flame |

(1)     Mean calculated (±°F)
(2)     Refurbished after recovery
(3)     Both primary and secondary O-rings affected

Examination of the records shows that if one defines any sort of damage around the O-ring as "distress", then there have been 10 "distressed" field joints, including the art field joint on the right-hand booster of 51-L. These data, which are tabulated above, show 10 instances of distress in a total of 150 flight exposures. One-half of the instances occurred in the aft joint, one-third in the forward joint, and one-fifth in the midjoint. Sixty percent of the distress occurred in the left Solid Rocket Motor.

(continues)

and/or penetrated the External Tank initiating vehicle structural breakup and loss of the Space Shuttle Challenger during STS Mission 51-L. . . .

2. The evidence shows that no other STS 51-L Shuttle element or the payload contributed to the causes of the right Solid Rocket Motor aft field joint combustion gas leak. Sabotage was not a factor.

3. Evidence examined in the review of Space Shuttle material, manufacturing, assembly, quality control, and processing of nonconformance reports found no flight hardware shipped to the launch site that fell outside the limits of Shuttle design specifications.

4. Launch site activities, including assembly and preparation, from receipt of the flight hardware to launch were generally in accord with established procedures and were not considered a factor in the accident.

5. Launch site records show that the right Solid Rocket Motor segments were assembled using approved procedures. However, significant out-of-round conditions existed between the two segments joined at the right Solid Rocket Motor aft field joint (the joint that failed).

6. The ambient temperature at time of launch was 36 degrees Fahrenheit, or 15 degrees lower than the next coldest previous launch. . . .

9. O-ring resiliency is directly related to its temperature.

   a. A warm O-ring that has been compressed will return to its original shape much quicker than will a cold O-ring when compression is relieved. Thus, a warm O-ring will follow the opening of the tang-to-clevis gap. A cold O-ring may not.

   b. A compressed O-ring at 75 degrees Fahrenheit is five times more responsive in returning to its uncompressed shape than a cold O-ring at 30 degrees Fahrenheit.

   c. As a result it is probable that the O-rings in the right solid booster aft field joint were

not following the opening of the gap between the tang and clevis at time of ignition.

10. Experiments indicate that the primary mechanism that actuates O-ring sealing is the application of gas pressure to the upstream (high-pressure) side of the O-ring as it sits in its groove or channel.

   a. For this pressure actuation to work most effectively, a space between the O-ring and its upstream channel wall should exist during pressurization.

   b. A tang-to-clevis gap of .004 inches, as probably existed in the failed joint, would have initially compressed the O-ring to the degree that no clearance existed between the O-ring and its upstream channel wall and the other two surfaces of the channel.

   c. At the cold launch temperature experienced, the O-ring would be very slow in returning to its normal rounded shape. It would not follow the opening of the tang-to-clevis gap. It would remain in its compressed position in the O-ring channel and not provide a space between itself and the upstream channel wall. Thus, it is probable the O-ring would not be pressure actuated to seal the gap in time to preclude joint failure due to blow-by and erosion from hot combustion gases.

11. The sealing characteristics of the Solid Rocket Booster O-rings are enhanced by timely application of motor pressure.

   a. Ideally, motor pressure should be applied to actuate the O-ring and seal the joint prior to significant opening of the tang-to-clevis gap (100 to 200 milliseconds after motor ignition).

   b. Experimental evidence indicates that temperature, humidity, and other variables in the putty compound used to seal the joint can delay pressure application to the joint by 500 milliseconds or more.

c. This delay in pressure could be a factor in initial joint failure.

12. Of 21 launches with ambient temperatures of 61 degrees Fahrenheit or greater, only four showed signs of O-ring thermal distress; i.e., erosion or blow-by and soot. Each of the launches below 61 degrees Fahrenheit resulted in one or more O-rings showing signs of thermal distress. . . .

a. Of these improper joint sealing actions, one-half occurred in the aft field joints, 20 percent in the center field joints, and 30 percent in the upper field joints. The division between left and right Solid Rocket Boosters was roughly equal.

b. Each instance of thermal O-ring distress was accompanied by a leak path in the insulating putty. The leak path connects the rocket's combustion chamber with the O-ring region of the tang and clevis. Joints that actuated without incident may also have had these leak paths. . . .

15. Smoke from the aft field joint at Shuttle lift off was the first sign of the failure of the Solid Rocket Booster O-ring seals on STS 51-L.

16. The leak was again clearly evident as a flame at approximately 58 seconds into the flight. It is possible that the leak was continuous but unobservable or non-existent in portions of the intervening period. It is possible in either case that thrust vectoring and normal vehicle response to wind shear as well as planned maneuvers reinitiated or magnified the leakage from a degraded seal in the period preceding the observed flames. The estimated position of the flame, centered at a point 307 degrees around the circumference of the aft field joint, was confirmed by the recovery of two fragments of the right Solid Rocket Booster.

a. A small leak could have been present that may have grown to breach the joint in flame at a time on the order of 58 to 60 seconds after lift off.

b. Alternatively, the O-ring gap could have been resealed by deposition of a fragile buildup of aluminum oxide and other combustion debris. This resealed section of the joint could have been disturbed by thrust vectoring, Space Shuttle motion and flight loads inducted by changing winds aloft.

c. The winds aloft caused control actions in the time interval of 32 seconds to 62 seconds into the flight that were typical of the largest values experienced on previous missions.

# Conclusion

*In view of the findings, the Commission concluded that the cause of the Challenger accident was the failure of the pressure seal in the aft field joint of the right Solid Rocket Motor.* The failure was due to a faulty design unacceptably sensitive to a number of factors. These factors were the effects of temperature, physical dimensions, the character of materials, the effects of reusability, processing, and the reaction of the joint to dynamic loading.

Source: Excerpt from the *Report of the Presidential Commission on the Space Shuttle Challenger Accident,* Washington, D.C., 1986.

# CONTEXT

As the story before the commission unfolded, it appeared that people associated with the project had written documents concerning problems with the O-rings several years prior to the accident. However, the messages had gone unheeded. Although engineers had become increasingly worried about the condition of the O-ring seal, managers never fully understood their concern. The presidential commission cited a failure to communicate effectively as a contributing factor in the disaster. The commission concluded, "There was a serious flaw in the decision-making process leading up to the launch of flight 51-L. A well structured and managed system . . . would have flagged the rising doubts about the Solid Rocket Booster (SRB) joint seal. Had these matters been clearly stated, . . . it seems likely that the launch of 51-L might not have occurred when it did" (*Report of the Presidential Commission* 1986).

In this chapter, as you study the various documents that were written and read by personnel involved in the manufacture and operation of the shuttle, you will have an opportunity to consider why the communication failures cited by the commission occurred.

## Organizational Structure of the Shuttle Project

The organizational structure in which the shuttle was designed, assembled, and operated determined lines of communication and influenced the way in which messages were written and read. The structure was complex, composed of different organizations with different goals, which often created conflicts between the structures as well as within them.

Three major organizations were involved in the shuttle program. The National Air and Space Administration (NASA) was responsible for the overall project. The design and manufacture of the shuttle was contracted to four private companies, including Morton Thiokol, Inc. (MTI), which was awarded the contract for the SRBs. The managerial responsibilities for the project were divided among NASA's three field centers. The Marshall Space Flight Center in Huntsville, Alabama, was delegated the responsibility for the SRBs (Figure 9.1). Conflicts relating to certain aspects of design arose periodically between the engineers at MTI and those at Marshall.

At both MTI and Marshall, the responsibilities for the shuttle project were divided between two divisions: engineering and managerial services. The Science and Engineering Division at Marshall and the Engineering Division at MTI were involved in the design and development aspects. The shuttle projects office at Marshall and the Space Division at MTI were responsible for support and managerial services. Communication between divisions within each organization occurred often. However, communication

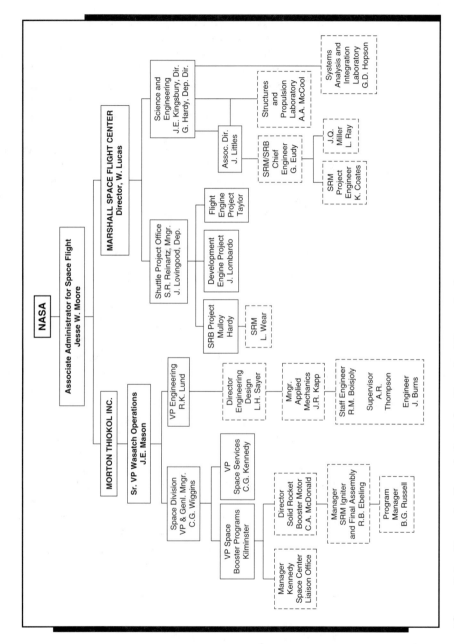

**Figure 9.1    Organizational chart for the shuttle project.**

between the two organizations, MTI and Marshall, was sporadic at best. Because of this lack of communication, engineers at Marshall often didn't have access to information that engineers at MTI had, including the information related to the test results indicating the second O-ring did not reseal when temperatures dropped below 50 degrees Fahrenheit.

Apparently there were also differences in point of view between engineers and higher level managers at both organizations. Although most managers had engineering backgrounds, their points of view often changed. Increasingly, they considered economic and political consequences as they moved up the organizational hierarchy. On the night before the Challenger took off on its last mission, engineers at MTI wanted to delay the launch until the temperature increased, but managers wanted to keep to the launch time to avoid the cost of a delay.

## Economic and Political Factors

Throughout its history the shuttle had been plagued by tight funds, a Congress increasingly set on cutting back NASA's budget, and a public that was no longer excited by space flights that had begun to appear routine. The companies that contracted with NASA to carry up their satellites placed pressure on the agency to meet deadlines. Delays were costly not only to the private companies but also to NASA. Long delays meant fewer flights per year and therefore less money from commercial customers to help defray the cost for the taxpayers. NASA managers continually felt pressured to cut costs, meet deadlines, and increase flights. It was under these conditions that managers from NASA, Marshall, and Morton Thiokol met to decide whether to launch the Challenger on January 27, 1986.

As of that date, the 51-L mission had been postponed three times and canceled once. On the night of January 27 the managers and engineers at Morton Thiokol huddled over a conference call with managers at Marshall and Kennedy to determine whether to launch the shuttle despite low temperatures and icy conditions. The managers at Marshall were pushing for a launch, but the Morton Thiokol participants were concerned with the cold temperatures. Eventually the Morton Thiokol managers capitulated to the desires of their customers at Marshall and agreed to a launch. The commission later found that the managers did not understand the criticality of the low temperature, despite various communications concerning it.

## Chronology of Events

MTI had begun testing the solid rocket motor in 1977. Results had indicated some problem with the O-rings in the joint. MTI engineers did not believe the results would cause a problem for the shuttle and stated so in a report, proposing to relax standards instead. However, Marshall engineers, who had disagreed with the MTI engineers over the type of O-ring being used since the inception of the project, believed that the shuttle could have a problem. One of the engineers, Leon Ray, was asked to review MTI's documents. Ray presented his evaluation in the memo in Figure 9.2, which he wrote to solid rocket motors chief engineer Glenn Eudy. Ray's immediate supervisor, John Miller, signed the memo to add authority to Ray's discussion. In a linear organizational hierarchy, employees tend to listen more to people who are higher in the hierarchy than they are. Ray sent copies of the memo to several other managers in both the Science and Engineering and the Space Project Divisions.

Eudy received the memo but took no special steps. It is quite possible that he simply scanned the long memo, never recognizing the importance of the point Ray was making. Rather, he continued to examine MTI's reports on additional flights as they came in.

Ray, however, was not satisfied, and a year he later wrote a second memo (Figure 9.3), which Miller also signed. The second memo is much shorter than the first; it is limited to a single page. Ray states the organizing idea in the first paragraph and then succinctly lists the claims to support his argument. He omits the details discussing the evidence. The document is much easier for the reader to scan and understand; the dense appearance of the first implies that the reader needs time to study the contents. A copy of this second memo was sent to the SRB project manager, one of the highest managers on the project.

As a result of this memo, Eudy joined Ray in a visit to two O-ring manufacturers to get their opinions on the MTI test results. Ray reported the results of that visit in the trip report in Figure 9.4. However, rather than summarize the results of the trip in relation to its purpose, Ray provides a chronological narration in which he includes the companies' opinions along with irrelevant information. A reader scanning the document could have easily missed the main points.

Morton Thiokol never received any written communication responding to the information or the point of view presented in these three memos.

In early 1981 the shuttle flights began. MTI conducted evaluations of the SRBs jettisoned from each flight and then disseminated the results to the company's divisions as well as to Marshall and NASA headquarters. Engineers evaluating the second flight found charring and erosion of the O-ring, but they believed it to be insufficient to delay launches until the problem could be solved. They also believed that the secondary O-ring provided sufficient backup. A number of subsequent flights also showed charring, but engineers continued to believe that the integrity of the joint was not threatened. However, in early 1985, a launch that resulted in erosion of the secondary as well as the primary O-ring led MTI engineers to question the

National Aeronautics and
Space Administration

**Lewis Research Center**
Cleveland, Ohio
44135

**NASA**

Reply to Attn of: **EP25 (78-1)**                    **January 9, 1978**

**TO:**      EE51/Mr. Eudy

**FROM:**    EP25/Mr. Miller

**SUBJECT:** Restatement of Position on SRM Clevis Joint O-Ring
Acceptance Criteria and Clevis Joint Shim Requirements

In view of recent events relating to proposals suggesting the relaxation
of standards for clevis joint O-ring acceptance and the use of a standard
shim thickness for clevis joints which allows O-ring compression to fall
below minimum industry accepted values, this office feels obligated to
restate its opposition to both proposals.  The following paragraphs
address each of the related subjects in terms of events leading to such
recommendations, risks involved by lowering standards, and recommendations
to resolve risks.

     a.  Relaxation of O-Ring Acceptance Standards - During the latter part
of November 1977, this office was requested by memorandum EE51 (77-291)
to review Thiokol documents STW7-2875, Standard Acceptance Criteria for
Preformed Packing (O-Rings) and 171-136, Standard Repair Instructions for
O-Rings (see enclosure 1).  Our response, which was documented in memorandum
EP25 (77-108) dated November 30, 1977 (see enclosure 2), recommended
rejection of both documents because of excessive deviations from MIL-STD-413
requirements, "visual inspection for rubber O-rings", and for lack of
clarification on several subjects.  Our memorandum also outlined recommended
allowable flaw sizes per MIL-STD-413 and allowables for other types of
defects which were not contained in MIL-STD-413.  On December 22, 1977, we
were provided with and asked to comment on a draft copy of memorandum EE51
(77-321) to program management (see enclosure 3) which contained EE51
comments and recommendations to Thiokol documents STW7-2375 and 171-136
which were not in agreement with our previous assessment.  Because of these
differences and to further amplify our position concerning O-ring defect
allowables, the following recommendations and justifications are restated:

     (1)  Inclusions - Remove all visible inclusions regardless of
size or type of included material.  The included material can be detached
during O-ring installation and use, creating debris and probable leakage.
Repair is required if the resulting void exceeds 0.025 inch diameter by
0.005 inch deep.  Deeper voids create a greater risk for leakage with low
compression (example: a void .015 inch deep reduces compression effect
by 5.5 percent).

**Figure 9.2    Marshall Space Flight Center internal memorandum.
First discussion regarding problems with SRB joint.** Source: *Report of the
Presidential Commission on the Space Shuttle Challenger Accident,* Washington, D.C., 1986.

---

*Margin annotations (left column):*

Long sentence

Purpose

Forecast

Background of O-ring
standards. Uses
chronological
organizational pattern.

Recommendations.
Indicates problem.

Claim
↓
Evidence

2

Generalization
↓
Details

(2) Mold Deposit Effects, Pits, and Voids - Each defect must be treated according to defect shape. Defects having sharp edges should be treated as a notch sensitive cut and repaired if the defect exceeds 0.025 inch diameter by 0.002 inch deep. Defects having smooth shapes should be repaired if either the diameter or depth exceeds 0.025 inch and 0.005 inch, respectively.

Claim
↓
Evidence

(3) Cuts - Radial cuts other than superficial cuts (cuts which cannot be felt with the thumbnail) are not allowed and must be repaired or dispositioned by splicing or rejection. The orientation of radial cuts is such that stretching of the O-ring can cause further tearing. Cuts parallel to the O-ring longitudinal axis must not exceed 0.002 inch deep by 0.060 inch long.

Generalization

(4) Repair Limitations - The limitations on maximum defect size acceptable for repair should be based on results of Thiokol's test program per TWR-11507. Deviations should be approved by EH01.

Background on O-ring compression. Uses chronological organizational pattern.

b. Below Minimum O-Ring Compression - Prior to the static firing of DM-1 in June 1977, shims were installed in the clevis joints to stop seal leakage caused by tang distortion. Shims of various thicknesses (0.010 to 0.031 in.) were placed around two of the joints according to gap width available (with some exceptions). No leaks were apparent during the test; however, the cavity pressure measurement on clevis joint number 5 (see enclosure 4) showed peculiar behavior (negative pressure to +8.3 psig). Calculations performed by MSFC and agreed to by Thiokol show that distortion of the clevis joint tang for any joint can be sufficient to cause O-ring/tang separation. Data from DM-1 shows that this condition can be created by joint movement (lowering of support chocks) and data from the hydroburst test shows the tang and clevis do not remain concentric during pressure cycling. All situations which could create tang distortion are not known, nor is the magnitude of movement known. Regardless of these unknowns, Thiokol then proposed to use a standard 0.020 inch thick shim for all SRM clevis joints including the STA-1 vehicle (see enclosure 5). Subsequent to arrival of the STA-1 vehicle at MSFC, Structures and Propulsion Laboratory was asked to assess the adequacy of the 0.020 inch shims which had been installed by Thiokol. The response, documented by memorandum EP01 (77-252) (see enclosure 5) recommended shim sizes ranging from 0.034 inch to 0.046 inch thick in order to maintain the industry recommended minimum compression value of 15 percent. It was, and still is, our desire to test with 15 percent minimum compression since this value is the industry wide minimum and was originally the minimum design value used by Thiokol prior to the tang distortion problem.

Claim
↓

Main organizing idea. Uses first person plural.

After issuance of the Structures and Propulsion Laboratory recommendations, an EH51 decision was made to use a 0.015 inch thick shim in the field joint of STA-1 which results in a minimum compression value of approximately 5.5 percent. This value assures no compression set. We strongly object to this proposal because it creates unacceptable risks which can and should be avoided.

(continues)

**Figure 9.2**    *(continued)*

**3**

Calculations conducted by this office show that in some instances, O-ring compression on flight vehicles has the potential of being <u>negative</u> by approximately 1.5 percent; these calculations included the effect of O-ring compression set. Thiokol test report dated August 15, 1977, per TWR-11507, "O-ring repair verification test plan" (see enclosure 7) shows that the parent O-ring material and splice joints exhibited maximum compression sets of 5.8 and 7.0 percent, respectively. Also, when considering that the SRM process demonstration segment O-ring suffered a compression set value of approximately 11.0 percent, one must treat these values as realistic and include their effects when calculating O-ring compression. It is recognized that O-rings will perform properly at lower values than the 15 to 25 percent range recommended; however, the higher values are used as a design point in order to account for losses such as O-ring compression set and defects in the hardware sealing surfaces and O-rings. Our recommendations to redesign on-coming hardware and custom shim each joint (with a range) on existing hardware as presented to you in October 1977, is still valid (see enclosure 8) The following recommendations and justifications are considered mandatory to provide adequate clevis joint sealing on all SRMs.

(1) Reshim STA-1 to obtain a minimum compression value of 15 percent in order to verify the design for flight.

(2) Redesign clevis joints on all on-coming hardware at the earliest possible effectivity to preclude unacceptable, high risk, O-ring compression values. This will eventually negate the use of shims, thereby reducing assembly time and eliminating shimming errors.

(3) Continue to use shims with existing and mixed hardware. Shims should be of sufficient thickness to provide a minimum O-ring compression of 15 percent. This value is used and recommended by Parker, Precision, CSD (Titan), Aerojet, and MSFC Science and Engineering Laboratories. We know of no instance where lower values are recommended.

(4) Direct the prime contractor and booster assembly contractor to <u>reinstate</u> the design requirements of 15 to 25 percent compression for clevis joint O-rings. We see no valid reason for not designing to accepted standards.

In summary, we believe that the facts presented in the preceding paragraphs should receive your most urgent attention. Proper shim sizing and high quality O-rings are mandatory to prevent hot gas leaks and resulting catastrophic failure. We will be pleased to provide assistance in any way possible.

Questions concerning the contents of this memorandum should be referred to Mr. W. L. Ray, 3-0459.

John Q. Miller
Chief, Solid Motor Branch

8 Enclosures

cc: w/o enc.
EP01/Mr. McCool
EP41/Mr. Hopson
EP21/Mr. Lombardo

**Figure 9.2**  *(continued)*

---

Indicates problem.

Evidence

Recommendations

Main organizing idea.

Conventional closing.

National Aeronautics and
Space Administration

**Lewis Research Center**
Cleveland, Ohio
44135

**NASA**

Reply to Attn of: **EP25 (79-13)**                                      **January 19, 1979**

TO:        EE51/Mr. Eudy

FROM:      EP25/Mr. Miller

SUBJECT:   Evaluation of SRM Clevis Joint Behavior

**Purpose**

**Finding**

**Evidence supporting finding.**

As requested by your memorandum, EE51 (79-10), Thiokol documents TWR-12019 and letter 7000/ED-78-484 have been reevaluated. We find the Thiokol position regarding design adequacy of the clevis joint to be completely unacceptable for the following reasons:

  a.  The large sealing surface gap created by excessive tang/clevis relative movement causes the primary O-ring seal to extrude into the gap, forcing the seal to function in a way which violates industry and Government O-ring application practices.

  b.  Excessive tang/clevis movement as explained above also allows the secondary O-ring seal to become completely disengaged from its sealing surface on the tang.

  c.  Contract End Item Specification, CPW1-2500D, page I-28, paragraph 3.2.1.2 requires that the integrity of all high pressure case seals be verifiable; the clevis joint secondary O-ring seal has been verified by tests to be unsatisfactory.

Questions or comments concerning this memorandum should be referred to Mr. William L. Ray, 3-0459.

John Q. Miller
Chief, Solid Motor Branch

CC:
SA41/Messrs. Hardy/Rice
EE51/Mr. Uptagrafft
EH02/Mr. Key
EP01/Mr. McCool
EP42/Mr. Bianca
EP21/Mr. Lombardo
EP25/Mr. Powers
EP25/Mr. Ray

**Figure 9.3   Shortened version of the January 1978 memo.**   Source: *Report of the Presidential Commission on the Space Shuttle Challenger Accident*, Washington, D.C., 1986.

National Aeronautics and
Space Administration

**Lewis Research Center**
Cleveland, Ohio
44135

**NASA**

Reply to Attn of:  **EP25 (79-23)**                                    **February 6, 1979**

TO:        Distribution

FROM:      EP25/Mr. Ray

SUBJECT:   Visit to Precision Rubber Products Corporation and
           Parker Seal Company

Purpose.

Refers to himself in
third person.

The purpose of this memorandum is to document the results of a visit
to Precision Rubber Products Corporation, Lebanon, TN, by Mr. Eudy, EE51 and
Mr. Ray, EP25, on February 1, 1979 and also to inform you of the visit
made to Parker Seal Company, Lexington, KY on February 2, 1979 by Mr. Ray.
The purpose of the visits was to present the O-ring seal manufacturers
with data concerning the large O-ring extrusion gaps being experienced on
the Space Shuttle Solid Rocket Motor clevis joints and to seek opinions
regarding potential risks involved.

Discussion of visit #1.

Recommendations.

Inappropriate
conventions for trip
report.

The visit on February 1, 1979, to Precision Rubber Products Corporation
by Mr. Eudy and Mr. Ray was very well received. Company officials, Mr.
Howard Gillette, Vice President for Technical Direction, Mr. John Hoover,
Vice President for Engineering, and Mr. Gene Hale, Design Engineer
attended the meeting and were presented with the SRM clevis joint seal
test data by Mr. Eudy and Mr. Ray. After considerable discussion,
company representatives declined to make immediate recommendations because
of the need for more time to study the data. They did; however, voice
concern for the design, stating that the SRM O-ring extrusion gap was
larger than that covered by their experience. They also stated that more
tests should be performed with the present design. Mr. Hoover promised
to contact MSFC for further discussions within a few days. Mr. Gillette
provided Mr. Eudy and Mr. Ray with the names of two consultants who may
be able to help. We are indebted to the Precision Rubber Products
Corporation for the time and effort being expended by their people in
support of this problem, especially since they have no connection with
the project.

Discussion of visit #2.

The visit to the Parker Seal Company on February 2, 1979, by Mr. Ray,
EP25, was also well received; Parker Seal Company supplies the O-rings
used in the SRM clevis joint design. Parker representatives, Mr. Bill
Collins, Vice President for Sales, Mr. W. B. Green, Manager for Technical
Services, Mr. J. W. Kosty, Chief Development Engineer for R&D, Mr.
D. P. Thalman, Territory Manager and Mr. Dutch Haddock, Technical
Services, met with Mr. Ray, EP25, and were provided with the identical

**Figure 9.4    Trip report in the form of a memo.**    Source: *Report of the*
*Presidential Commission on the Space Shuttle Challenger Accident,* Washington, D.C., 1986.

**2**

Recommendations.

SRM clevis joint data as was presented to the Precision Rubber Products Company on February 1, 1979.  Reaction to the data by Parker officials was essentially the same as that by Precision; the SRM O-ring extrusion gap is larger than they have previously experienced.  They also expressed surprise that the seal had performed so well in the present application. Parker experts would make no official statements concerning reliability and potential risk factors associated with the present design; however, their first thought was that the O-ring was being asked to perform beyond its intended design and that a different type of seal should be considered. The need for additional testing of the present design was also discussed and it was agreed that tests which more closely simulate actual conditions should be done.  Parker officials will study the data in more detail with other Company experts and contact MSFC for further discussions in approximately one week.  Parker Seal has shown a serious interest in assisting MSFC with this problem and their efforts are very much appreciated.

*William L. Ray*
William L. Ray
Solid Motor Branch, EP25

Distribution:
SA41/Messrs. Hardy/Rice
EE51/Mr. Eudy
EP01/Mr. McCool

**Figure 9.4**   *(continued)*

effectiveness of the secondary O-ring as a backup in cold weather. They initiated a series of tests to examine the effects of low temperatures on the O-ring and found that at temperatures below 50 degrees Fahrenheit, the secondary O-ring did not reseal. At this point MTI engineers recognized that the O-ring problem needed to be solved quickly.

When NASA received the report on the erosion of the secondary O-ring, it also became concerned. NASA officials requested an engineer with their Shuttle Propulsion Division make a visit to Marshall to meet with personnel to "see what kind of problem [they] were having" so that top management could clear up the problem quickly. The memo in Figure 9.5, which was distributed internally at NASA headquarters in Washington, is a report of that trip.

Engineers at MTI were becoming increasingly concerned with the problem. In his progress report, Roger Boisjoly attempted to persuade MTI to concentrate its resources on solving the problem. He presented the following argument based on management's values (Figure 9.6).

Boisjoly addressed the problem again a week later in a memo to one of the top people in the hierarchy, the vice president of engineering. Rather

National Aeronautics and
Space Administration

**Lewis Research Center**
Cleveland, Ohio
44135

NASA

JUL 17 1985

Reply to Attn of   **MPS**

**TO:**        M/Associate Administrator for Space Flight

**FROM:**      MPS/Irv Davids

**SUBJECT:**   Case to Case and Nozzle to Case "O" Ring Seal Erosion
             Problems

**Purpose**

As a result of the problems being incurred during flight on both
case to case and nozzle to case "O" ring erosion, Mr. Hamby and I
visited MSFC on July 11, 1985, to discuss this issue with both
project and S&E personnel. Following are some important factors
concerning these problems:

**Subheads**

**A.   Nozzle to Case "O" ring erosion**

**Discussion of problem**

There have been twelve (12) instances during flight where there
have been some primary "O" ring erosion. In one specific case
there was also erosion of the secondary "O" ring seal. There
were two (2) primary "O" ring seals that were heat affected (no
erosion) and two (2) cases in which soot blew by the primary
seals.

**Cause of problem**

The prime suspect as the cause for the erosion on the primary "O"
ring seals is the type of putty used. It is Thiokol's position
that during assembly, leak check, or ign'tion, a hole can be
formed through the putty which initiates "O" ring erosion due to
a jetting effect. It is important to note that after STS-10, the
manufacturer of the putty went out of business and a new putty
manufacturer was contracted. The new putty is believed to be
more susceptible to environmental effects such as moisture which
makes the putty more tacky.

**Alternative solutions**

There are various options being considered such as removal of
putty, varying the putty configuration to prevent the jetting
effect, use of a putty made by a Canadian Manufacturer which
includes asbestos, and various combination of putty and grease.
Thermal analysis and/or tests are underway to assess these
options.

**Personal response to
solutions**

Thiokol is seriously considering the deletion of putty on the QM-
5 nozzle/case joint since they believe the putty is the prime
cause of the erosion. A decision on this change is planned to be
made this week. I have reservations about doing it, considering
the significance of the QM-5 firing in qualifying the FWC for
flight.

**Figure 9.5    A trip report.**   Source: *Report of the Presidential Commission on the
Space Shuttle Challenger Accident,* Washington, D.C., 1986.

It is important to note that the cause and effect of the putty varies. There are some MSFC personnel who are not convinced that the holes in the putty are the source of the problem but feel that it may be a reverse effect in that the hot gases may be leaking through the seal and causing the hole track in the putty.

Considering the fact that there doesn't appear to be a validated resolution as to the effect of putty, I would certainly question the wisdom in removing it on QM-5.

**B.   Case to Case "O" Ring Erosion**

There have been five (5) occurrences during flight where there was primary field joint "O" ring erosion.  There was one case where the secondary "O" ring was heat affected with no erosion. The erosion with the field joint primary "O" rings is considered by some to be more critical than the nozzle joint due to the fact that during the pressure build up on the primary "O" ring the unpressurized field joint secondary seal unseats due to joint rotation.

The problem with the unseating of the secondary "O" ring during joint rotation has been known for quite some time.  In order to eliminate this problem on the FWC field joints a capture feature was designed which prevents the secondary seal from lifting off.  During our discussions on this issue with MSFC, an action was assigned for them to identify the timing associated with the unseating of the secondary "O" ring and the seating of the primary "O" ring during rotation.  How long it takes the secondary "O" ring to lift off during rotation and when in the pressure cycle it lifts are key factors in the determination of its criticality.

The present consensus is that if the primary "O" ring seats during ignition, and subsequently fails, the unseated secondary "O" ring will not serve its intended purpose as a redundant seal.  However, redundancy does exist during the ignition cycle, which is the most critical time.

It is recommended that we arrange for MSFC to provide an overall briefing to you on the SRM "O" rings, including failure history, current status, and options for correcting the problem.

*Irving Davids*
Irving Davids

cc:
M/Mr. Weeks
M/Mr. Hamby
ML/Mr. Harrington
MP/Mr. Winterhalter

**Figure 9.5**    *(continued)*

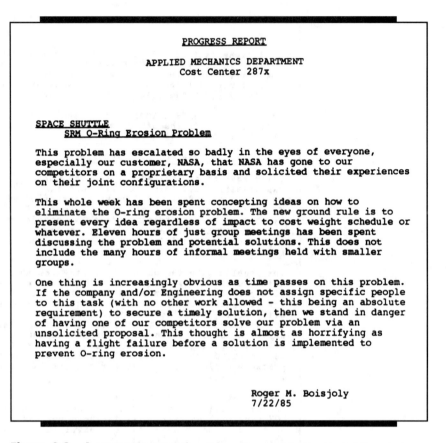

**Figure 9.6  A progress report.**  Source: *Report of the Presidential Commission on the Space Shuttle Challenger Accident,* Washington, D.C., 1986.

than repeat the facts, he interprets them, warning of the consequences of continuing to launch the shuttle (Figure 9.7). Rather than skirt the issue, or build his case before making his recommendation, he specifies his purpose at the very beginning of the message. His tone is direct; he states unequivocally that the problem can result in "loss of human life." The tone is also authoritative—for example, he states that "the team *must* (emphasis is author's) be given the responsibility." There are no hedge words. Furthermore, his tone is emotional (e.g., "The result would be a catastrophe of the highest order"). He packs the memo with adjectives (e.g., "*mistakenly* accepted position," "*drastically* changed") and uses jargon to emphasize a point (e.g., "jump ball").

## MORTON THIOKOL, INC.
### Wasatch Division

COMPANY PRIVATE

**Interoffice Memo**

31 July 1985
2870:FY86:073

TO:        R. K. Lund
           Vice President, Engineering

CC:        B. C. Brinton, A. J. McDonald, L. H. Sayer, J. R. Kapp

FROM:      R. M. Boisjoly
           Applied Mechanics – Ext. 3525

SUBJECT:   SRM O-Ring Erosion/Potential Failure Criticality

**Purpose**

This letter is written to insure that management is fully aware of the seriousness of the current O-Ring erosion problem in the SRM joints from an engineering standpoint.

**Problem**

The mistakenly accepted position on the joint problem was to fly without fear of failure and to run a series of design evaluations which would ultimately lead to a solution or at least a significant reduction of the erosion problem. This position is now drastically changed as a result of the SRM 16A nozzle joint erosion which eroded a secondary O-Ring with the primary O-Ring never sealing.

**Use of modifiers**

**Consequences**

**Personal response**

If the same scenario should occur in a field joint (and it could), then it is a jump ball as to the success or failure of the joint because the secondary O-Ring cannot respond to the clevis opening rate and may not be capable of pressurization. The result would be a catastrophe of the highest order – loss of human life.

**Recommendations**

An unofficial team (a memo defining the team and its purpose was never published) with leader was formed on 19 July 1985 and was tasked with solving the problem for both the short and long term. This unofficial team is essentially nonexistent at this time. In my opinion, the team must be officially given the responsibility and the authority to execute the work that needs to be done on a non-interference basis (full time assignment until completed).

(continues)

**Figure 9.7    Confidential memo.**    Source: *Report of the Presidential Commission on the Space Shuttle Challenger Accident,* Washington, D.C., 1986.

```
R. K. Lund                                                    31 July 1985

It is my honest and very real fear that if we do not take immediate action to
dedicate a team to solve the problem with the field joint having the number
one priority, then we stand in jeopardy of losing a flight along with all the
launch pad facilities.


R. M. Boisjoly


Concurred by:


J. R. Kapp, Manger
Applied Mechanics
                                            COMPANY PRIVATE
```

**Figure 9.7**   *(continued)*

The O-ring problem was also a major item on the agenda for the problem review board teleconference between Marshall and MTI managers in July. In response to two questions originating in that conference, an engineer at MTI sent a letter (Figure 9.8) to one of the quality assurance people at Marshall. The letter, in a question and answer format, is factual and informative but offers no interpretation of the data. When Thomas received the memo, he made copies of it and sent it to his supervisor, Lawrence Mulloy, so that Mulloy could send it to NASA. However, by the time Mulloy saw it, MTI had made a presentation on the problem to NASA. Mulloy thought the information was old news and sent the memo back to Thomas. The official to whom it would have been sent has since commented that even if he had received the memo, he might not have understood the implications of it. And Mulloy has echoed that sentiment: "There were a whole lot of people who weren't smart enough to look behind the veil and say, 'Gee, I wonder what that means'" (Presidential Commission 1986). However, it appears less a veil than a failure of the writer to understand what readers needed to know. Because readers either were in a different organization from the writer or were not experts in the writer's topic, readers needed to know the facts and needed to understand the relationship between the facts and the safe operation of the shuttle. Furthermore, because readers at Marshall had not been kept informed of the O-ring issue, they needed background information.

Windsor suggests that Boisjoly and Russell failed because of the culture of their organization, which was proprietary by nature and did not want its failures or problems known, especially by those outside the company (Windsor 1988). Thus, in writing to an external audience, Russell is

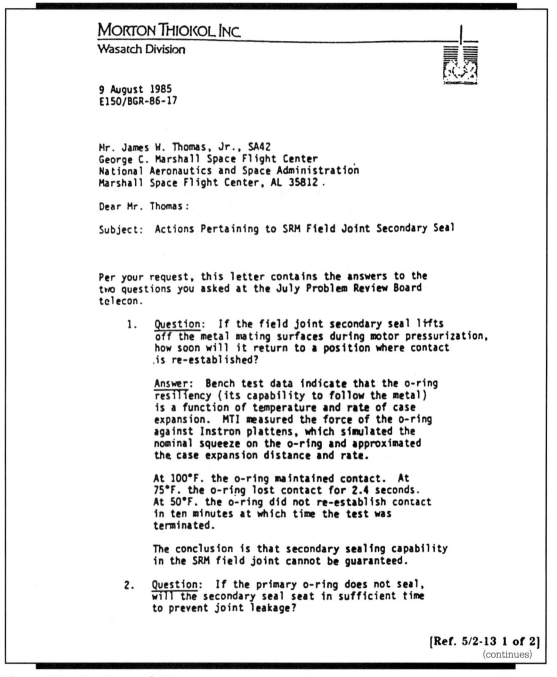

## MORTON THIOKOL, INC.
### Wasatch Division

9 August 1985
E150/BGR-86-17

Mr. James W. Thomas, Jr., SA42
George C. Marshall Space Flight Center
National Aeronautics and Space Administration
Marshall Space Flight Center, AL 35812.

Dear Mr. Thomas:

Subject: Actions Pertaining to SRM Field Joint Secondary Seal

Per your request, this letter contains the answers to the
two questions you asked at the July Problem Review Board
telecon.

1.  Question: If the field joint secondary seal lifts
    off the metal mating surfaces during motor pressurization,
    how soon will it return to a position where contact
    is re-established?

    Answer: Bench test data indicate that the o-ring
    resiliency (its capability to follow the metal)
    is a function of temperature and rate of case
    expansion. MTI measured the force of the o-ring
    against Instron plattens, which simulated the
    nominal squeeze on the o-ring and approximated
    the case expansion distance and rate.

    At 100°F. the o-ring maintained contact. At
    75°F. the o-ring lost contact for 2.4 seconds.
    At 50°F. the o-ring did not re-establish contact
    in ten minutes at which time the test was
    terminated.

    The conclusion is that secondary sealing capability
    in the SRM field joint cannot be guaranteed.

2.  Question: If the primary o-ring does not seal,
    will the secondary seal seat in sufficient time
    to prevent joint leakage?

**[Ref. 5/2-13 1 of 2]**
(continues)

**Figure 9.8  A response letter.**   Source: *Report of the Presidential Commission on the Space Shuttle Challenger Accident*, Washington, D.C., 1986.

Mr. James W. Thomas, Jr., SA42
9 August 1985
E150/BGR-86-17
Page 2

<u>Answer</u>: MTI has no reason to suspect that the
primary seal would ever fail after pressure
equilibrium is reached, i.e., after the ignition
transient. If the primary o-ring were to fail
from 0 to 170 milliseconds, there is a very high
probability that the secondary o-ring would hold
pressure since the case has not expanded appreciably
at this point. If the primary seal were to fail
from 170 to 330 milliseconds, the probability
of the secondary seal holding is reduced. From
330 to 600 milliseconds the chance of the secondary
seal holding is small. This is a direct result
of the o-ring's slow response compared to the
metal case segments as the joint rotates.

Please call me or Mr. Roger Boisjoly if you have additional
questions concerning this issue.

Very truly yours,

Brian G. Russell, Manager
SRM Ignition System

BGR/co

cc: L. Wear, SA42
    E. Skrobiszewski, SA49
    I. Adams, MTI/MSFC

bcc: J. Kilminster
     A. McDonald
     R. Ebeling
     J. Elwell
     B. Brinton
     A. MacBeth
     R. Boisjoly
     A. Thompson
     S. Stein

[Ref. 5/2-13 2 of 2]

**Figure 9.8** *(continued)*

careful not to state anything negative about his organization (see Figure 9.8). Notice that Boisjoly in his memo in Figure 9.7 is careful to type *COMPANY PRIVATE* on the bottom of the page to ensure that the information remain confidential within the organization.

On August 19, 1985, MTI and Marshall program managers briefed NASA headquarters on the problem of the seals. NASA applied pressure to solve the problem, and on August 20 Thiokol's vice president for engineering announced the establishment of a task force to recommend short- and long-term solutions. But problems with the task force existed from the start.

Because of the importance of the task force's work and because time was critical, the task force communicated its progress to MTI management in weekly activity reports. The activity report contained in the memo in Figure 9.9 is as direct in its approach as the memo written by Boisjoly. The very first word carries a personal and emotional tone that is not found in most business correspondence. Although the report deviates from the conventional style of an activity report, it adheres to the conventional format. It is divided according to the various activities involved, with headings introducing each activity. The report is sent to the director of the solid rocket motor project, who will then report to his supervisors on the progress of the task force. Copies are sent to those associated with the O-ring problem as well as those on the task force. Oddly, Ebeling deviates from the conventional format in that he signs the memo at the end, as if it were a letter, rather than putting his name next to the *FROM* subcategory in the heading.

Shortly after this memo went out, Boisjoly wrote one echoing Ebeling's complaints (Figure 9.10). His style continues to be personal and direct. Note, however, that the activity report is not formatted for easy reading as Ebeling's is; Boisjoly's report is an attachment to a memo rather than a part of one.

Boisjoly's and Ebeling's direct and emotional memos, which not only interpreted the facts for the readers but also provided them with an understanding of the consequences, were apparently successful in communicating to their readers the significance of the O-ring problem. During that fateful night of January 27, 1986, MTI managers initially supported their engineers' decision to halt the launch. It was only after Marshall asked for a management decision that they capitulated to what they believed to be their customers' demands. On January 28, when temperatures were well below 50 degrees Fahrenheit, the Challenger was sent on its final fatal mission.

MORTON THIOKOL INC
Wasatch Division

Interoffice Memo

1 October 1985
E150/RVE-86-47

TO:                A. J. McDonald, Director
                   Solid Rocket Motor Project

FROM:              Manager, SRM Ignition System, Final Assembly, Special
                   Projects and Ground Test

CC:                B. McDougall, B. Russell, J. McCluskey, D. Cooper,
                   J. Kilminster, B. Brinton, T. O'Grady, B. MacBeth,
                   J. Sutton, J. Elwell, I. Adams, F. Call, J. Lamere,
                   P. Ross, D. Fullmer, E. Bailey, D. Smith, L. Bailey,
                   B. Kuchek, Q. Eskelsen, P. Petty, J. McCall

SUBJECT:           Weekly Activity Report
                   1 October 1985


EXECUTIVE SUMMARY

HELP! The seal task force is constantly being delayed by every possible
means. People are quoting policy and systems without work-around. MSFC
is correct in stating that we do not know how to run a development
program.

GROUND TEST

1.  The two (2) GTM center segments were received at T-24 last week.
Optical measurements are being taken. Significant work has to be done
to clean up the joints. It should be noted that when necessary SICBM
takes priority.

2.  The DM-6 test report less composite section was released last week.

ELECTRICAL

As a result of the latest engineering analysis of the V-1 case it
appears that high stress risers to the case are created by the phenolic
DFI housings and fairings. As it presently stands, these will probably
have to be modified or removed and if removed will have to be replaced.
This could have an impact on the launch schedule.

**Figure 9.9   A memo containing an activity report.**   Source: *Report of the Presidential Commission on the Space Shuttle Challenger Accident*, Washington, D.C., 1986.

A. J. McDonald, Director
1 October 1985
E150/RVE-86-47
Page 2

FINAL ASSEMBLY

One SRM 25 and two SRM 26 segments along with two SRM 24 exit cones were
completed during this period. Only three segments are presently in
work. Availability of igniter components, nozzles and systems tunnel
tooling are the present constraining factors in the final assembly area.

IGNITION SYSTEM

1. Engineering is currently rewriting igniter gask-o-seal coating
requirements to allow minor flaws and scratches. Bare metal areas will
be coated with a thin film of HD-2 grease. Approval is expected within
the week.

2. Safe and Arm Device component deliveries is beginning to cause
concern. There are five S&A's at KSC on the shelf. Procurement,
Program Office representatives visited Consolidated Controls to discuss
accelerating scheduled deliveries. CCC has promised 10 A&M's and 30
B-B's no later than 31 October 1985.

O-RINGS AND PUTTY

1. The short stack finally went together after repeated attempts, but
one of the o-rings was cut. Efforts to separate the joint were stopped
because some do not think they will work. Engineering is designing
tools to separate the pieces. The prints should be released tomorrow.

2. The inert segments are at T-24 and are undergoing inspection.

3. The hot flow test rig is in design, which is proving to be
difficult. Engineering is planning release of these prints Wednesday or
Thursday.

4. Various potential filler materials are on order such as carbon,
graphite, quartz, and silica fiber braids; and different putties. They
will all be tried in hot flow tests and full scale assembly tests.

5. The allegiance to the o-ring investigation task force is very
limited to a group of engineers numbering 8-10. Our assigned people in
manufacturing and quality have the desire, but are encumbered with other
significant work. Others in manufacturing, quality, procurement who are
not involved directly, but whose help we need, are generating plenty of
resistance. We are creating more instructional paper than engineering
data. We wish we could get action by verbal request but such is not the
case. This is a red flag.

R. V. Ebeling

**Figure 9.9**   *(continued)*

```
                            ACTIVITY REPORT




        The team generally has been experiencing trouble from the business

    as usual attitude from supporting organizations.  Part of this is due to

    lack of understanding of how important this task team activity is and

    the rest is due to pure operating procedure inertia which prevents

    timely results to a specific request.

        The team met with Joe Kilminster on 10/3/85 to discuss this

    problem.  He wanted specific examples which he was given and he simply

    concluded that it was every team members responsibility to flag problems

    that occurred to organizational supervision and work to remove the road

    block by getting the required support to solve the problem.  The problem

    was further explained to require almost full time nursing of each task

    to insure it is taken to completion by a support group.  Joe simply

    agreed and said we should then nurse every task we have.

        He plain doesn't understand that there are not enough people to do

    that kind of nursing of each task, but he doesn't seem to mind directing

    that the task never-the-less gets done.  For example, the team just

    found out that when we submit a request to purchase an item, that it

    goes through approximately 6 to 8 people before a purchase order is

    written and the item actually ordered.

        The vendors we are working with on seals and spacer rings have

    responded to our requests in a timely manner yet we (MTI) cannot get a

    purchase order to them in a timely manner.  Our lab has been waiting for

    a function generator since 9-25-85.  The paperwork authorizing the

    purchase was finished by engineering on 9-24-85 and placed into the

    system.  We have yet to receive the requested item.  This type of
```

**Figure 9.10  Activity report.**  Source: *Report of the Presidential Commission on the Space Shuttle Challenger Accident*, Washington, D.C., 1986.

example is typical and results in lost resources that had been planned to do test work for us in a timely manner.

   I for one resent working at full capacity all week long and then being required to support activity on the weekend that could have been accomplished during the week. I might add that even NASA perceives that the team is being blocked in its engineering efforts to accomplish its tasks. NASA is sending an engineering representative to stay with us starting Oct 14th. We feel that this is the direct result of their feeling that we (MTI) are not responding quickly enough on the seal problem.

   I should add that several of the team members requested that we be given a specific manufacturing engineer, quality engineer, safety engineer and 4 to 6 technicians to allow us to do our tests on a non-interference basis with the rest of the system. This request was deemed not necessary when Joe decided that the nursing of the task approach was directed.

   Finally, the basic problem boils down to the fact that ALL MTI problems have #1 priority and that upper management apparently feels that the SRM program is ours for sure and the customer be damned.

Roger Boisjoly 10/4/85

**Figure 9.10**   *(continued)*

## CHAPTER SUGGESTIONS FOR DISCUSSION AND WRITING

1. Write a comparison of the two trip reports in Figures 9.4 and 9.5. Examine the respective focus, style, and format of each. Which one is easier for the reader to comprehend? Why? Which provides an appropriate focus? Why?

2. In both the January 1979 and the July 31, 1985, memos, the writers want the readers to do something. Compare the two. What does each one want the readers to do? Do you think the 1979 memo would persuade you to take immediate action? Would the 1985 memo? Revise one or both of the two memos so that readers are more likely to be persuaded to follow the writers' recommendations.

3. Compare the tones of the July 17, the July 31, and the October 1 memos. How would you characterize the tone in each? What key words, phrases, and rhetorical devices alter the tone in each? To which memo do you think the reader is most likely to make the desired response? Why?

4. Compare the formats of the July 17, July 31, and October 1 memos. Why do you think the writers use such different formats?

5. Boisjoly specifies *Company Private* on his July 31 memo in an effort to keep the information confidential. Why do you think he wishes confidentiality? Do you believe this is ethical, or do you think that because of the significance of the problem, Boisjoly should send this information to others? Would sending on the information be ethical? Would he jeopardize his job or his employer's trust if he did so?

   If, prior to writing this memo, Boisjoly had written to you, as an employee of the company, and asked your opinion about the confidentiality of the information, what advice would you have given him? Assume the role of an engineer at MTI and write a memo in response to Boisjoly, presenting your view on whether the information should be kept confidential.

6. Consider the July 17 memo. The writer uses technical terms and acronyms without defining them. Highlight several. Bracket the ones you think readers will have difficulty understanding. Examine the syntax. Do the sentences flow smoothly? Bracket sentences you think readers will have difficulty reading.

7. The reports in Figures 9.9 and 9.10 cite a problem with getting work done. Is an activity report an appropriate subgenre for communicating this information? Can you suggest a better subgenre form? What other actions might the writers take to get these problems solved?

8. Examine the July 22, the July 31, and the October 1 memos. Compare the claims the writers make. Which claims in each of the memos are related to readers' attitudes, values, and needs? What evidence do the writers use to support their claims? Which claims in each of the

memos will persuade the readers? Can you think of other claims the writers might make to persuade the readers?

9. *The Report of the Presidential Commission* is written for the president, Congress, persons associated with the space industry, and interested citizens. Thus, the audience ranges from experts to novices. You fit somewhere in this range. Were you able to understand the description of the O-rings on page 380? If not, what did you have difficulty understanding? What strategies might the writers have used to help you better understand the section?

10. Compare the description of the O-rings on pages 380 with the description of a nuclear reaction on pages 359. Both deal with complicated subjects. Both should be clear enough for novices to understand. Which one do you think is easier for a novice to understand? Why? What strategies does the writer use to help readers understand? How could the writers of the other text use these strategies? Try to rewrite a passage of the more difficult text to make it easier for a novice to understand.

11. Compare the memos in Figures 1.1, 8.2, and 9.8. All of them failed to persuade their readers to act as the writers wanted. What do they have in common, if anything? Why did they all fail?

12. During the investigation into the Challenger disaster, a manager at the Marshall Space Flight Center who received the MTI letter in Figure 9.8 from his engineer admitted that he was "not smart enough to look behind the veil and say, 'Gee, I wonder what this [letter] means.'" Furthermore, the manager at NASA who should have received the letter indicated that he might not have understood the implications of the temperature data even if he had received it. Why do you think these two managers were unable to understand the letter? Try to revise the letter so that it could be easily comprehended by the two managers.

## Notes

Presidential Commission on the Space Shuttle Challenger Accident. 1986. *Report of the Presidential Commission on the Space Shuttle Challenger Accident.* Vol. 1–5. Washington, D.C.: Government Printing Office.

Windsor, Dorothy A. 1988. Communication failures contributing to the Challenger accident: An example for technical communicators. *IEEE Transactions in Technical Communication* 31 (September): 101–7.

# CHAPTER 10

# The African Slave Burial Ground

## FOLEY SQUARE PROPOSED FEDERAL COURTHOUSE AND FEDERAL/MUNICIPAL OFFICE BUILDING

At the end of the 1980s, the federal government proposed to build a federal courthouse and federal/municipal office building in downtown Manhattan. To comply with the National Environmental Protection Act (NEPA), discussed in Chapter 7, the government had to write an Environmental Impact Statement (EIS). The General Services Administration's (GSA) New York planning staff contracted with a private engineering consulting firm to study the proposed site and to write the EIS. In investigating the archaeological and historical value of the area, consultants discovered that a portion of the site would be located over a former African slave burial ground. However, after studying maps of the area, the consultants concluded that, because buildings with deep basements had previously been built over the site, the remains would have already been destroyed, leaving little of significance.

## RESEARCH DESIGN FOR ARCHAEOLOGICAL, HISTORICAL, AND BIOANTHROPOLOGICAL INVESTIGATIONS OF THE AFRICAN BURIAL GROUND

The EIS was accepted, and construction on the project begun. However, as contractors began to clear the site, they discovered human remains. Previous construction had not destroyed the burial site. In all, over 400 bodies were found. The site contained the largest extant slave burial ground in the north. News reached the public, which demanded a stop to the construction. Many people objected to digging up the bodies; others simply wanted the bodies dug up in such as way as to preserve them for study. Excavation was halted at a cost of $100,000 per day for construction delays that would be paid by the federal government. One government official estimated that the cost of dealing with the site would be approximately $16 million.

The government was faced with the question of what to do with the site and the remains on it. They requested another consulting firm to study the problem and develop a research design which, according to NEPA requirements, should be presented to the public for input in the same way in which an EIS is.

The excerpt starting on the next page is part of the research design. The segment is an information report that bears close resemblance to the academic reports you have written throughout your academic career. It is based largely on written sources and includes in-text citation, using an author/date style. Section 2.2.1 is organized chronologically. Although the narration is formal and impersonal, the writers are aware of the economic, political, and social conflicts surrounding the topic, and they adopt a viewpoint that appeals to their audience's values and attitudes. As you read this section, consider the writers' bias toward the subject.

Section 2.2.2 is organized analytically according to the specific resources derived from the burial ground site. Consider the writers' point of view and focus as you read this section also.

# RESPONSE LETTERS

Reaction to the research design was mainly split between responses by academicians, historians, and anthropologists to the technical aspects of the design, and the responses of the African-American community to the humane and ethnic aspects of the situation. The letters in Figures 10.1, 10.2, 10.3, and 10.4 are indicative of the types of responses the GSA received from people representing their respective professional and cultural communities. As a consequence of these responses, the federal government permanently halted construction on the site, agreeing instead to re-inter the bodies after they had been studied and to declare the area a historical landmark. The government also agreed that the remains should be studied at Howard University.

FOLEY SQUARE FEDERAL COURTHOUSE
AND OFFICE BUILDING
NEW YORK, NEW YORK

RESEARCH DESIGN
FOR
ARCHEOLOGICAL, HISTORICAL, AND
BIOANTHROPOLOGICAL INVESTIGATIONS
OF THE AFRICAN BURIAL GROUND
AND FIVE POINTS AREA
NEW YORK, NEW YORK

PREPARED FOR

EDWARDS AND KELCEY ENGINEERS, INC.
70 SOUTH ORANGE AVENUE
LIVINGSTON, NJ 07039

AND

GENERAL SERVICES ADMINISTRATION
REGION 2
JACOB K. JAVITS FEDERAL BUILDING
NEW YORK, NY 10278

PREPARED BY

HOWARD UNIVERSITY
WASHINGTON, DC 20059

AND

JOHN MILNER ASSOCIATES, INC.
309 NORTH MATLACK STREET
WEST CHESTER, PA 19380

OCTOBER 15, 1992

## 2.0 EVALUATION OF SIGNIFICANCE

### 2.1 Introduction

This section of the Research Design presents an evaluation of the significance of human remains and archeological resources present in the project areas in terms of the National Register of Historic Places. Cultural resources are afforded certain protections from federally funded, licensed, or approved undertakings under the provisions of Section 106 of the National Historic Preservation Act, as amended, provided that the resources are listed on or eligible for the National Register. The Criteria for Evaluation of the National Register (36 CFR 60.4) are as follows:

> The quality of significance in American history, architecture, archeology, engineering and culture is present in districts, sites, buildings, structures and objects that possess integrity of location, design, setting, materials, workmanship, feeling, and association and
>
> (a) that are associated with events that have made a significant contribution to the broad patterns of our history; or
>
> (b) that are associated with the lives of persons significant in our past; or
>
> (c) that embody the distinctive characteristics of a type, period, or method of construction, or that represent the work of a master, or that possess high artistic values, or that represent a significant and distinguishable entity whose components may lack individual distinction; or
>
> (d) that have yielded, or may be likely to yield, information important in prehistory or history.

These criteria are applied to each project area's cultural resources in turn, beginning with the Broadway Block.

(continues)

2.2 Broadway Block

The National Register eligibility of human remains and archeological resources on the Broadway Block are evaluated in this portion of the
Research Design. A summary history of the block is followed by a summary
of the archeological resources found on the block. Finally, the National
Register's Criteria for Evaluation are applied to the block's cultural resources.

2.2.1 Summary History

The streets bounding the Broadway Block were not laid out until ca.
1784-1795. Before that, the southern portion of the block, which was outside the city's palisade, had been used by New York City's African community as a burial ground. The "Negros Burial Ground" is clearly marked
on a historic map dating to the mid-1700s (Maerschalk 1754).

By most accounts the burial ground was being used as a final resting place
for Africans, slaves and free people at least as early as 1712 (Stokes
1915-1928; Valentine 1847). Valentine's *Manual* provides one of the few
known descriptions of the burial ground:

> Beyond the commons lay what in the earliest settlement of the town had
> been appropriated as a burial place for Negroes, slave and free. It was
> a desolate, unappropriated spot, descending with a gentle declivity to-
> wards a ravine which led to the Kalkhook pond. The Negroes in this city
> were, both in the Dutch and English colonial times, a proscribed and de-
> tested race. . . . Many of them were native Africans, imported hither in

slave ships, and retaining their native superstitions and burial cus-
toms, mummeries and outcries. . . . So little seems to have been thought
of the race that not even a dedication of their burial-place was made
by the church authorities, or any others who might reasonably be supposed to have any interest in such a matter. The lands were unappropri-
ated, and though within convenient distance from the city, the locality
was unattractive and desolate, so that by permission the slave population were allowed to inter their dead there (Valentine 1847:567). . . .

In 1628 a vessel of the Dutch company brought three enslaved African women
to Manhattan. Although there is no explicit evidence that suggests the
company intended to enter the business of breeding slaves, its records indicate that these women had been purchased, as the company's clerk put
it, "for the comfort of the company's Negro men" (McManus 1966). Over the
years, some enslaved Africans managed to establish stable families whose
conjugal and consanguineal bonds the Dutch Reformed Church sanctioned
through its marriage and baptismal ceremonies (see Baptisms in the Dutch
Reformed Church of New Amsterdam and New York City December 25, 1639 to
December 27, 1730, published in 1890). This is not to say that the Dutch
created a benign slave regime. However much the Dutch clerics professed
their concern for the fate of the souls of Africans in the afterlife, they
did little to ameliorate the slaves' conditions on earth. Although a few
free blacks and free mulattoes of mixed African and Dutch ancestry inhabited the region, the vast majority of the people of African descent
who resided on Manhattan Island during Dutch rule were slaves, whom the

(continues)

Dutch company and the free inhabitants held in the abject condition of perpetual bondage (Swan 1990:1-6). . . . .

From 1626 to 1659 shipments of human cargoes arrived at New Amsterdam, adding to the Dutch settlement's numerically modest population. Slave laborers, mostly young men from the Congo-Angola region of Africa, were put to work on the construction of the fort and other public works projects that were indispensable to the success of the Dutch enterprise at Manhattan. Other enslaved Africans became agricultural workers who cultivated the company's farms; the largest farm was located north of Wall Street and extended to Hudson Street. The products of the enslaved Africans' labor on this unit of agricultural production fed the inhabitants of the fledgling Dutch settlement on the island (Rink 1986:100-101).

By the eve of the English conquest of 1664, no fewer than 700 enslaved Africans inhabited the island of Manhattan—that is, approximately 40 percent of the Dutch settlement's total population at that time (Brodhead 1859). As already noted these uprooted laborers of African ancestry made indispensable contributions to the project of colony-building under Dutch rule.

Although the proportion of enslaved Africans in the colonial settlement's total population declined during English rule, the colony of New York, which was named for the new English proprietor, the Duke of York later

King James II, supplemented its inadequate white labor force with enslaved Africans (Table 1). . . .

TABLE 1

Population of Manhattan or the County
of New York, 1698-1786

| Census Year | Total | European | African |
|---|---|---|---|
| 1698 | 4,937 | 4,237 | 700 |
| 1703 | 4,375 | 3,645 | 630 |
| 1712[a] | 5,841 | 4,886 | 975 |
| 1723 | 7,248 | 5,886 | 1,362 |
| 1731[b] | 8,622 | 7,045 | 1,577 |
| 1737[c] | 10,664 | 8,945 | 1,719 |
| 1746 | 11,717 | 9,273 | 2,444 |
| 1749 | 13,249 | 10,926 | 2,368 |
| 1756 | 13,046 | 10,768 | 2,278 |
| 1771 | 21,863 | 18,726 | 3,137 |
| 1786 | 23,614 | 21,507 | 2,107 |

Source:

Ira Rosenwaike, *Population History of New York City* (1972), 8.

[a] Returns for the Census of 1712 were incomplete.

[b] Robert V. Wells, "The Census of 1731," *NYHSQ,* 57(1973), 255-259.

[c] Gary B. Nash, "The New York Census of 1737: A Critical Note on the Integration of Statistical and Literary Sources," *WMQ,* 3d. ser., 36(1979), 428-435.

The African Burial Ground was the Africans' only autonomous social space, the only place where they were allowed to congregate with regularity, in large numbers, and beyond the purview of the authorities. This cemetery

(continues)

was located outside the city's limit, north of the palisades that was erected in 1745, and adjacent to the Collect Pond. The earliest surviving reference to the burial ground can be found in a letter Chaplain John Sharpe wrote to an English colleague. In that letter, dated 1712, Sharpe complained that even Christianized slaves "are buried in the Common by those of their country and complexion without office; on the contrary the Heathenish rites are performed at the grave by their countrymen" (New York Historical Society Collections 1880:341). Later in the century poor Europeans, who inhabited the northern fringes of the city's limit which now extended near the cemetery, submitted to the city council a petition that reported loud drumming and chanting emanating from the burial ground. The petition complained that the ecstatic ceremonies held at the cemetery sometimes lasted late into the night and disturbed the sleep of the residents who lived nearby. The city council took no prohibitive action except to reiterate the sunset curfew already placed on African funerals and to restate its injunction against gatherings of more than twelve mourners at these funerals. The Africans ignored these city ordinances, and the English authorities seldom demonstrated a desire to enforce the laws pertaining to the Africans' activities at the cemetery (Minutes of the Common Council of the City of New York 1901:III, 277–278).

Unmolested by outside interference, the Africans venerated their deceased family and friends at the African Burial Ground from roughly 1712 to 1790.

Enslaved Africans perhaps led other slaves in traditional rites intended to memorialize the dead. Through the performance of these rituals African ways were, perhaps, transmitted from one generation to the next. For more than two generations the African Burial Ground provided a precious space in which a unique African American society and culture took shape. . . .

2.2.2 Summary of Archeological Resources

2.2.2.1 African Burial Ground Resources

Archeological resources associated with the African Burial Ground consist of an unknown number of burials, of which 420 have been carefully excavated and removed. The majority of excavated burials are of individuals of African descent, although approximately seven percent appear to be of European descent. The depth of the burials below grade is variable, but ranges from 16 to 28 feet. The disparity in relative depth reflects the historic slope of the terrain in this area, which is now masked by fill introduced in the nineteenth century.

The condition of the human remains is quite variable, ranging from excellent (in a very few cases) to extremely poor. Many burials show evidence of coffin wood remains (mostly in the form of organic wood stains), coffin nails, and shroud pins. A few graves also contain buttons from clothing, beads, copper alloy finger rings, and other small items.

Other cemetery-related features consist of a row of postholes aligned in an east-west orientation which may delineate the original northern boundary of the African Burial Ground. . . .

2.2.3 Statement of Significance

2.2.3.1 African Burial Ground Resources

The histories, narratives, and records depicting African life in New Amsterdam and early New York in life and in death have been written from the perspectives of neither the enslaved nor the freed African New Yorkers. What is known today as "the history" of Africans in colonial New Amsterdam and New York was written by the enslavers, many of whom commonly viewed Africans as beasts of burden. The role of Africans in early New York was that of supplying a sorely needed work force. There is almost nothing that presents a glimpse of the lives of seventeenth and eighteenth century African New Yorkers from their own perspectives, in their own words.

Toni Morrison (1987, *New York Times*) has remarked that this nation has erected no monument honoring the memory of enslaved Africans. Furthermore, M.A. (Spike) Harris, an avocational historian, notes in *A Negro History Tour of Manhattan* (1968:xii) that "a stroll along Manhattan's streets reveals almost nothing except dark faces to connect Negroes with the history of New York City." Although a great deal of scholarship has been devoted to the study of the African/African American experience and a wealth of publications present facts and figures, the history of New York City's African community has not been made easily accessible. The African Burial Ground provides a focal point for bringing to life a past that has long been hidden. Historically, the significance of the site lies as much in the present as in the past. It is an opportunity to make available information that already exists on the history of New York City's African population and to generate new information through scholarly research.

The African Burial Ground is significant for the focus it brings to professional studies relating to the particular experience of New York City's enslaved population. The interdisciplinary nature of the project team will allow an informed approach to such challenging questions as whether the burials are examples of Christian or African religious practice or a complex fusion of both traditions. The African groups making up New York City's eighteenth century population may be better defined and their particular contributions to the city's growth and culture identified. The analysis of human skeletal remains provides data that are complimentary to historical and archeological research, and often yields information that is unavailable through any other means (Handler and Lange 1978; Rathbun 1987). Parrington (1987:57) notes that:

(continues)

. . . burial remains and associated artifacts are direct and conscious manifestations of ideological beliefs and practices and can potentially provide more explicit information about the cultural standards of the society being studied.

Biological studies of past human populations have significant applications in medicine, epidemiology, environmental toxicology, forensic science, DNA and genetics research, and demography. The discovery and excavation of over 400 burials at the African Burial Ground provides a unique opportunity to collect biological data on the earliest African skeletal sample in the United States, as well as for a significant sample of colonial Europeans from the lowest socioeconomic levels. Framed within a bio-cultural context that draws on historical, archeological, and comparable biological data, the analysis of the osteological remains from the African Burial Ground will yield crucial information on colonial Africans and Europeans unavailable from any other source.

Samples of over 20 individuals from a single colonial cemetery are rare, and those that have been recovered most often represent short-term cemetery use (Owsley 1990). The considerable number of remains recovered from the African Burial Ground is advantageous in that it will provide an adequate sample for a range of statistical analyses, and will also yield meaningful demographic data. . . .

Source: Excerpts from the *Research Design for Archeological, Historical, and Bioanthropological Investigations of the African Burial Ground and Five Points Area, New York, New York.* Prepared for Edwards and Kelcey Engineers, Inc., and General Services Administration, Region 2. Prepared by Howard University and John Milner Associates, Inc., October 15, 1992.

Unfortunately, in describing the historical background of the lots impacted by the GSA construction and in describing both through historical maps and legal descriptions of present day insurance maps, there are no illustrations in the text to assist in orienting the reader to the project area. The assumption is that the reader is already familiar with the documents described and knows the location of the project area. While this might be in part true, there are times in the text where the legal insurance map description is exclusively used, for example Lot 160. Without an illustration clarify the part of the project area being described, the reader is at best confused.

Other comments from our Southeast Regional Office are

o Further emphasizing the previous point, a serious flaw exists that the report lacks illustrations. This inhibits the readers, especially those unfamiliar with the project area, from grasping the spatial and directional relationships alluded to in the text. For the record, and to assist readers at all levels to follow the discussions, this must be corrected. For example, while most of the maps, drawings, and other illustrations in the Historic Conservation and Interpretation, Inc (Ingle et al 1990) Stage 1A report need not be repeated, several key maps, such as those showing historic street orientations and the projected boundaries of the 18th century "Negro Burying Ground" (page 7) are essential.
o Page 49, last paragraph: The report is vague on the future of unexcavated burials and other archeological resources in the pavilion area. Can we assume that the remaining unexcavated burials are protected and will be preserved in place? How many remaining burials will be excavated? How does the future treatment of these burials affect the design of the pavilion? How will GSA assure the public that unexcavated burials and other archeological resources will not be affected by construction activities?

o Page 52, first paragraph: What is meant by the term "anthropological archeologist"? In the U.S., most archeologist receive advanced degrees in anthropology. The usual parlance specifies a "prehistoric" vs. "historic" specialization or a chronological or regional specialization such as colonial America, the Caribbean, etc.

o Page 56, last paragraph: Are the Broadway Block specimens seen as comprising a sufficient sample for "baseline biology" of the entire African American population in the U.S.? Why is this so? How could this be? What about regional and cultural variations? Further explanation is needed here.

**Figure 10.1   A response to the research design.**   Source: *Comments on the Draft Research Design for the Archeological, Historical, and Bioanthropological Investigations of the African Burial Ground and Five Points Area, New York, New York.* Published by the General Services Administration, Region 2, 1993.

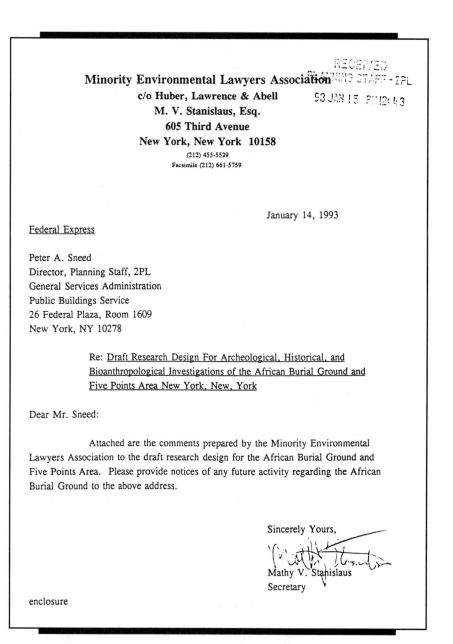

**Figure 10.2  A response to the research design.**  Source: *Comments on the Draft Research Design for the Archeological, Historical, and Bioanthropological Investigations of the African Burial Ground and Five Points Area, New York, New York,* Published by the General Services Administration, Region 2, 1993.

## INTRODUCTION

The Minority Environmental Lawyers Association ("MELA") is an organization comprised of minority environmental lawyers, primarily in the New York metropolitan area. Among MELA's objectives is to participate in activities affecting minority communities. As an advocate of minority environmental issues, MELA provides these comments to the Draft Research Design. In addition to our participation in the Draft Research Design, we request that MELA be involved in all decisions regarding the African Burial Ground (hereinafter, the "Burial Ground"), also referred to as being part of the Broadway Block in the Draft Research Design, including decisions regarding leaving in situ the unexcavated portion of the Burial Ground in the pavilion area, reinterment of excavated remains, a memorial and public education.

## SUMMARY OF COMMENTS

First, and foremost, it must be noted clearly in the Research Design that it was prepared subsequent to when required under the National Historic Preservation Act (hereinafter, the "NHPA") and its implementing regulations. Specifically, it must be noted that the preparation of the Draft Research Design was initiated after excavation of the majority of the graves at the Burial Ground - approximately 400.

The public outreach regarding the Draft Research Design and proposed project as a whole and has been dismal. Any future activities should be include the participation of all interested parties, including representatives of the descendant communities.

2

(continues)

**Figure 10.2** *(continued)*

In terms of the Draft Research Design itself, it fails to consider whether the Burial Ground is significant because of its contributions to American history and culture, as required under the Advisory Council on Historic Preservation (hereinafter, the "ACHP") regulations for significance of the Burial Ground and effect of the proposed project on the Burial Ground. Moreover, if such an analysis was conducted and if it had concluded that the Burial Ground was significant because of its contribution to American history and culture, the proposed project would be required to be modified.

The ACHP recognizes that while graves and their contents are important sources of information when studied by archaeologists and other specialists, they also represent deceased human beings, whose remains should be treated with respect[1]. The remains often have powerful emotional importance for their descendants. In the ACHP's policy statement regarding the treatment of human remains and grave goods, the importance of consultation with the descendants of the dead is stressed. The concerns of descendants should be weighed equally against the scientific interests in studying the remains.[2]

The ACHP's policy statement indicates a clear preference for leaving graves undisturbed. It also states that for activities subject to review under NHPA § 106, ACHP will seek preservation in place of sites known or likely to contain graves

---

[1] Memo to Federal Agency Preservation Officers Re: *Treatment of Human Remains Under Section 106*, October 12, 1988.

[2] Id.

3

**Figure 10.2**   *(continued)*

whenever this is feasible and prudent[3]. This requirement should have been applied to the entire Burial Ground prior to any excavation activities.

The fact that approximately 400 graves were excavated prior to any significant consultation with representatives of descendant communities, would appear to be violative of the ACHP policy statement and NHPA. The Draft Research Design should have been completed prior to any excavation - assuming such excavation is appropriate after application of the NHPA and its implementation regulations - particularly since information identifying the existence of the Burial Ground was available to the General Services Administration (hereinafter, "GSA") prior to the initiation of excavation activities.

Since it is impossible to apply the ACHP's policy statement's preference for preservation to the excavated portion of the Burial Ground, it should be applied to the unexcavated graves located in the pavilion area. Therefore, GSA must comply with this requirement by preventing any further excavation of graves for scientific studies and redesign the pavilion area to preserve the graves undisturbed. To resume excavation would be a second violation of NHPA and ACHP's regulations and policy. Additionally, the excavation of the remaining graves would exacerbate the disrespect, both actual and spiritual, already shown to not only the forgotten people buried there but to their descendants. . . .

**Figure 10.2**   *(continued)*

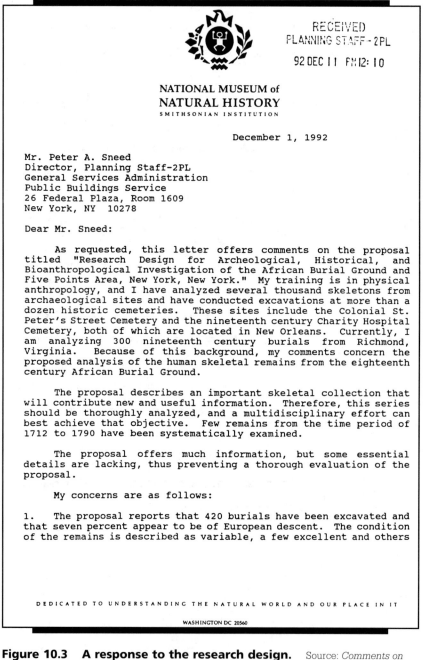

RECEIVED
PLANNING STAFF - 2PL

92 DEC 11 PM 12: 10

**NATIONAL MUSEUM of**
**NATURAL HISTORY**
SMITHSONIAN INSTITUTION

December 1, 1992

Mr. Peter A. Sneed
Director, Planning Staff-2PL
General Services Administration
Public Buildings Service
26 Federal Plaza, Room 1609
New York, NY  10278

Dear Mr. Sneed:

As requested, this letter offers comments on the proposal titled "Research Design for Archeological, Historical, and Bioanthropological Investigation of the African Burial Ground and Five Points Area, New York, New York." My training is in physical anthropology, and I have analyzed several thousand skeletons from archaeological sites and have conducted excavations at more than a dozen historic cemeteries. These sites include the Colonial St. Peter's Street Cemetery and the nineteenth century Charity Hospital Cemetery, both of which are located in New Orleans. Currently, I am analyzing 300 nineteenth century burials from Richmond, Virginia. Because of this background, my comments concern the proposed analysis of the human skeletal remains from the eighteenth century African Burial Ground.

The proposal describes an important skeletal collection that will contribute new and useful information. Therefore, this series should be thoroughly analyzed, and a multidisciplinary effort can best achieve that objective. Few remains from the time period of 1712 to 1790 have been systematically examined.

The proposal offers much information, but some essential details are lacking, thus preventing a thorough evaluation of the proposal.

My concerns are as follows:

1.  The proposal reports that 420 burials have been excavated and that seven percent appear to be of European descent. The condition of the remains is described as variable, a few excellent and others

DEDICATED TO UNDERSTANDING THE NATURAL WORLD AND OUR PLACE IN IT

WASHINGTON DC 20560

**Figure 10.3    A response to the research design.**    Source: *Comments on the Draft Research Design for the Archeological, Historical, and Bioanthropological Investigations of the African Burial Ground and Five Points Area, New York, New York,* Published by the General Services Administration, Region 2, 1993.

Peter A. Sneed
December 1, 1992
Page 2

extremely poor. Page 59 mentions that the soil is highly acidic, and elsewhere the proposal states that the remains were deeply buried by overburden. Page 60 notes the relatively poor condition of much of the bone. These descriptions suggest generally poor preservation, which will limit the information that can be recovered.

Consequently, more specific information about the condition of the remains is essential for determining the types of analyses that will be feasible for this collection. The research design mentions research on DNA and chemical analysis, but it does not indicate whether the remains are sufficiently well preserved to support these kinds of studies. How many skulls are complete enough to justify digitized computer imaging (page 101)?

2.    The proposal preface lists a large number of individuals who appear to be involved in this project, but the responsibilities of the different team members are not specified? Why is the MFAT team doing "preliminary assessments of age, sex, and ancestry of each individual, a preliminary description of any pathological conditions observed, and preliminary measurements of specific bones (page 60)"?. Page 94 indicates that the first step in the investigation is preliminary analysis to determine age, sex, and stature. What about determination of race? Why is the work classified as "preliminary"? Who is responsible for the final determinations? This exercise suggests duplication of effort, thus unnecessary expense. Accurate assessments and measurements are needed, but they should be taken once by someone who knows what he or she is doing.

The data recovery format and analytical procedures are not clearly defined.

(Under separate cover, I was sent additional information from the Metropolitan Forensic Anthropology Team. That information indicates thorough awareness of the literature and standards used to determine age, sex, and race by the members of this team.)

Page 94 states that the synthetic work will be conducted at Howard University. The rationale justifying this transfer should be documented, because it is not apparent. What is the basis for this determination in terms of facilities and personnel?

The proposal states that the materials have been removed from the site to a laboratory at Lehman College. It is an unnecessary

expense to establish museum-quality storage facilities at Lehman, with the required environmental controls, and then to establish a duplicate set of conditions at Howard University.

Sincerely yours,

*Douglas W Owsley*

Douglas W. Owsley, Ph.D.
Research Scientist/Curator

**Figure 10.3**    *(continued)*

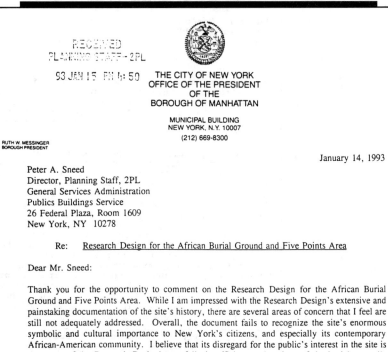

RECEIVED
PLANNING STAFF - 2PL

93 JAN 15 PM 4: 50     THE CITY OF NEW YORK
OFFICE OF THE PRESIDENT
OF THE
BOROUGH OF MANHATTAN

MUNICIPAL BUILDING
NEW YORK, N.Y. 10007
(212) 669-8300

RUTH W. MESSINGER
BOROUGH PRESIDENT

January 14, 1993

Peter A. Sneed
Director, Planning Staff, 2PL
General Services Administration
Publics Buildings Service
26 Federal Plaza, Room 1609
New York, NY  10278

Re:     Research Design for the African Burial Ground and Five Points Area

Dear Mr. Sneed:

Thank you for the opportunity to comment on the Research Design for the African Burial Ground and Five Points Area. While I am impressed with the Research Design's extensive and painstaking documentation of the site's history, there are several areas of concern that I feel are still not adequately addressed. Overall, the document fails to recognize the site's enormous symbolic and cultural importance to New York's citizens, and especially its contemporary African-American community. I believe that its disregard for the public's interest in the site is the cause of the Research Design's woefully insufficient explanations of the burial ground's future. Issues requiring more comprehensive and focused attention include: continued archeology and research; the eventual reinterment of the remains; curation and exhibition of the artifacts that have been recovered; education; and plans to establish a suitable memorial at the site.

3.2.1     The Research Design does not present a coherent plan to prevent further desecration of this sacred site, even indicating that the destruction of the burial site could be mitigated by archeological work in advance of construction. While acknowledging that "the destruction of significant human remains and archeological resources constitutes an adverse effect," it goes on to state that "a research program of archeological data recovery and analysis mitigates the adverse nature of the effect, provided that the resources are significant for their potential to provide information concerning the past." The Research Design should explain further how it is intended to "limit effects on the human remains and archeological resources present" during further construction. Additionally, it should make clear what would be considered "unavoidable effects to human remains and archeological resources" resulting from future construction.

5.3.2     I am concerned that the document does not mandate the reinterment of the African-American skeletal remains on-site. While the document does state that "following the

**Figure 10.4     A response to the research design.**     Source: *Comments on the Draft Research Design for the Archeological, Historical, and Bioanthropological Investigations of the African Burial Ground and Five Points Area, New York, New York.* Published by the General Services Administration, Region 2, 1993.

completion of analysis, the human remains will be reinterred," it neither specifies possible locations, nor addresses the feasibility of returning the human remains to their original resting place within the Broadway block.

5.3.2.1        I also have questions regarding the establishment of an appropriate repository for the long-term curation of the artifacts that were uncovered in the course of the archeological excavations. Given my commitment to the incorporation of a suitable and significant memorial and museum space within the new Federal Office Building, I am distressed by the Research Design's failure to consider the future location of this repository. It should be more explicit about the plans for this facility, including a consideration of the feasibility of locating it on-site. It must also acknowledge the responsibility of the Federal Government to establish this museum, including an appropriate financial commitment. Finally, I wish to address the criteria and characteristics that will be used to choose a repository for the artifacts and records. Although appropriate from an archeological standpoint, these criteria do not take into account the social historical context in which the museum is being created. The museum is required because of the tremendous cultural and symbolic import of the burial ground to the African-American community, and the overwhelming public interest that it has generated. But it is also important to remember and respect the fact that the subject matter of the museum is human remains and personal artifacts. Sensitivity to these important issues must be included in the criteria used to develop this facility.

6.2.2        The Research Design fails to adequately treat plans for the former pavilion portion of the Burial Ground, plans for the reinterment, plans for a memorial, plans for an exhibit, and plans for a museum, in stating that the National Steering Committee will address these issues. It is within the purview of the Research Design to make recommendations for these future issues. Given the importance of the issues that are addressed by the National Steering Committee, the document should at the very least be more explicit about the process in which these decisions will be made. Must the Federal Government follow the recommendations of the National Steering Committee? If not necessarily, then what other criteria will be used to make these determinations?

Please feel free to contact David Freudenthal (669-2226) in my office if you have any questions regarding my comments on the Research Design. I look forward to continuing our dialogue on the preservation of this national treasure.

Sincerely,

Ruth Messinger

cc:     Hon. Barbara J. Fife
        Hon. Laurie Beckleman
        Community Board #1
        Municipal Art Society
        Historic Districts Council
        Landmarks Conservancy
        National Steering Committee

**Figure 10.4**   *(continued)*

---

## CHAPTER SUGGESTIONS FOR DISCUSSION AND WRITING

1. Figure 10.1 is an excerpt from a letter from the U.S. Department of the Interior responding to the research design. The excerpt is specifically concerned with the readability of the document. Compare the letter writer's points with strategies discussed in Chapter 2. Although the letter does not include any criticism of the pages included in this excerpt, the writer may have missed the problems in readability and comprehension readers may have. Re-examine the excerpt and determine whether such problems exist on these pages. Mark the problems in the margin, then revise them.

2. What do the writers of each letter attempt to persuade the GSA to do? What claims do the writers make in their arguments? On what values are these claims based? Why do you think some readers share the values on which these claims are based? Why do you think readers may be persuaded to act as the writers want? What other claims do you think the writers might make to persuade the readers?

3. Assume the role of Peter Sneed, the planning director of the Foley Square project, and write a letter to one of the people who has responded to the research design. You may want to respond to the person's specific criticisms and recommendations, or you may simply want to advise the person that you have received the letter and will consider it as you make your final decision about the site. Select one of the letters in Figures 10.1, 10.2, 10.3, or 10.4, and write your letter of response.

4. What is the writers' viewpoint in the research design? What do you think are the writers' biases toward the topic? Give examples of passages that indicates the writers' biases.

# Index